391
AF654443

WATER RESOURCES, SUSTAINABLE LIVELIHOODS AND ECO-SYSTEM SERVICES

About the Editors

Kanchan Chopra is Professor at the Institute of Economic Growth, Delhi. Her research interests span the areas of development, agricultural economics, natural resource management and sustainable development strategies. She is currently Associate Editor, *Environment and Development Economics* and Member, Assessment Panel, Millennium Ecosystem Assessment. She has authored and co-authored several books, in addition to having published extensively in national and international journals. Her books include *Operationalising Sustainable Development: Economic-Ecological Modelling for Developing Countries* (with G.K. Kadekodi) 1999 and *Migration of Common Property Resources and Environmental Degradation* (with S.C. Gulati) 2001.

C. H. Hanumantha Rao formerly Director and Professor of the Institute of Economic Growth, Delhi is now an Honorary Fellow and Chairman, Board of Governors atf this Institute and at the Centre for Economic and Social Studies, Hyderabad. He was Member, Planning Commission, 1982-86, member, Seventh and Eighth Finance Commissions,Chairman, National Commission on Rural Labour, 1990-91, and Chairman, High Powered Committee on Fertilizer Pricing Policy, 1997-98. His publications include *Technological Change and Distribution of Gains in Indian Agriculture*, 1975 and *Agricultural Growth, Rural Poverty and Environmental Degradation in India*, 1994.

Ramprasad Sengupta is Professor of Economics at the Jawaharlal Nehru University, New Delhi. He has been in the teaching and research in Economics for the last 33 years. He was for 3 years Adviser in the Planning Commission, Government of India for Energy Policy Modelling. He has recently been Professor at the Indian Institute of Management, Calcutta for more than two years. His area of research interest is energy, environment and development economics. His latest published work in this area has been *Ecology and Economics: An Approach to Sustainable Development*, Oxford University Press, India, 2001.

WATER RESOURCES, SUSTAINABLE LIVELIHOODS AND ECO-SYSTEM SERVICES

Edited by
Kanchan Chopra
C. H. Hanumantha Rao
Ramprasad Sengupta

Published for
INDIAN SOCIETY FOR ECOLOGICAL ECONOMICS
by
CONCEPT PUBLISHING COMPANY, NEW DELHI-110059

All rights reserved. No part of this work may be reproduced, stored in a retrieval system, or transmitted in any form or by any means electronic, mechanical, photocopying, micro-filming recording or otherwise, without the prior written permission of the copyright owner and the publisher.

Indian Society for Ecological Economics
Registered office: Institute of Economic Growth, University Enclave, Delhi-110007
Email: insee@ieg.ernet.in
Website: www.indev.nic.in/insee.html.

ISBN 81-8069-059-8

First Published 2003

© Institute of Economic Growth, New Delhi-110007

Published and Printed by
Ashok Kumar Mittal
Concept Publishing Company
A/15-16, Commercial Block, Mohan Garden
New Delhi-110059 (INDIA)

Phones : 25351460, 25351794
Fax : 091-11-25357103
Email : publishing@conceptpub.com.

Preface

It is with a sense of satisfaction that we see this volume go into publication. The Indian Society for Ecological Economics (INSEE) selected *Water Resources, Sustainable Livelihoods and Eco-system Services* as the theme for its Second Biennial Conference way back in 2000. Following on the conference held at the Institute for Forest Management, Bhopal in December, 2001, authors of papers presented there were asked to send papers for the proposed volume. We received a very enthusiastic response from paper writers and it became evidence that all contributed papers could not be accommodated in the proposed volume, given the exigencies of time, space and resources involved. However, the issues discussed were wide-ranging and we did want to give the reader of the volume an insight into the range and depth of their coverage. Chapter 1 entitled "Introduction" aims to do this. It encompasses all papers presented at the conference, bringing them into the ambit of one or other of the sub-themes that the volume is divided into.

The four sections of the volume are to an extent, self-explanatory. To begin with, the theme of the volume is examined from the macro perspective and from the viewpoint of specific problem-related or regional issues. A large part of the focus in the second section is on emerging issues of urban and industrial pollution. The third section extends to institutions for water management, while the fourth relates to aquatic eco-systems and the services that they offer. The volume ends with a section which brings together papers that indicate future directions.

A volume such as this constitutes a reflection of the state of research in the country. This perhaps accounts for the somewhat disproportional emphasis on "water resources" at the cost of "eco-system services" and perhaps even "livelihoods". It is our hope that sometime in the future, when INSEE decides to come back to a theme such as this, there shall be a far more in-depth and interdisciplinary discussion of issues relating ecosystem services and human wellbeing. That in fact should be the direction for research in this area in the future.

A large number of individuals provided inputs during the two years leading to the publication of the volume. We wish to thank members of the successive Executive Committees of INSEE and members of the Conference Programme Committee for the Second Biennial Conference for their inputs into the selection of papers for the conference and finally for the volume. Authors of papers appearing in the volume were prompt with making revisions in the light of comments whenever asked to do so. The conference and the publication of the volume received generous financial support provided by the Swedish International Development Agency (SIDA), Winrock International India and by the Ministry of Finance, Government of India and we are thankful to them on our behalf as well as on behalf of INSEE.

We also wish to thank Shri Ashok Kumar Mittal of Concept Publishing Company who took interest in the speedy publication of the volume, Mrs. Meera Juneja, who did a thorough copy editing task on the manuscript and the INSEE Secretariat which provided efficient support.

March 18, 2003

Kanchan Chopra
C. H. Hanumantha Rao
Ramprasad Sengupta

Contents

SECTION-IV
AQUATIC ECOSYSTEMS AND SERVICES

SECTION-V
DIRECTION FOR THE FUTURE

List of Tables

List of Figures

Contributors to the Volume

Alagh, Y.K., Professor & Vice Chairman, Sardar Patel Institute of Economics and Social Research, Thaltej Road, Ahmedabad 380054, Gujarat, Formerly Minister of Science and Technology, Government of India.

Babu, C.R., Director, Centre for Environmental Management of Degraded Ecosystems, and Pro-Vice Chancellor, University of Delhi, Delhi 110007.

Bandyopadhyay, Jayanta, Research Professor, Centre for Development and Environment Policy, Indian Institute of Management, Joka Diamond Harbour Road, Kolkata, 700104, West Bengal.

Brij Gopal, Professor, School of Environmental Sciences, Jawaharlal Nehru University, New Delhi 110067.

Chopra, Kanchan, Professor, Institute of Economic Growth, University Enclave, Delhi 110007.

Dasgupta, Purnamita, Research Associate, Institute of Economic Growth, University Enclave, Delhi 110007.

Goldar, Bishwanath, Professor, Institute of Economic Growth, University Enclave, Delhi 110007.

Iyer, Ramaswamy R., Visiting Professor, Centre for Policy Research, Chanakyapuri, New Delhi 110044, and Formerly Secretary, Government of India.

Jani, B.M., Professor, Department of Economics, Saurashtra University, Rajkot, 360001, Gujarat.

Kateja, Alpana, Assistant Professor, Department of Economics, JN University, Jodhpur, Rajasthan.

Kumar, M. Dinesh, Consultant and Project Director, North Gujarat Groundwater Institute, International Water Management Institute, India Project Office, Vallabh Vidyanagar 388120, Anand.

Kumar, Pushpam, Reader, Institute of Economic Growth, University Enclave, Delhi 110007.

Love, Amit, Scientist, Centre for Environmental Management of Degraded Ecosystems, University of Delhi, Delhi 110007.

Mallik, Bidisha, Research Associate, Centre for Development and Environment Policy, Indian Institute of Management, Joka Diamond Harbour Road, Kolkata 700104, West Bengal.

Marvania, P.G., Associate Professor, Department of Economics, Saurashtra University, Rajkot, Gujarat.

Nikku, Bala Raju, Ph.D. student, Irrigation and Water Engineering Group, Wageningen University, The Netherlands & Gujarat Institute of Development Research, Ahmedabad.

Prasad, M.S.V., Assistant Professor, College of Management Studies, Gitam, Visakhapatnam, 533045.

Rao, C.H.Hanumantha, Honorary Professor & Chairman, Institute of Economic Growth, Delhi, Chairman, Centre for Economic & Social Studies, Hyderabad, Former Member of Planning Commission, New Delhi 110001.

Sekar, C., Associate Professor, Department of Bio Energy, College of Agricultural Engineering, Tamil Nadu Agricultural University, Coimbatore-3, Tamil Nadu.

Sengupta, Nirmal, Professor, Madras Institute of Development Studies, 79, Second Main Road, Gandhi Nagar, Adyar, Chennai 600020.

Sengupta, Ramprasad, Professor, School of Social Sciences, Jawaharlal Nehru University, New Delhi 110067.

Sharma, Radhey Shyam, Scientist, Centre for Environmental Management, Department of Botany, University of Delhi, Delhi 110007.

Singh, C.J., Senior Scientist, Punjab Agriculture University, Regional Research Station, Dabwali Road, Bathinda 151001, Punjab.

Singh, Dalbir, Director, Centre for Human Development, Chittorgarh 312 025, Rajasthan.

Singh, Ksh. Jhaljit, Assistant Professor, Department of Economics, Manipur University, Chanchipur 795003, Imphal.

1

Introduction

KANCHAN CHOPRA, C.H. HANUMANTHA RAO
& RAMPRASAD SENGUPTA

Water is arguably the resource which presents the most intractable challenge for scientists, planners and policy makers. The issues that arise in its use are spread over a range of areas impinging on human well-being. Water features in the production of food, and is therefore an essential ingredient of food security. This is perhaps any aspect that captures the major policy focus in developing countries such as India. It is also the best known catalyst for initiating development interventions as any development practitioner will testify. People rally around water-related issues very easily, be it in urban environments or in far-flung rural areas. For, water sustains livelihoods and indeed life. Preservation of water sources in their pristine and unpolluted state is at the core of a large number of enriching cultural practices of the major cultures of the world.

It seemed appropriate, therefore, that the Indian Society for Ecological Economics should focus on water, sustainable livelihoods and eco-system services for its second biennial conference. A range of issues and problems impinging on water were discussed at this conference. This volume brings together some papers presented under a set of schematic themes[1] providing thereby an overview of the dimensions explored, at both the plenary session and in the contributed papers.

The Macro Perspective

A number of the papers placed water-related scientific and policy issues in the broader perspective of macro-level as well as sectoral-level water balances. This larger picture was viewed by the contributors from the multiple perspectives of hydrology, economics, political economy and water-related institutions. The Presidential address by C.H. Hanumantha Rao[2] sets the stage by examining the sustainable use of water in Indian agriculture. Considering that more than 80 per cent of the total future water demand projections in India are for use by the agricultural sector, this issue is critical to overall sustainable development. Drawing upon studies both international [such as those of the International Water Management Institute, (IWMI)] and national [the National Commission for Integrated Water Resource Development Plan (NCIWRDP)], he illustrates the criticality of improving irrigation use efficiency in the efforts at bridging gaps between demand and supply of water. He points out that improvement in irrigation use efficiency had been very low in India due to the continued focus on land use efficiency. Water saving defined as "volume or value of crop output per unit of water" is not, in general, accorded priority by planners. He reviews different methods for raising water use efficiency, among which technological changes (such as biotechnology-related changes) to improve crop yields, institutional changes to motivate water savings on the part of different stakeholders and selective dependence on the rest of the world for food are important. He ends by reiterating that appropriate institutions that bridge the gap between knowledge and its application are of the essence in this context.

Jayanta Bandyopadhyaya's paper[3] provides a hydrological perspective of the water balance. Reviewing recent attempts at forecasting water demand and supply for the agricultural, industry, power and household sectors, he comments on official strategies for water resource development which have focused on the development of large storages. He concludes that an appropriate water policy needs to be based on an understanding of the hydrological cycle and the eco-system services that water performs prior to and after being used for direct consumptive purposes by humans.

Amongst papers focusing on the broader picture, Ramaswamy Iyer's[4] is couched in a political economy framework. Reviewing the stance taken by government as also other initiatives for providing a long-term view of the water problem in India, he re-examines critically the notion of "developing" water resources. Most future scenarios are drawn with the underlying assumption that the gaps between demand and supply need to be bridged through development by using technological interventions of the usual kind, i.e. building storages and using water through a system of canals. This raises, according to him, a number of dilemmas. These are described as the money dilemma, the markets dilemma, the environmental dilemma and the human rights dilemma. In each of these contexts, the questions raised are fundamental and the state of knowledge inadequate. How can the functioning of markets be regulated so as to ensure that they operate where they can in an efficient manner. Enough is not known about the environmental impacts to say whether they are significant or not in particular cases. Substantial differences are also present in approaches to ethical questions about the rights of competing stakeholders. This perceptive treatment of a range of issues leaves one asking the question: is this not a situation where the "precautionary approach" (as expounded in the policy literature with regard to ecosystems) is the best one to follow?

Another paper that begins with an examination of the macro situation with respect to water demand and supply is provided by Dinesh Kumar.[5] This paper also focuses on the institutional changes needed for appropriate demand management. It argues for multi-tier organizations of water-users and a critical role for property rights which enable markets to function.

Specific Problem-related or Regional Issues

A second set of papers presented at the conference discussed specific problems or policy-related issues such as salinity of water, urban water management, technological interventions to improve availability as well as efficiency of water, rainwater harvesting, industrial water pollution, valuation of water quality, environmental regulation of water quality, etc. Specific problems related to water management were focused on, either from the national or the regional perspective.

Among these, Nirmal Sengupta's paper[6], titled, "Salinization and the importance of rainwater harvesting" adopts a multi-region approach. Dwelling at length on the issue of salinization in the context of different regions in the country, it combines a historical perspective with careful analysis to establish the role that rainwater harvesting can play within the repository of salinity control techniques. It draws on evidence from traditional knowledge and argues forcefully for learning from this to manage our water resources.

Urban water management issues and industrial pollution of water bodies are becoming increasingly prominent and these issues are addressed in a large number of papers. Among the papers that dwelt on problems of urban water management and waste water disposal[7], Choudhary has focused on the demand for and differential access to water of different classes of households in Berhampur in Orissa while Fadnavis studies the deteriorating water quality in Bhandara district of Maharashtra. B.M. Jani has made a comparative analysis of Class 1 and Class 2 cities of Gujarat with respect to waste water disposal while K.J. Singh has examined the salient features of the water supply in Imphal city. Chakravarty and Sharma have studied aspects of water demand and supply in Kolkata and Gaya, respectively. Both of them discover discrepancies between demand and supply. Purnamita Dasgupta addresses the problem of water stress viewed as water scarcity and water pollution in Delhi and uses an econometric approach to estimate willingness of households to pay for a minimum assured water supply. It is found that the total willingness of the surveyed households to pay can meet the long run cost of production and distribution of water. Other aspects of household behaviour are also studied.

Water pollution caused by industry has been studied in a large number of papers. It is interesting to note that in most cases these studies have gone beyond just trying to estimate the nature and extent of industrial pollution. They have also addressed issues regarding the cross-sectoral impact of this pollution on agriculture or on drinking water. The notion of water as a resource with multiple uses and hence interconnections both positive and negative between these uses is the central point of this research.

C. Sekar examines the effect of pollution on agriculture caused by tanneries in Tamil Nadu. Using a hedonic value approach, he finds that value of farmland is impacted by the water quality index and the nature of forest resources in the vicinity.

Amarnath similarly examines the influence of pollution caused by tanneries on water quality in agricultural farms. Datta *et al.* aim at setting up water quality indices for industry. Mishra and Sahu value the damage to drinking water due to industrial pollution in Orissa. They find that rural women need to spend an additional two hours each day to collect water due to this pollution.

Water quality impacts, industrial water use and pollution are addressed in methodologically innovative fashion by another set of papers. Goldar asks questions relating to the value of water for India's industrial sector and the price elasticity of the demand for water. He finds a low price elasticity of demand using a water demand function. An increase in water price for industries shall therefore increase revenue for water supplying agencies, though it will not reduce consumption of water by industry. Vasisht examines econometric problems linked with the use of static models to estimate yield losses associated with salt concentrations in irrigated water. Murty and Kumar study the effect of environmental regulation relating to water pollution on the productive efficiency of firms. They find that the technical efficiency of firms increases with the intensity of environmental regulation.

Region-specific issues in water management both in rural and urban contexts, are studied in a large number of contributed papers. These include, for instance, salinity in coastal Saurashtra (Marvania), and efficient use of canal water in Punjab (C.J. Singh), water management in Haryana (Attri and Verma) declining groundwater levels in Rajasthan (A. Kateja), the potential for solar photo-voltaic pumping in Gujarat (Navadkar and Birari), local harvesting technology in the dry zone of Maharashtra (Jugale) and conservation of water in Pondicherry (Velmurgan).

Institutions for Water Management and Conflicts Over Water

Formal and informal institutions at different levels of management are critical for water allocation. This issue received due attention in the papers presented at the conference. Bala Raju Nikku dwells at length on the new "participatory irrigation management" introduced by law in Andhra Pradesh. The paper succeeds in drawing attention to the pitfalls that government-sponsored "participation" is likely to encounter. Dalbir Singh, in an in-depth study of some villages in Rajasthan, points out how water markets in the current legal context (where groundwater access is linked to ownership of land) treat groundwater as a readily accessible resource and exacerbate unequal access. This needs to be complemented by informal distributional arrangements.

Institutions can be interpreted to mean different kinds of organizations as well. Arguing in this vein, Vishwanathan examines the impact of major and minor irrigation projects in Kerala. While many may not agree with this interpretation of "institutions" the main contention of the paper that the present distribution arrangements with respect to canal water result in inefficiency is established in other regions and contexts as well.

The political economy of water and legal frameworks for solving conflict were studied by two interesting papers. Chaube and Chaube examine inter-state water disputes in India and the different frameworks within which solutions have been sought for them. Girdner points out in his paper, that water in the Middle East is not only a resource and technology question but a political-strategic question. He warns that the potential exists for a war between Turkey, Syria and Iraq over the use of water in the Euphrates-Tigris river basin. To us in India, that sounds familiar in the context of our own conflicts, both past and present in South Asia and within India. The paper by Bon Emmanuel points towards new possibilities of managing water using the French approach rooted in "patrimonial strategies". A patrimonial audit of water resources keeps in mind the varying claims of stakeholders and holds the water resources of the nation (in terms of both quantity and quality) in trust for the future.

Aquatic Eco-system Services

Another focal point of the conference was the eco-system services that water bodies provide. This perspective goes beyond the traditional use-oriented treatment of water and extends to an analysis of the contribution of water-bodies such as lakes and wetlands to human well-being. Brij Gopal's paper[8] provides the starting point into this subtheme. Describing aquatic eco-systems in some detail, he describes services such as treatment of human wastes, flood control and water recharge which rivers, floodplains and wetlands have rendered for humans. Diversion of water flow from these bodies for so-called "economic uses" has a cost that needs to be valued in terms of the long-run loss of these services.

Kumar[9] *et al.* in a study of floodplains in a twenty-five kilometre stretch of the Yamuna river in Delhi estimate the economic impact of bio-physical changes resulting from urbanization, embankment and encroachment. Using the production function approach, it is estimated that the mean value of the water recharge function for farmers in the region is Rs. 5.36 lacs per annum for the area under cultivation. Additional benefits accrue to the residents of Delhi in the form of augmented water supply.

Dikshit and Kulkarni highlight the role played by forests in the catchment of the five lakes from which Mumbai gets its supply of water. Verma *et al.* in a study of the Bhoj wetlands in Bhopal draw current and future scenarios of levels of pollution of the lakes with and without intervention. Different interventions to ensure sustainable use are suggested while alternative valuation techniques are utilized to determine its value to residents. Gopikuttan examines the impact of a conversion of paddy fields to non-agricultural use in Kerala and finds that the economic value of the services lost due to such conversion are substantial; almost as high as the value of the agricultural output. Jha and Saha examine the role of wetlands in providing livelihoods through traditional agriculure and the consequent resistance to declaring some of these wetlands as protected areas. Sahu and Pradhan study marine fisheries in Orissa as a source of sustainable livelihoods. Their contention is that mechanization of boats and trawlers results in over-extraction of fish and that regulatory institutions are needed to prevent this.

This set of papers provides pointers to the trade-offs between developmental gains and gains from eco-system services that emerge when the ecological significance of fresh water eco-systems is recognized in making the choice among alternative land uses. Social decision processes need to be evolved for choosing between the provisioning (food and water), regulating (life-supporting) and enriching (cultural) functions of these eco-systems. Research needs to support such decision-making by providing clearer pictures of the elements in this trade-off, both in the long and short-run.

Directions for the Future

The conference was also enriched by a set of papers that focused on providing directions for the future. The valedictory address by Y.K. Alagh[10] does this in an incisive manner, linking development issues with micro level initiatives in an era of globalization and corporatization. He argues that community-based institutions shall have to be central to the next phase of development in India. Three critical areas need to be examined in this context: how can communities and individuals at risk in fragile regions be turned around? What are the policy and macro implications of supporting such strategies? And thirdly, how can partnerships be fostered between participatory transparent institutions of people living in different socio-ecological systems? This indeed is the essence of the development problem itself and Alagh argues forcefully for innovative organizational and financial rules that make strategic alliances across different kinds of institutions possible. Such an approach shall assist the mainstreaming of the better amongst the land and water-based initiatives.

Some other papers also provide insights which could chart future directions. Prasad argues for a systems approach to water resource management, based on an inter-disciplinary perspective. Bhavanishankar and Shivashankar, in a similar vein, point towards general principles for water management and Desarda argues for watershed development based on traditional knowledge.

Issues relating to attitudes towards the natural environment, appropriate institutional framework for decision-making and innovative technologies are critical for directions that the future shall take. In an invited presentation, John Proops[11] studied environmental values and modes of introducing change in them using market research techniques. Purporting that peoples' attitudes towards the environment should constitute the starting point of environmental policy, he proposed the Q methodology as a technique to span the divide between qualitative and quantitative approaches to attitudinal research. Chachadi and Ferriera,[12] discuss the relevance of new mapping techniques for potential problem areas with respect to seawater intrusion. Raina dwells on an important issue, namely the decision processes for technology generation and utilization. The paper presents a case for agro-ecological methods to replace conventional decision-making which is often highly compartmentalized. She illustrates her point using the case of decision-making with respect to the increased sodicity of water in Haryana. The point at issue is:

will such a change come from a state fiat or will it have to be built into the existing provisions for decentralized governance such as the Panchayats Act?

Conclusions

The wide range of issues covered in the papers indicate that serious and interesting research initiatives in this emerging area of ecological economics are emerging. They point to some of the policy and research challenges in the area. Traditionally studied areas such as increasing water use efficiency in agriculture continue to be significant since technology adoption and institution identification are by no means settled issues. Meanwhile, structural changes in the economy such as high rates of growth and urbanization with concentration of population have created a new set of water-related issues. Increased recognition of the role of ecological services provided by eco-systems provides a challenge at the frontiers of research. The findings of this research need to be integrated into policy-making processes using basic economic principles

Finally, land and water are best managed locally. In an era of globalization, new satellite-based technology and corporate forms of governance, any innovative institutions that link the local to the global need to be encouraged and strengthened. And this is a challenge that presents itself in relation to an array of resources, among them water. We feel that the papers in this volume put together these issues in a careful, well-researched manner for the discerning reader. We do hope that they shall be of interest to a wider audience.

NOTES

1. It has, for different reasons not been possible to include all papers presented at the conference in this volume. however, this introduction endeavors to give a flavor of all kinds of issues and themes covered by the papers.
2. The Presidential address at the conference.
3. Invited paper.
4. Invited paper.

5. "Demand Management in the face of growing water scarcity and conflicts in India" Contributed Paper.
6. Invited Paper.
7. These are all contributed papers
8. Invited Paper.
9. Contributed Papers
10. Entitled "Emerging Institutions for Rural Development".
11. The presentation was entitled, "Attitudes Towards the Environment: Informing and Reforming Economic Policy"
12. Chachadi *et al*. and Raina are contributed papers.

SECTION-I

THE MACRO PERSPECTIVE

2

Sustainable Use of Water for Irrigation in Indian Agriculture*

C.H. HANUMANTHA RAO

Growing Water Scarcity

Water resources are becoming extremely scarce. According to the projections made by the National Commission for Integrated Water Resource Development Plan, the requirement of water for irrigation in India will grow by more than 50 per cent in the next 50 years. The water requirements for household consumption and for industry would rise even faster. In view of this, even after fully exploiting the usable water resources, the balance between the supply and demand for irrigation water can be achieved only by improving the level of irrigation efficiency in a big way from about 36 per cent efficiency in 1993-94 to 60 per cent in the year 2050 (Government of India, 1999). A ten per cent improvement in the efficiency of water use would be equivalent to adding some 14 million hectares of gross irrigated area (Saleth, 1996). A model of water supply and demand for 118 countries accounting for 93 per cent of world population, built by the International Water Management Institute (IWMI), shows that around 50 per cent of the increase in demand for water by the year 2025 can be met by increasing the effectiveness of irrigation. Most of this gain in irrigation effectiveness would be in countries with a high percentage of

* The author is thankful to S. Mahendra Dev, V. Ratna Reddy and S. Subrahmanyam for their comments and suggestions.

irrigated rice. India and China together would account for as much as one-half of the world's total estimated water savings from increased irrigation effectiveness. Therefore, the capacity of large countries like India and China to efficiently develop and manage water resources is likely to be a key determinant of global food security in the twenty-first century (Seckler, *et al.*, 1998).

Usable water supplies are highly unevenly distributed across the globe and across regions within a large country like India. Therefore, notwithstanding the balance achieved at the national level between the demand and supply of water, shortage of drinking water would pose a serious problem in the drought-prone regions of the country, especially because over 90 per cent of rural population depends on groundwater for drinking purposes. The poor who depend upon dug wells, which dry up fast, are the first to suffer. Under such circumstances, the only way to ensure local food security and water security to meet the basic needs for drinking and cooking, is by undertaking rainwater harvesting and groundwater recharge on a large scale through the participation of the community, including especially the women and the landless. This should be possible because, unlike in regions of high rainfall, land area per head is larger in the low rainfall or drought-prone areas where 10 tiny dams with a catchment of one ha. each can collect more water than one larger dam with a catchment of 10 ha. (Agarwal, 2000).

Variations in Water Productivity

In the last two decades, we have been faltering in our efforts to augment water resources, even as the use of the available water resources has become increasingly unsustainable. Public investment in major and medium irrigation schemes has been declining, in real terms, throughout the 1980s and the 1990s, even though as much as 40 per cent of the potential still remains to be exploited. We have not taken significant steps so far for improving water-use-efficiency through modernization/ renovation of existing systems which have deteriorated over the years. According to the Mid-Term Appraisal of Ninth Five Year Plan, the progress achieved so far in Participatory Irrigation Management (PIM), designed to improve water-use-efficiency, is rather slow. The irrigated area transferred to Water Users Associations (WUAs) in India is only about 7 per cent as against the 45 per cent in Indonesia, 66 per cent in Philippines, and 22 per cent in Thailand (Government of India, Planning Commission, 2000). As regards groundwater, the rate of extraction has been far above the rate of recharge. As a result, the number of over-exploited and dark blocks has increased by nearly 70 per cent over the last 14 years (Government of India, Planning Commission, 2000).

Ultimately, savings in irrigation water can be achieved only by raising the productivity of water, defined broadly as the volume or value of crop output per unit of water used. The definition of water productivity varies in the literature depending on how the denominator in this ratio is specified. When, at one extreme, water released from the system is used as the denominator, water productivity becomes all-inclusive subsuming water-use-efficiency, that is, the ratio of consumptive use of water to the water released. At the other extreme, when the denominator consists of water lost as evapo-transpiration by plants in any particular season, then an improvement in water productivity can arise basically from the improvement in yields.

Interestingly, water savings in global crop production in the last few decades have accrued indirectly from a rise in crop yields and very little directly from improvement in water-use-efficiency (CGIAR, 2001*a*). This is particularly so in developing countries like India where the improvement in irrigation efficiency has been very slow. For one thing, a clear focus on water productivity, that is, productivity per unit of water used is a new concept for the international agricultural research system (CGIAR, 2001*b*). Nearly two hundred years ago, Thomas Malthus focused on the fixity or scarcity of land in relation to population growth as the cause for recurring famines. This perception prompted the evolution and use of land-saving technologies and practices which substantially raised productivity per unit of land. Until recently, irrigation water has seldom been regarded as a scarce resource in the favourable rainfed areas of Europe, North America and Japan, where water losses on account of evaporation and evapotranspiration are also lower because of a cooler climate. In the last century, major breakthroughs in crop yields per unit of land were witnessed in these countries, from where such technologies were transferred to the developing countries. Recently, however, the growing competition for water demand among industry, by urban areas for household consumption as also for environmental purposes in these developed countries has been exerting pressure to save the water allocated to irrigation. Besides, the rising level of water pollution is accentuating the scarcity of fresh water in all sectors.

The Green Revolution has substantially raised yields in the irrigated areas of India, resulting in a rise in the productivity of irrigation water. Such an increase in water productivity on account of land-saving technological changes can eventually ease the pressure on water for irrigation, even when there is no change in water-use-efficiency. Of the area under foodgrains since irrigated area accounts for the bulk of the increase in foodgrains output, growth of foodgrains output in relation to the growth of irrigated area under foodgrains would give a good measure of the changes in the productivity of irrigation water. Despite the Green Revolution, which was confined largely to wheat in the north-western region in the 1970s, the foodgrains output grew at about the same rate as gross irrigated area. Thus, there was no growth in productivity of irrigation water for foodgrains in the country in this decade (Table 2.1). During the 1980s, however, the Green Revolution was broad-based covering rice in the central and eastern regions of the country where high rainfall supplements irrigation. Thus, foodgrains output grew at a higher rate than irrigated area resulting in a positive growth in the productivity of irrigation water. However, in the post-reform decade of the 1990s, there was a deceleration in the growth of foodgrains output leading to a significant decline in the productivity growth of irrigation.

Table 2.1: Productivity of irrigation for foodgrains in Indian agriculture

(*Growth Rates*)

	1970-71 to 1980-81	*1980-81 to 1990-91*	*1990-91 to 1996-97*
Gross Irrigated Area	2.31	1.72	1.70
Output	2.30	2.90	2.00
Productivity of Irrigation	–0.01	1.18	0.30

Sources: (1) Government of India, Ministry of Agriculture, *Indian Agricultural Statistics*, 1992-1997.
(2) Government of India, Central Statistical Organisation, *Statistical Abstract* — India, 2000-2001.

The decline in water productivity growth in the foodgrains sector of Indian agriculture in the last decade may be attributed partly to the loss of momentum in the development of yield-increasing technologies and partly to the political economy of irrigation from groundwater sources. For example, there was a sharp decline in agricultural growth in East U.P. on account of severe cuts in the supply of power for pumping water, which was diverted to West U.P. to satisfy the powerful farm lobby (Shah, 2001). The known potential from Green Revolution is yet to be fully realized in the eastern states where the productivity of irrigation water can be raised significantly both on account of high precipitation which supplements irrigation water and the controlled nature of irrigation from groundwater sources which are abundant in this region: Whereas eastern India has over one-fourth of India's usable groundwater resources, less than one-fifth of this is developed.

Sources for Raising Water Productivity

Although research focused on water productivity is relatively new, the gap between the available knowledge to increase water productivity and its beneficial application appears quite large (CGIAR, 2001*a*). In the case of India at present, such a gap between knowledge and its application is greater in the case of water productivity as compared to land productivity. While property rights for land are well-defined and protected being therefore the single most important factor in providing incentives for farmers to successfully participate in the Green Revolution the same is not true of irrigation. Secondly, the Green Revolution has been facilitated by institutional support such as the provision of extension services, credit and marketing. Thirdly, agricultural price policy, whether for output or for input, has provided the necessary stimulus for stepping up output per unit of land through input intensification. As far as irrigation is concerned, the failures have been massive on all the three fronts, namely, property rights, institutions and public policy. It would be instructive, therefore, to identify the major problems afflicting irrigation in each of these areas.

But, what are the major theoretical possibilities for raising the productivity of irrigation water? An understanding of these would help to place the problems afflicting irrigation in a proper perspective. We have seen that technological changes which raised crop yields per unit of land have so far been the major source of the rise in productivity per unit of irrigation water. Since genetic improvements for raising the productivity of land further are very much in the realm of possibility, they can, at the same time, significantly raise water productivity, especially if such genetic improvements are targeted towards saving water. These improvements include the development of crop varieties with better tolerance to drought, cool seasons which reduce evaporation and evapo-transpiration, and saline conditions. The potential of biotechnology to raise yields and impart stability in adverse environments is well known (CGIAR, 2001*b*). In India, technological changes to improve water productivity by raising crop yields seem to hold a better promise in the short and medium-term than the attempts to improve water-use-efficiency, which can be expected to yield significant results only in the long-run, owing to the severe political constraints, and managerial and institutional bottlenecks besetting these efforts.

Secondly, the productivity of irrigation water can be raised by skilfully supplementing it with rain water, especially in the high rainfall areas, despite the possible losses on account of evaporation and evapo-transpiration. Because of this, agricultural development strategy that accords greater priority to high rainfall areas like the central and eastern regions in the country may result in significant savings on high-cost surface irrigation.

Third, water losses may occur between the point of delivery from the system to the farmer's field due to evaporation, flow of usable water to sinks, pollution, salinity and waterlogging. Such losses cannot eventually be recovered at the basin level. These can be minimized with appropriate management practices, provided there are adequate incentives to farmers for adopting water-saving practices.

Fourth, farmers are known to reallocate land and water to high value crops in response to the changing demand. As incomes rise, consumer demand shifts away from some of the water-intensive crops like rice to water-saving horticultural products. However, policies for trade and price as well those for input subsidies, including policies on irrigation water, would determine whether farmers would be induced to switch over to water-conserving enterprises.

Finally, globalization opens up opportunities for countries faced with water scarcities to specialise in the production and export of water-saving crops and import water-intensive ones. Favourable rainfed areas of Europe, North America and parts of South America may gain a comparative advantage in the export of water-intensive crops (Seckler, 1996). A country like India, faced with water scarcity, can safeguard its food security by entering into collaborative arrangements with certain countries, which have abundant land and water resources but are short of labour and capital and do not have an excessive domestic demand, for the production and exchange of foodgrains and other water-intensive products like sugar. Such arrangements are already working in the case of fertilizers and petrochemicals (Hashim, 2000).

Government versus Market Failure in Water Management

Well-defined and secure or enforceable property rights on irrigation water; a volumetric pricing of surface water as well as electricity for pumping water, are both necessary to induce farmers to adopt water-saving practices. However, under the present system of public management of irrigation, we have failed to ensure these pre-conditions for an efficient use of water. The well-known explanation for government failure in respect of irrigation is a lack of accountability in its management to the user-farmers, and its growing politicization. The most worrisome manifestation of this politicization in the recent period is the political rent-seeking by offering free electricity for pumping water which has led to an over-exploitation of groundwater resources.

Therefore, in this context in considering whether or not privatization of irrigation can provide the necessary correctives it should be kept in mind that market failure is as endemic in irrigation and water management as government failure. Market failure is caused, in the first place, by the widespread prevalence of externalities defined as uncompensated costs or benefits accruing to some owing to the activities of some others. Then, there are high transaction costs, for example, costs of infrastructure for setting volumetric pricing and its enforcement. Thirdly, property rights for irrigation water are insecure and ineffective. This is commonly illustrated by the unequal access of farmers at the head reaches of the system who appropriate much more water than they are entitled to, at the expense of those at the tail end (Perry *et al.*, 1997). Full-scale privatization of irrigation under such circumstances, in the absence of necessary institutional mechanisms for correcting the distortions arising from market failure, can lead to serious inequities and inefficiencies in the use of water.

Irrigation Management Transfer (IMT) to the user-farmers is being increasingly advocated and practiced the world over to provide correctives to the distortions arising from the failure of the market as well as the state. India has embarked upon Participatory Irrigation Management (PIM), under which the management of some of the systems is being turned over to the Water Users' Associations (WUAs). Conservation of land and groundwater resources through rain water harvesting is another area where market failure as well as government failure have been glaring. As a solution to this, Participatory Watershed Development is being evolved and practiced in many countries, including India, for the protection and sustainable use of land and water resources.

Performance of Participatory Irrigation Management

The evidence so far on Irrigation Management Transfer (IMT) in several countries, and on Participatory Irrigation Management (PIM) in India, shows some positive results in terms of an increase in yields, water productivity and farm incomes. Incentives to the farmers to participate and the quality of leadership seem to be the key for the success of these ventures (Guerra *et al.*, 1998). In India, the performance of PIM so far indicates that the unwillingness of the lower levels of bureaucracy to fully transfer powers and functions to the user associations proves to be a major bottleneck in the successful working of these institutions (Brewer *et al.*, 1999). Such a reluctance to delegate powers and functions, by stifling local initiative, prevents the emergence and development of leadership at the grassroots — an essential pre-requisite for the success of these institutions.

Similarly, the experience with watershed development under the new participatory approach in India shows that the immediate results are positive in terms of a rise in the water table, in crop yields, and in the employment and incomes of the participants (Rao, 2000). And, where property or use rights on common property resources are well-defined and protected, there has been a significant decline in the distress migration of labour (Chopra and Gulati, 1997;2001). But there is still considerable uncertainty regarding the long-run sustainability of these institutions, that is, their ability to hold the different social groups together and mobilize resources from the community for the maintenance of structures already created and for further development when the assistance under the current officially sponsored programmes ceases (Rao, 2000; Reddy *et al.*, 2001). This holds true even for a relatively promising experiment towards Participatory Irrigation Management in Andhra Pradesh (Raju, 2000). The performance of Irrigation Management Transfer (IMT) so far in several other countries points to similar uncertainties (Perry *et al.*, 1997).

The present obsession of bureaucracy in India with the fulfilment of physical and financial targets has to give place, firstly, to greater devolution of powers and functions to the user-associations, including their involvement in the renovation and modernization of the systems as well as management by them right from the outlet. Secondly, much greater stress needs to be laid on building up social capital through social mobilization of different groups of user-farmers for a sustainable use of land and water through conflict resolution among them and their capacity building by putting in place an autonomous and credible infrastructure for training. The institutions developed in this process need to be integrated with more formal legal or market institutions for sustaining them (Chopra, 2001). Thirdly, independent and professional evaluations of the on-going programmes can help mid-course corrections that would greatly enhance the fulfilment of programme objectives.

Public Policy

In the area of irrigation and water management, the state should increasingly concern itself with overall public policy to create a conducive socio-economic environment for the efficient and productive use of water. The state should divest itself of the task of *managing* the systems, because its failures have been conspicuous in this area which accounts for the major opportunities available for improving water productivity (CIGAR, 2001*a*).

Governments in the states and at the centre need to accord the highest priority to the renovation/modernization of the existing systems which account for nearly 40 per cent of the irrigated area from the major and medium irrigation projects. The rate of return from such investments would be very high when compared to the projects that create new irrigation potential. Modernization of the delivery systems and the distribution channels for the existing projects would have a high pay-off, as they would facilitate a clear definition of property rights or entitlements of farmers and their effective enforcement. They would also facilitate the adoption of measures by the water users to improve water-use-efficiency and productivity.

The price and procurement policy of the government has an important bearing on the efficiency and productivity of water use through its impact on the cropping pattern. A glaring illustration of this is the price policy for rice and wheat in the recent period which has resulted in the accumulation of over 60 million tonnes of foodgrains — nearly thrice the actual requirement — with the Food Corporation of India (FCI). It has been estimated that for each hectare of paddy raised under irrigated conditions, as much as 16,000 tonnes of irrigation water is applied in the semi-arid North-Western Indian states (Dhawan, 2001), from where a bulk of the rice is procured by the FCI. One can thus imagine how many billion tonnes of scarce water has been wastefully used over the years in the semi-arid plains of India, which could have been saved or diverted to higher productivity uses if the price and procurement policies were better aligned with the emerging domestic and export demand for foodgrains.

The experimental work backed by the ICAR found that 25-40 per cent savings in irrigation water could be achieved through intermittent submergence and transplanting paddy seedlings at about the time of onset of monsoon rains. Similarly, water savings of the order of 50 per cent in the case of drip irrigation and 25 per cent in the case of sprinkler irrigation can be realised (Dhawan, 2001). But, only about one per cent of irrigated area in the country is presently covered by the drip and sprinkler methods of irrigation. These technologies are not adopted by the farmers because water is available at a very low cost or is even free. The incentives are in fact perverse inasmuch as the capital cost of drip and sprinkler irrigation is prohibitively high for the Indian farmer due to various taxes and a high interest rate (Dhawan, 2001).

The debate on water rates in the country so far has centred round the revenue accruing to the government for covering the operation and maintenance charges. The objective of conserving water needs to be brought into the forefront in these discussions. Given that water rates have depreciated very much in real terms, it is not certain that a significant rise in these rates within the feasible range would induce the farmers to save water under the existing system of management. On the other hand, a system of rationing water supplies in water-scarce regions even with moderate charges can induce farmers to increase water productivity, provided the supplies are reliable in terms of quantity and timings as in the case of the Warabandi system in North-Western India (Berkoff, 1990; IWMI, 2001). In any case, raising water rates significantly would not be politically feasible unless it is linked with the performance of irrigation services and made explicit that revenues so raised would be reinvested locally to upgrade the services.

Water Users' Associations and Panchayats should be fully empowered and induced to charge and collect water and electricity rates from the farmers on the basis of the volume of actual consumption where metering is possible, or, on the basis of the quantities of water received or electricity consumed as estimated by the farmers' associations where metering is uneconomical or otherwise infeasible. The revenues so collected should be allowed to be retained entirely by the Water Users' Associations under the aegis of Panchayats for maintenance and development of irrigation services, with possible supplementation from the state governments.

It is clear from the foregoing discussion that given the technology and public policy, institutions concerning water use hold the key to raising water productivity by bridging the vast gap that now exists between knowledge and its application. Water institutions are a relatively new and challenging area of research for social scientists. Also, research in this area is highly inter-disciplinary.

References

Agarwal, Anil (2000): 'Drought? Try Capturing the Rain', *The Asian Journal*, Vol.7, No.3.

Berkoff, D.J.W. (1990): *Irrigation Management on the Indo-Gangetic Plain*, World Bank Technical Paper No.129, The World Bank.

Brewer, J.S., Kolavalli *et al.*, (1999): *Irrigation Management Transfer in India: Policies, Processes and Performance*, IIMI, Ahmedabad & IIMI, Oxford & IBH Publishers: Colombo.

Chada, D.K. (2000): 'Groundwater Recharge: Option for Groundwater Management', *The Asian Journal*, Vol. 7, No. 4.

Chopra, Kanchan and S.C. Gulati (1997): 'Population, Poverty and Environmental Degradation: The Role of Property Rights', in, *The Challenge of the Balance-Environmental Economics in India*, Anil Agarwal (ed.), Centre for Science and Environment, New Delhi.

————(2001): *Migration, Common Property Resources and Environmental Degradation – Inter-linkages in India's Arid and Semi-arid Regions*, Sage Publications: New Delhi.

Chopra, Kanchan (2001): "Social Capital and Development Processes: The Role of Formal and Informal Institutions," Mimeo, Institute of Economic Growth, Delhi.

Consultative Committee for International Agricultural Research (CGIAR), Technical Advisory Committee (2001*a*): "Water and the CGIAR," Discussion Paper.

————(2001*b*): "Water Productivity Research in the Context of CGIAR," Discussion Paper.

Dhawan, B.D. (2001): "Technological Change in Indian Irrigated Agriculture: Diffusion of Water Economising Technologies and Practices," Mimeo, Institute of Economic Growth, New Delhi.

Government of India, Ministry of Water Resources (1999): *Report of the National Commission for Integrated Water Resources Development Plan*, Vol. 1, New Delhi.

Government of India (2001): *Economic Survey, 2000-2001*, New Delhi.

————Planning Commission (2000): *Mid-Term Appraisal of Ninth Five Year Plan (1997-2002)*, New Delhi.

Guerra, L.C., S.I. Bhuiyan, T.P. Tuong and R. Barker (1998): "Producing More Rice with Less Water from Irrigation Systems," SWIM Paper 5, International Water Management Institute (IWMI).

Hashim, S.R. (2000): "From Food Security to Water Security", *The Asian Journal*, Vol. 7, No. 4.

International Water Management Institute (IWMI) (2001): *Basin-Level Use and Productivity of Water: Examples from South Asia*, Research Report 49.

Perry, C.J., Michael Rock, and D. Seckler (1997): *Water as an Economic Good: A solution, or a Problem?*, Research Report 14, IIMI.

Raju, K.V. (2000): *Participatory Irrigation Management in Andhra Pradesh: Promise, Practice and a Way Forward*, Working Paper 65, Institute for Social and Economic Change.

Rao, C.H.H (2000): 'Watershed Development in India — Recent Experience and Emerging Issues', *Economic and Political Weekly*, Vol. 35, No. 45, November 4.

Reddy, V. Ratna, M. Gopinath Reddy, S. Galab and Oliver Springate-Baginski (2001): "Watershed Development and Livelihood Security: An Assessment of Linkages and Impact in Andhra Pradesh, India," (Mimeo), Centre for Economic and Social Studies, Hyderabad and School of Geography, University of Leeds, U.K.

Saleth, R.M. (1996): *Water Institutions in India: Economics, Law and Policy*, Commonwealth Publishers, New Delhi.

Seckler, David (1996): *The New Era of Water Management: From "Dry" to "Wet" Water Savings*, Issues in Agriculture 8, CGIAR, Washington, D.C.

Seckler, David, Upali Amarasinghe, David Molden, Radhika de Silva and Randolph Barker (1998): *World Water Demand and Supply, 1990 to 2025: Scenarios and Issues*, Research Report 19, International Water Management Institute (IWMI), Colombo: Sri Lanka.

Shah, Tushar (2001): *Wells and Welfare in the Ganga Basin: Public Policy and Private Initiative in Eastern Uttar Pradesh, India*, Research Report 54, IWMI.

3

The Dilemmas of 'Water Resources Development'*

RAMASWAMY R. IYER

I

At the outset an explanation is necessary for the inverted commas within which the title "water resources development" has been enclosed. This is to indicate that the term "water resources development" (WRD) has acquired a special meaning among those concerned with water resources in this country and perhaps elsewhere. In the case of products such as steel or chemicals or soap, additions to the supply-capacity are made to meet future demand; by analogy the same approach is extended to water and we talk about "developing water resources", meaning adding to the availability of water. But how can we add to the availability of water in nature? Is it not finite? Yes, indeed. Here we must take note of two other technical terms that the water people use, namely, "available" and "usable" water resources. By "available" water resources they mean the annual flows as measured in rivers, preferably close to the terminal point. Only a part of this is considered actually "usable"—particularly the quantum that has been stored in reservoirs behind dams or diverted through barrages or other similar structures. "Developing" water resources in this context therefore means bringing more of the "available"

* In writing this I have made some use of material from earlier papers, but the theme and perspective of the present piece of writing are different.

quantum into the "usable" category. The National Water Policy 1987 has a proposition urging that this be done[1].

I have reservations regarding the notions of "developing" water resources, as well as the term, "development" as generally used with a strong positive resonance; and I think that the terms 'available' and 'usable' water resources can be used in a much wider sense than we tend to do. However, those points will emerge in due course as we proceed with expounding the generally prevalent ideas. In our current usage, then, WRD generally means large supply-side storage/diversion projects. (It also means groundwater extraction and use, but that is a separate and extensive subject, which involves a whole range of issues. Here we are concerned mainly with surface water and only a brief reference to groundwater will be made.)

II

I used the term "supply-side" because what is involved is a supply-side response to projected demand. Let us therefore take a quick look at the demand-supply calculus. The average annual precipitation by way of rain and snow over India's landmass is 4000 cu km, but the annual water resources of the country, measured in terms of flows in the river systems, have been estimated by the National Commission on Integrated Water Resources Development Plan (NCIWRDP)[2] as 1953 cu km. The groundwater availability has been put at 432 km.

As against those "availability" figures, the NCIWRDP[3] estimates that the annual 'usable' water resources of the country are 690 cu km of surface water and 396 cu km of groundwater, making a total of 1086 cu km. The present quantum of use is put at around 600 cu km. It follows that in national terms the position is not uncomfortable at the moment. However, this will obviously change with the growth of population and the processes of urbanization and "development". The NCIWRDP has made various assumptions with regard to these matters (high, medium and low rates of change), and come to the conclusion that by the year 2050 the total water requirement of the country will be 973 to 1180 cu km under "low" and "high" demand projections, which means that supply will barely match demand.[4] It is the Commission's view that there will be a difficult situation but no crisis, *provided* that a number of measures on both the demand side and the supply side are taken in time.

That precarious balance between supply and demand can, of course, tip over into a crisis if the actual developments fail to conform to the assumptions. Moreover, apart from demand putting pressure on the available supplies, the supplies themselves may also be seriously affected by the growing incidence of pollution and contamination of water sources.[5] Besides, national aggregates and averages are very misleading, as there are wide variations, both temporal and spatial, in the availability of water in the country. Much of the rainfall occurs within a period of a few months during the year, while during that period the intensity is concentrated within a few weeks. Spatially, there is a wide range in precipitation — from 200mm or less in parts of Rajasthan to 11000mm in Cherrapunji. The standard engineering response to these temporal and spatial variations is to propose: *(i)* the storing of river waters in reservoirs behind large dams to transfer water from the season of abundance to that of scarcity as well as from good years to bad, and *(ii)* long-distance water transfers from 'surplus' areas to water-short areas.

Water resource planning in India as in others countries therefore focuses on the setting up a number of large dam-and-reservoir projects. In its response to the Report of the World Commission on Dams, the Government of India, Ministry of Water Resources, stated: "Having made impressive strides since independence in developing our water resources, India proposes to continue with its programme of dam construction to create another 200 billion cu metres of storage in the next 25 years or so to ensure continued self-sufficiency in foodgrain production and to meet the energy and drinking water needs of a growing population."[6]

However, as the title of my talk indicates there are four

main dilemmas that such policy projections give rise to (with connections among them): the money dilemma, the markets dilemma, the environmental dilemma and the human rights dilemma.

III

Let us consider the "*money dilemma*" first. In the late nineties the India Water Vision 2025 exercise (under the auspices of the Global Water Partnership) for presentation at the Hague Forum of March 2000, postulated an investment of Rs. 5000 billion in 25 years, or Rs. 200 billion per year.[7] The NCIWRDP's rough estimate of amount needed for completing spill-over projects are Rs. 70,000 crores in the Tenth Plan and Rs. 110,000 crores in the Eleventh Plan.[8] Not only does this leave no scope for an major new projects, but the difficulty of finding funds of this order necessitates a severe selectivity even with regard to the continuance of what are called 'on-going projects'. The Report devotes a whole chapter to the 'prioritization' of these.[9] Some very hard choices are called for: some projects may have to be accelerated, others restructured and drastically pruned, and yet others abandoned.

What we must accept in a clear-headed manner is that the actual availability of investment funds in the public sector is likely to be no more than a small fraction of the projections made. Nor should we delude ourselves into thinking that the answer lies in private sector investment. National private sector investment in this sector, if forthcoming, is likely to be marginal at best, and we have to be very wary of possible investments by foreign or multi-national corporations; but we shall come to that later. Thus, the pursuit of WRD in terms of large projects is likely to be a frustrating exercise financially and much time will be lost in this process with uncertain prospects of success. The "money dilemma" in other words is that though we seem convinced that a massive construction programme is urgently necessary, the financial resources needed are nowhere in sight: how then are we going to meet future needs as we see them?

IV

Let me turn now to what I have referred to as "the markets dilemma". If engineers and administrators tend to argue for supply-side projects, "liberal" economists and officials of the multilateral financial institutions tend to argue for water markets. To them water is a commodity like any other, governed by the laws of supply and demand: if the state stops meddling and leaves it to the private sector and to the operation of market forces, then supply will meet demand, prices will be right, economy and conservation will be ensured, and conflicts will be automatically resolved by the market. This route is also expected to resolve the financing dilemma that was mentioned earlier. The slogan is: "water is an economic and social good". (I suspect that the preferred formulation would have been simply that "water is an economic good"; the addition of the word "social" seems to be a bit of a compromise, a genuflection in the right direction. Perhaps it can also be regarded as a "sweetener" for making the formulation acceptable.) Yes, water is an "economic good" when it is used for industry or agriculture, and perhaps a 'social good' when used for sanitation or in hospitals or for fire-fighting; but can this be adequate description of water which is basic requirement for human and animal survival? Can water in that basic aspect be reduced to a commodity like cement or steel or fertilizers or soap? Is it not more akin to the air?

The doctrinaire call for 'privatisation' includes allowing the corporate private sector to build and operate dams across rivers for hydro-electric power and/or for irrigation. Assuming that the private sector is interested in investing in such capital-intensive, long-gestation, modest-return projects, how are the environmental and social impacts (which have presented serious difficulties to the state in the past) to be handled by the private entrepreneur and manager? Supply may match demand but resource conservation may receive scant consideration; resettlement and rehabilitation aspects are likely to be given grudging attention only to the extent that resistance by those affected and public opinion compel such attention; and it is naïve to imagine that market forces will obviate conflicts or provide a magical route to their resolution. Whatever the position may be in relation to industry, water cannot be left to market forces. One is not arguing for a dominant role for the state, but the alternative player is civil society, not the corporate sector.

Moreover, in the case of other commodities, properly so described, the expression "market" often refers to the global market. The argument is that everything need not be produced domestically, and that imports should be freely allowed. What application does this have to water? Water is currently not an internationally traded commodity to the extent that oil is — this is subject to a point that I am going to make shortly—and the availability of water in a distant part of the world is of no great relevance to a country or region with water scarcity. People need sources of water close to their homes and lands; inter-country sharing or exchanges in relation to water can at best happen between neighbouring countries.

There is indeed a notion of trading in "virtual water", that is, in commodities, say, foodgrains, grown in water-abundant countries, but this has serious implications. A critical dependence on imports of foodgrains, even if the balance of payments position allows this, may introduce a degree of vulnerability that may not be desirable for a country like India. However, this complex subject cannot be adequately discussed here: the issue is merely flagged for attention.

In the context of the advocacy of water markets ("define water rights and allow trading") it may be asked if water is either a state resource (as governments tend to argue) or a community resource (as many believe), and if what a user (an individual or an institution) gets is a user-right, how can he (or she or it) have a surplus for sale, except temporarily and under special circumstances? This may seem an odd conundrum to raise, because we know that water markets do exist and serve some useful purposes; but there *are* difficult issues involved.

A reference must be made in this context to a serious development. The doctrine that has been prevalent for a long time that rights over water are only use-rights and not ownership rights seems to be giving way to the view that water is a tradable commodity. The latter view is now enshrined in several international agreements, and a handful of giant corporations are trying to gain control over the world's water resources. Doubtless they will receive the full protection of the WTO system, and attempts by member countries to preserve their natural resources from the onslaught of these giants may fall foul of the WTO regime. There seems to be considerable danger of developing countries losing control over their own natural resources including water. Vandana Shiva and Vinod Raina have been speaking and writing about this issue and hoisting danger signals.[10] If this issue is not tackled with determination right now, it is capable of assuming menacing proportions within a short period of time. Vinod Raina will probably argue that it has done so already.

Returning to the domestic sphere, water markets often tend to emerge in the context of the extraction and use of groundwater, and they serve some useful purposes. For instance, they make it possible for irrigated agriculture to be undertaken by poorer farmers who cannot afford to invest in tubewells or borewells. However, there are dangers of unsustainable extraction as also of inequitable relationships between sellers and buyers. Under Indian law, the ownership of land carries with it the ownership of the groundwater under it, subject to regulation and control by the state. To quote Chhatrapati Singh, "groundwater is attached, like a chattel, to land property". Again, "There is no limitation on how much groundwater a particular landowner may draw".[11] It follows that only those owning land can have rights over groundwater; the landless (including communities, tribal and other, who may have been using certain natural resources for centuries) can have no such rights. Further, this legal position leads to inequities of various kinds: a rich farmer can install power-driven tubewells or borewells in his land and their operation can make dugwells in the neighbourhood run dry; he can sell water so extracted to his poorer neighbours even though the water may come from a common aquifer running under their lands; and he can deplete the aquifer through excessive exploitation. The easement right makes regulation difficult.

In the light of the above analysis, the dilemma here is a multiple one:

(*i*) In the domestic context, how to allow limited and regulated water markets to function without inequity and injustice or danger to the resource, and without transferring rights and control substantially away from the community to the state; and how to accept (and indeed insist on) a stringent commercial approach in the context of supply for industrial use, for commercial agriculture, and for luxury consumption by the affluent, without carrying the "commodification"[12] of water too far; and

(*ii*) In the international sphere, how to ensure that the few instances of international trade in water that now occur do not burgeon into a massive bulk trade in water as in the case of oil; and how to protect the rights of the poorer and weaker countries over their own natural resources from predatory corporate giants.

V

Let us proceed to the "environmental dilemma". As we saw earlier, the Ministry of Water Resources has put before the nation an ambitious target of construction of 200 BCM of storage capacity in the next 25 years. It is clear that even if a part of that target is attempted, the impacts on the natural environment and the ecological system will be very substantial. The environmental impacts will of course vary in range and severity from case to case, but most such projects have some common and inescapable consequences, such as:

- Violent disturbance of pristine areas; varying degrees of submergence of land including forests in some cases;
- Impacts on flora and fauna, leading to a reduction in bio-diversity;
- In particular, severe impacts on the fish population in the river;

- The stilling of flowing waters leading to temperature stratification as well as variations in nutrient content and dissolved oxygen, rendering the water inhospitable to aquatic life;
- Drastic changes in the river regime downstream of the dam (reduced flows affecting aquatic life and riparian communities, reduced capacity for self-regeneration, increased pollution levels, reduced recharge of groundwater aquifers, adverse impact on estuarine conditions); and so on.

Some of these effects cannot be remedied or even mitigated; and in some cases, efforts at the mitigation of or compensation for environmental impacts may in turn create further problems. Further, it is clear from past experience that all the consequences and ramifications arising from the damming of a river cannot really be fully foreseen and planned for.

The usual answer to these criticisms is to say that all this can be taken care of through comprehensive environmental impact assessment (EIA) studies; proper cost-benefit analyses based on such EIA studies, and investment decisions with reference to such cost-benefit analyses; and remedial and/or compensatory measures to counter the expected impacts.

Unfortunately, EIAs are notoriously undependable. When they are undertaken in-house by the project planners, the desire to get the project approved may influence the EIA and render it suspect. Even when a reputed external consultancy firm is engaged (as is often the practice), the thoroughness and objectivity of the study cannot be taken for granted. The insidious pressure on the consultant to be "positive" about the project could be very strong: to say this is not to imply that there is collusion between the project-planner and the consultant. The latter has an interest (not necessarily conscious) in coming to the conclusion that the adverse impacts of the project can be remedied or mitigated or compensated for; that the project will still remain viable; and that the overall balance of costs and benefits will be favourable. A consultant who says: "The impacts of this project are too grave to be mitigated or offset: the project should not be undertaken" is unlikely to secure many assignments. It is only a disinterested examination by an independent appraisal agency, say, the Ministry of Environment and Forests or an agency appointed by it, that could be expected to be truly neutral and objective. Even that agency could come under strong pressure from other agencies *within* the government to be "positive" and supportive of 'development'. Further, as already mentioned, it is impossible to foresee all the consequences of such major interventions in nature; nor is it always possible to remedy or mitigate or compensate for the ill effects of such projects.

As for a cost-benefit analysis, it is a flawed basis for decision-making because: (*i*) it is susceptible to manipulation (costs are usually understated and benefits overstated); *(ii)* it is necessarily incomplete and inadequate (not every aspect or dimension can be brought within the ambit of the calculus); and *(iii)* it is morally blind (the infliction of misery on some people is often sought to be justified on the ground that a larger number elsewhere will be benefited; but we shall revert to that later). Moreover, the rationale of a cost-benefit analysis is that the benefits (direct and indirect) justify the costs (financial and social); and that justification tends to get undermined by the fact that the costs are certain to be incurred and are almost always higher than projected, whereas the claimed benefits are often problematic and may not be fully realized. (This is not an argument against cost-benefit analyses; they are necessary; this is merely a plea that their limitations as the basis for investment decisions be recognized. They need to be widened in scope, converted into multi-criteria analyses and improved in quality; and they need to be supplemented by other things.)

The dilemma here, that is, in the environmental context, is that we think that we can "harness" natural resources for human benefit while limiting the harm that we do, and that the "trade-off" is in most cases positive; but this is open to serious question. We may in fact be sitting on the edge of a branch and axing it at the junction to the tree, or perhaps cutting down the tree itself. If, in undertaking major interventions in nature to increase the availability of water for use, we do not know exactly what we are doing and put the ecological system at risk, what will happen eventually to the water? If we forget for a moment the questionable calculus of supply and demand and look at "security" from the point of view of protecting the ecological system and planet earth, we begin to realize that by building a series of large WRD projects we may not be *ensuring* security but *endangering* it.

At this juncture, and by way of digression, two questions posed by a distinguished professor at a conference on the eastern Himalayan rivers at Kathmandu a few years ago may be relevant. Referring to the sustainability debate, he said that the concerns could be captured by the following two questions: (*i*) How far can we go in our draft on natural resources before we cause harm to future generations? (*ii*) How far can we go in dumping pollutants and contaminants in water, soil and the atmosphere before we cause irreversible damage? At first sight these questions seem well formulated, but if we consider them more carefully, we realize that they ask not a *positive* question ("how can we live in harmony with nature?") but a *negative* one ("how much harm and damage can we afford to inflict?"). Behind this formulation lies a certain view of the relationship between humanity and nature, and the philosophy (still prevalent, though no longer stridently voiced) of "conquest of nature". This is the legacy of the western legend of Prometheus who is said to have brought fire to earth in defiance of the gods. Under the influence of that legacy, we are driven by technological hubris to undertake the "harnessing" of nature for "development". That equestrian metaphor is an indicator of an essentially adversarial relationship with nature. In contrast, we have the Indian legend of Bhagiratha who brought water — the river Ganga — to earth in a spirit of prayer. In our tradition, rivers have been regarded as deities to be worshipped, not as horses to be "harnessed" and ridden. (That of course has not prevented us from polluting and contaminating our sacred rivers!)

VI

The fourth and final dilemma is the "human rights dilemma" or simply the human dilemma. The projected benefits of a project have a cost: not merely a financial or an environmental cost, but also a human cost; and while the benefits accrue to one set of people, the "social costs" — a euphemism for suffering — are imposed on another and usually poorer set of people. In the case of most projects, there will be varying degrees of displacement of human settlements, with the attendant problems of resettlement and rehabilitation; this impact often falls on poor and disadvantaged sections, particularly on tribal communities. There are inherent difficulties in resettlement and rehabilitation: a lack of full knowledge of the numbers and categories of people likely to be affected; separation of communities from the natural resource base on which they are dependent; inadequacy of land for land-based rehabilitation; scattering of well-knit communities; resettlement in distant and unfamiliar areas; difficulties with the host communities in the resettlement areas; major transformations in the ways of living, loss of old coping capabilities and the need to learn new skills and ways of living; and so on. However good and enlightened are the rehabilitation policies and "packages", there will inevitably be great hardship and suffering, to which the response of the governmental machinery is rarely adequate, much less imaginative.

It is fashionable to refer to project-affected people (PAPs) as "partners in development", but that sanctimonious formulation bears little resemblance to reality. Efforts to involve them in decision-making and to give them their rightful share in the benefits of the projects that impose hardships on them have either not been seriously pursued or have been unsuccessful.

It must be noted that there is no necessary or inescapable dilemma here. There is no reason why the people likely to be adversely affected by a project cannot in fact be treated as "partners in development" and made to benefit from the project; but the sad fact is that this rarely happens. A draft

National Rehabilitation Policy has been under consideration for over fifteen years, but despite repeated revisions (doubtless to make it blander and blander) it has not yet received the official imprimatur.

VII

Having set forth these dilemmas, we must look for answers. One possible answer would be to say, "No more big projects". Many do say this. That answer seems to get rid of all the dilemmas at one stroke: no need for huge investments, no induction of the corporate private sector, no environmental impacts, no displacement. (The market dilemma will not entirely disappear, but let us ignore that for the present.) However, that simple-seeming recommendation, if carefully thought through, will take us into deep waters. Large dams are not isolated phenomena; they are parts of a complex system. It is possible to marshal an impressive array of evidence against dams; but it is equally possible to build up a strong case against other symbols of "development": coal-burning and nuclear power plants; metallurgical, chemical, hydrocarbon and petrochemical industries and mining complexes; monstrous megalopolises; the exploding automobile population; vast networks of railways and highways built by trenching into flood-plains, drainage channels, fields, forests and wildlife habitats, and by blasting hillsides and tunnelling through mountains; the onslaught on aquatic life by giant trawlers and whaling vessels; the staggering global trade in oil and the ever-present threat of oil-spills; and so on. All these are manifestations of a certain conception of "development" and a related attitude to nature. Is it possible to change these?

All of us are familiar with Mahatma Gandhi's famous remark about the world having enough for everyone's need but not enough for anyone's greed. His idea of "need" was austere; and by "greed" he meant consumption in excess of need so defined. It seems to me that if we are concerned about the sustainability of human life on planet earth and indeed that of the planet itself, the sanest voice that we can listen to is that of Gandhi. What we call "development" is "greed" in his sense, and with the triumph of the 'market forces' ideology, the march of globalization and the irresistible tide of consumerism all over the world, greed seems to have come to stay. At the moment it is difficult to see how a change of direction is going to be brought about and doom averted.

If that is in fact the world we live in and radical change seems unlikely, then perhaps we shall have to be content with the limited, flawed and as yet unimplemented concept of "sustainable development" as generally understood, which implies the postponement rather than the elimination of doom. Even this will call for a moderation of lifestyles from what we might describe as American standards of living (using America as a symbol) not perhaps to Gandhian austerity but to a level of modest prosperity: a difficult enough task, and one that the world as a whole cannot achieve unless America and the West in general are willing to moderate their own lifestyles.

Keeping in view "sustainable development" so defined we should perhaps consider what *can* be done in practical terms and in limited contexts. From that point of view, and in the context of water resources, it seems very necessary to explore all other possibilities of meeting future needs before we consider recourse to large-dam projects. In other words, we may not be able to say "No" to big-dam projects, but we can minimize their number by treating these as projects of the last resort. Such an approach would greatly reduce the severity of the dilemmas mentioned earlier, even if it does not eliminate them altogether.

VIII

That seems a good point at which to end this paper, but some difficult questions remain:

(*i*) What does the expression "projects of the last resort" mean in practical terms? Can we really cut down the number of big projects drastically? How then will our future needs be met?

(*ii*) If some big projects still have to be undertaken, some of the dilemmas mentioned earlier will arise: how should we deal with them?

For answering the first set of questions, we need to examine: *(a)* whether the demand for water will really be of the magnitude estimated by the NCIWRDP, and *(b)* whether, on the supply side, there are other alternatives or options.

Taking "Demand" projections first these are generally based on current patterns of water-use with some adjustments for improvements in efficiency and resource-conservation, and on prevailing notions of "development". It is self-evident that with a growing population, an increasing pace of urbanization and the processes of "development", the demand for water must necessarily go up very sharply, and that the needed supplies must somehow be found. If in fact water is a scarce resource that is becoming scarcer because of increase in population, should not the consciousness of this scarcity and the impending crisis guide our planning? I suggest that the approach common in the case of other consumer or industrial goods, of projecting demand and providing the supply through production, is inappropriate in the case of water. Here we need to start from the recognition of finite availability and learn to live with it. With that kind of reversal of approach, the "demand" projections may undergo drastic changes.

Regrettably, even the NCIWRDP does not attempt such a radical re-examination. It assumes slow improvements in irrigation efficiency (from the existing level of 35-40 per cent to 50 per cent by 2025 and 60 per cent by 2050), and a very modest increase in yields from irrigated agriculture (from 3 tonnes per hectare in 2010 to 4 by 2050).[13] It also expects a saving of no more than 20 per cent over current rates of use in the industrial demand for water.[14] As for drinking water and domestic uses, the Commission would like to bring the norms for rural and urban areas closer, and projects an increase over the present norms, but does not ask whether these norms can be scaled down. That may seem a shocking suggestion, but the Commission itself recognizes that actual water supply varies widely from 50 lpcd to 800 lpcd.[15] When we talk about the existing average supply of 182 and 103 lpcd, respectively for Class I and Class II cities, we must not forget that those average supplies are not equitably distributed: there is minimal consumption by the poorer sections, fairly heavy use by the middle classes (perhaps in the region of four to five times the average figure), and profligate use (much in excess of the 800 lpcd mentioned above) by the affluent. Should we aim at *improving the norms for supply* from the current figures to an average of 200 lpcd for urban areas and 150 lpcd for rural areas as proposed by the NCIWRDP, or at maintaining or even reducing current norms and *enforcing economies on the middle and upper classes, whether rural or urban*? Doubtless the latter would be an extremely difficult undertaking. However, should we attempt this, or should we forthwith seek supply-side solutions? Again, what will the demand projections look like if irrigation efficiency improves from the current level of 35-40 per cent to say 65 or 70 per cent instead of the 60 per cent assumed by the NCIWRDP, and earlier than the year 2050? Is that utterly impossible? Further, should we be content with a yield of 4 tonnes per hectare from irrigated agriculture? Is it possible to raise this to 6 or 7 tonnes? Should we not insist on the multiple recycling of water by industry, allowing no more than 10 per cent make-up water? Should we not strenuously promote improvements in efficiency and technological innovations in every kind of water-use to maximize what we get out of each drop of water? Apart from minimizing waste, should we not recognize that domestic and municipal waste is also a source from which water for some uses needs to be extracted? And if we do all this, will the demand picture remain the same?

In this context, I must mention a study by Kanchan Chopra and Biswanath Goldar entitled, *Sustainable Development Framework for India: The Case of Water Resources.*[16] This makes projections of future water needs under three different "scenarios", namely, "BAU" or Business As Usual (meaning 'carrying on as we have been doing'); 'High Growth'; and 'Sustainable Development'. Under the Sustainable Development scenario, the study takes into account possibilities of demand management, resource conservation, economies in water use, technological improvements, etc. and water requirements for environmental protection purposes. The study is indeed a commendable effort, and in my view it is better than the India Water Vision 2025 of the India Water Partnership and even the work of the NCIWRDP. However, while the study is very good as far as it goes, it is (quite understandably) not a radical enough departure from the conventional approach. It still begins with economic growth (as commonly understood) at certain desiderated rates, projects water demands on that basis, tries then to bring in the "sustainability" criterion, and applies corrections that range from limited to modest to significant.

That comment is likely to be countered with the question: "Is that not the right thing to do? How else would you have proceeded?" I would have liked to envision a future on lines very different from the prevailing notions of "economic growth" and "development", but that was not the task that the authors had undertaken, and we cannot fault them for not doing something that they did not set out to do. Moreover, any radical departure from current thinking will undoubtedly be fraught with serious difficulty and will pose formidable challenges. We could refuse to face them; we could assume ("realistically", as some may say) that no more than moderate improvements in efficiency and economy in water-use are likely to happen; and we could assume further (even more "realistically") that there will be no change in our ideas of "development" (that is, the pursuit of higher and higher "standards of living", with all countries aspiring to the condition of America); under those assumptions there will indeed be a horrendous water scarcity necessitating massive supply-side projects. We must then also "realistically" acknowledge that sustainable development is an impossibility, and that planet earth (and along with it humanity) is doomed. Instead of perplexing ourselves with such apocalyptic forecasts, will it not in fact be *more realistic* to recognize that given the precious nature of this life-sustaining element and its finite supply on this planet, a tremendous effort needs to be made not merely at efficiency, economy and conservation, but also at a re-ordering of our lives?

Turning to the supply side, large-dam projects are not the

only answer; alternatives and options are indeed available. From the perspective of national planning of large-dam projects it is not that some other possibilities can be regarded as alternative routes which we must choose in an 'either/or' sense. Both options may need to be used. However, in a given case there may be a choice of going in for a big dam or preferring other possibilities; and a number of choices of that kind, exercised in several cases, may mean that fewer dams are undertaken. That was what I meant by the expression 'projects of the last resort'.

We must also shake ourselves free of the usual engineering conventions of defining "available water resources" in terms of flows in rivers, and "usable water resources" in terms of what is stored behind a dam. What is available in nature is rainfall, not just river-flows; and while storing river waters behind a dam doubtless converts "available" water into "usable" water, so does *in situ* rainwater harvesting (that is, "catching the raindrop as it falls") and local watershed development. These are also part of the supply-side answers to the demand. (The Chopra-Goldar Study does take this into account?)

Fortunately, many successful examples of such initiatives are available. Achievements such as those of Anna Hazare (Ralegan Siddhi village in Maharashtra), the late P.R. Mishra (Sukhomajri in Haryana), Rajendra Singh and the NGO Tarun Bharat Sangh (several hundred villages in Rajasthan), and governmental efforts on a much larger scale in Madhya Pradesh under the leadership of Chief Minister Digvijay Singh, have become well known. These are not "small" instances but significant developments. If these examples could be replicated in thousands across the country (wherever feasible), they could be far more significant components in national water planning than we can now imagine. A veritable transformation of the water scene may result.

Instead of long-distance water transfers, whether inter-basin or intra-basin, we need to look at the various drought-prone and arid areas of the country severally and in each case explore local possibilities of water-harvesting and conservation, keeping in mind what has been achieved in some places (for example, Rajasthan, Madhya Pradesh, Gujarat), and thinking of recourse to external water only as a last resort. The same approach applies to urban centres. They should not be helplessly dependent on water from distant sources but must seriously explore possibilities of local augmentation.

In the case of electric power also, all options including demand management, energy-saving, increasing output from capacities already installed, minimizing energy inputs through technological improvements and innovations, extensive decentralized generation through biomass (integrating agriculture), wind/solar/tidal energy, etc., need to be explored, minimizing the need for large centralized generation. The approaches advocated by Dr. A.K.N. Reddy, Girish Sant and K.R. Datye[17] deserve more careful consideration than they have so far received.

It must be mentioned however that even small and modest water-harvesting or watershed-development activities are not without their effects on the environment (whether benign or adverse). These will probably be relatively minor, but they do have to be looked at. Similarly, even such local, limited initiatives could lead to conflicts, though they may not be as bitter and divisive as the conflicts generated by big projects and may be more amenable to resolution. Mechanisms for the resolution of conflicts must be worked out and made part of the institutional arrangements relating to the initiatives.

IX

How should we proceed with the planning and implementation of those large projects that are still felt to be necessary? I do not need to say much on the techno-economic aspects. All of us know the ills that have plagued such projects in the past: poor planning; slipshod, unprofessional processes of evaluation and investment decision; endless "time and cost over-runs"; the perpetual "spill-over" problem, with several projects continuing from Plan to Plan and never reaching closure; a thin spreading of limited resources over far too many projects; the absence of any post-project evaluation; the non-generation of revenues even to cover the running costs (O&M), much less capital-related costs or further investments; and so on. It is clear enough that we need far more professional planning; that it should be inter-disciplinary, integrated and holistic, and should include a consideration of options and alternatives; that the processes of examination and investment decision should be far more stringent; that there should be a willingness to say "No" to those projects that fail to meet the prescribed requirements; that in addition to an economic IRR, a financial return criterion should be added; that there should be machinery for proper monitoring and for post-completion evaluation; and so on. All this is familiar ground. Many Committees and Commissions have said these things. We can only hope that the future will be better than the past in this regard. However, something more needs to be said regarding the environmental and human concerns, which even now receive only grudging acceptance as externally imposed constraints.

I have already said that project-planning should be inter-disciplinary, integrated and holistic. Assuming that that prescription is followed, what is needed thereafter is a thoroughly reliable, rigorous EIA as part of the investment decision processes. I had explained earlier why EIAs are often unsatisfactory. To make these truly professional they should be independent both of the people preparing the projects and of those approving them. EIAs should be made a professional undertaking akin to the practice of law or medicine or audit, with a statutory charter under the Environment Protection Act, a professional council, codes of conduct, disbarment provisions, etc. Such institutional arrangements are no guarantors of quality or integrity, as we know from our experience in relation to the other professions mentioned; but they will mark a great advance over the present situation.

There should also be an independent, statutory Environmental Regulatory Authority established under the Environment Protection Act. Many of the functions now performed by the ministry should be transferred to the Regulatory Authority.

What is particularly important is that the project proposers should not be the paymasters for EIAs; the payments should be made by the Regulatory Authority (or an office established under it) and recovered from the project sponsors, or alternatively, the latter may be required to pay on the recommendations of the former. A further possibility is to provide that the EIA agency in each case will be nominated by the Regulatory Authority from panels maintained by it. (This would be similar to the maintenance of panels of auditors by the CAG for the purpose of the audit of public enterprises.)

It is of course necessary to ensure that the regulatory system does not degenerate into one of bureaucratic control. It is of special importance to this Conference, to examine further how can we widen the scope of EIAs so as to cover not merely a few "environmental impacts" in a narrow sense but the implications for the ecological system as a whole? INSEE needs to spell this out and see that its concerns are included in the charter that is written for EIA professionals.

X

For such projects, public hearings have now become a statutory requirement under the Environment Protection Act, but this has not become fully effective yet. When it does, the affected people will undoubtedly make use of the opportunity to make their concerns known. "People's participation" and "consultation" of affected people are principles subscribed to by all, and are part of the prevailing policies and guidelines, but these practices do not come easily to project planners and implementers. It is fashionable to talk about "stakeholder consultation (or participation)". This forms part of the Dublin-Rio principles, and has gained greatly in currency in recent years. Unfortunately, both the terms of that phrase are dubious.

"Participation" can vary from the full involvement of the people from the earliest stages of planning (putting people at the centre) to the mere formality of asking for comments on a plan, programme or project prepared entirely within the governmental machinery, with no serious intentions of making any significant changes. (A similar point could be made with reference to the implementation stage as well.)

The term "stakeholder" is flawed and has great potential for misuse. First, it is a notion drawn by analogy from prospecting for oil or minerals and carries the connotation of an individualistic claim with the underlying implication of contestation. Secondly, it is an ethically neutral concept that lumps together every person or party having any kind of connection or concern with the project. Not only those who are likely to be adversely affected by the project or expect to enjoy the benefits that it will bring, but a wide range of others who are concerned with it in one form or other come within the ambit of the term. Thus, politicians, bureaucrats, engineers, donor agencies, consultants and contractors are all "stakeholders". The interests and concerns of these diverse categories may not in all cases be benign and legitimate, and some may have a more vital "stake" than others, but the term "stakeholder" makes no distinctions: it legitimizes and levels all kinds of "stakeholding". Everyone is a stakeholder, and the primacy of those whose lands and habitats are taken away and who suffer traumatic uprooting is not recognized by the term.

Even taking only two categories, namely, project-affected people and prospective beneficiaries, the vital difference between the two tends to get blurred by the bland assimilating term "stakeholders". There is a cruel irony in describing the involuntary and helpless victims of a project as "stakeholders", and this is compounded when they are put on the same footing as those who stand to benefit from the project. Let us not forget that while in the case of the former *existing* rights (that is, natural and often centuries-old rights of access and livelihoods) are taken away, in the case of the latter the project, by diverting river waters through canals, confers *new* rights not earlier enjoyed. The former are *stake-losers*, whereas the latter are *stake-gainers*.

A standard response to the hardships inflicted by such projects on the affected people is to say that while everything must be done to mitigate their hardship, development does involve costs and that some groups may have to accept a measure of hardship ("sacrifice") in the larger interests of the nation. Without entering into a detailed discussion of the fallacies involved in this line of argument, let us merely note that 'sacrifice' is the wrong word to use for an involuntary displacement from land and homestead. Gandhiji, in whose view the Benthamite doctrine of the maximum good of the greatest number was immoral, would surely have refused to countenance the 'sacrifice for development' argument. Nor is Pareto optimality an adequate answer to this: it is not enough to say that while some are enabled by state action to acquire wealth others must be at least not worse off. That too is injustice. The project-affected persons (PAPs) must actually benefit from the project.

XI

In regard to both the environmental and the human aspects, there are very useful recommendations in the *Report of the World Commission on Dams*[18], but the Ministry of Water Resources have comprehensively rejected those recommendations. They feel that the requirements are too stringent and that no project will go through if they are adopted; in particular, they are very apprehensive of the principle of "free, informed prior consent" by the affected people and the "rights and risks" approach that the WCD has recommended. The pity is that much of what the Commission has recommended is already part of our own policies and guidelines; and where specific procedural recommendations are likely to cause practical difficulties, suitable adjustments could have been made. The summary rejection of the WCD's recommendations is a seriously retrogressive step. I have dealt with this in detail in an article in the *Economic and Political Weekly*[19], and shall not cover that ground here. The Ministry's critique of the WCD Report contains some points that need careful answers. It is sad to see that having taken several steps forward during the last decade or two, the Ministry (and others of like opinion) have now rejected even accepted principles such as equity, consultation, participation, consideration of options and alternatives, sustainability, etc. Efforts need to be made to rectify that grievous error. If the Ministry of Water Resources and other ministries concerned at the centre, the state governments, the Indian engineering community and their professional *confrères* abroad could be persuaded to look upon the WCD's Report with less jaundiced eyes, they will find much wise counsel in it.

XII

Summing up, the dilemmas I started with can be escaped, or at least made more manageable, by the following steps[20]:

- Reverse the usual approach of proceeding from projections of demand to supply-side answers in the form of 'water-development' projects; start with a recognition of finite supply and learn to live with it; shift the focus from 'water resources development' to 'water resources management';
- On the supply side, reverse the ranking of big 'WRD' projects as primary and local water-harvesting and watershed-development programmes as secondary and supplementary; treat the latter as primary and the former as projects of the last resort;
- Where a big project is put forward as necessary, make the planning inter-disciplinary from the start, with all environmental, ecological and human concerns fully internalized; and put it through a stringent process of comprehensive and rigorous evaluation to make sure: *(a)* that in itself it is a good proposition, and *(b)* that in comparative terms it represents the best choice out of the available options and alternatives[21];
- Make the EIA a truly independent and professional activity backed by a statute, a professional code, disbarment provisions, etc., under the supervision of a statutory Environmental Regulatory Authority; distance the EIA (and payments for it) from the project planners and approvers;

- Transform the Cost-Benefit Analysis into a careful, comprehensive and sensitive multi-criteria analysis;
- Give primacy to the affected people, make them part of the planning and decision-making process from the start, and give them the first rights over the benefits of the project; and
- Adopt the 'rights and risks' approach[22] and the principle of 'free, informed prior consent' recommended by the WCD, with such practical modifications as may be found necessary.[23]

The approach and procedures suggested above will of course apply to future decisions, but there is an existing backlog of completed and current projects where the environmental and human impacts have not been fully recognized or properly dealt with. The clearance of that backlog of deficiencies should be the first priority, and should take precedence over new project decisions.

Notes

1. Government of India, Ministry of Water Resources, National Water Policy 1987, paragraph 3.1.
2. *Integrated Water Resource Development – A Plan of Action, Report of the National Commission on Integrated Water Resources Development*, Government of India, Ministry of Water Resources, September 1999 (hereinafter referred to as "*Report*"). For the figures cited, *see* pp. 23, 31, 37, 40.
3. Treating those two figures as distinct could conceivably involve some double counting in view of the complex interactions between surface water and groundwater; but we need not go into that question here.
4. *Report*, p. 69.
5. The question of "water quality" is another important subject that falls outside the purview of this paper.
6. Government of India, Ministry of Water Resources, Letter No. 2/WCD/2001/DT (PR) Vol. III dated 1 February 2001 to the Secretary General of the World Commission on Dams.
7. *India Water Vision 2025 – Report of the Vision Development Consultation*, India Water Partnership and Institute for Human Development,

published by the Institute for Human Development, New Delhi, 2000, pp. 2, 70.

8. *Report*, p. 242.
9. *Report*, Chapter 10.
10. Presentations at: (*i*) International Conference on Globalization, Environment and People's Survival organized by the Research Foundation for Science, Technology and Ecology, New Delhi, 29 September-1 October, 2001; and *(ii)* WISCOMP Symposium on 'Conflict Resolution: Trends and Prospects', New Delhi, 3-7 October 2001.
11. *Water Rights and Principles of Water Resources Management*, Chhatrapati Singh, Indian Law Institute, N.M. Tripathi Pvt. Ltd. Bombay, 1991.
12. A term that has come into use, though (if a special term is considered necessary) I would prefer 'commoditization'.
13. *Report*, pp. 57, 58.
14. *Report*, p. 62.
15. *Report*, p. 60.
16. *Sustainable Development Framework for India: The Case of Water Resources* by Kanchan Chopra and Biswanath Goldar, Institute of Economic Growth, New Delhi, for The UN University, Tokyo, October 2000.
17. Amulya K.N. Reddy and Girish Sant; and K.R. Datye: Submissions to the Five-Member Group on the Sardar Sarovar Project, *Report of the FMG*, Vol. II, Appendices, Ministry of Water Resources, 1994.
18. *Dams and Development – A New Framework for Decision-Making*, The *Report of the World Commission on Dams*, Earthscan Publications Ltd. London, November, 2000.
19. "World Commission on Dams and India: Analysis of a Relationship", *Economic and Political Weekly*, Bombay, 23 June, 2001.
20. In this summing up I have not referred to the water markets dilemma, because the formulation of that dilemma at the end of Section IV is also a statement of the solution.
21. In such a comparison, the minimization of environmental and human impacts should be an important consideration; it may be useful to include "least displacement" and "minimum environmental impact" among the criteria for selection.
22. WCD suggests that we should ask ourselves "Whose rights are affected?" and "Who bears the risks?"
23. In addition to the rights and risks of the people who are affected, we need to consider also the rights of flora and fauna, of nature in general, and of the river itself. The word "rights" is of course a figure of speech here, as these mute entities cannot speak for themselves or give their "free, informed prior consent". Those concerned with environmental issues will have to speak for them in the processes of environmental appraisal and clearance.

4

Ecology and Economics in Sustainable Water Resource Development in India

JAYANTA BANDYOPADHYAY AND BIDISHA MALLIK

The survival of all forms of life on earth depends on the availability of water. Supply of this key natural resource in acceptable quality, in adequate quantity, and at the required place and time, is of utmost importance, whether in the world's megacities or in a tiny human settlement in a remote mountain area. With a rapid growth in the demand for water in all parts of the world, knowledge of the ecology of water resources became popular in the 1970s. In the decades that followed, the issue of water security attracted greater significance in research and global policy platforms (Gleick, 1993; Cosgrove and Rijsberman, 2000) along with a growing interest in the economics of this natural resource (Briscoe, 1996; Gibbons, 1986). Not only has the direct economic contribution made by water to human economic activities of agriculture and industry been highlighted but an attempt to evaluate, the less obvious contributions provided by water in the form of ecosystems services have also attracted a great amount of research interest. Over the last few decades, water indeed has evolved as an important object of ecological economic research. In India too, there is an emerging interest in water resources stemming from the interdiscipline of ecological economics. In this paper, some of the possible roles that ecology and economics can jointly play in the promotion of sustainable development of water resources in India, will be addressed. However, before going into the details of such an agenda, it is important to present a description of the quantitative availability and requirements of water resources of India. In this context, the areas where ecology and economics can jointly play useful roles in guiding policy for the sustainable development of this resource will be explored.

The Hydrological Cycle and Availability of Freshwater

The earth is a planet rich in water and its hydrosphere contains about 1,386 million cu km of water. Some parts of this continuously move and this movement takes place along paths described by the global hydrological cycle. All the processes of renewal of water resources, used by all plants and animals, including humans, are governed by this cycle, which is shown schematically in Fig. 4.1. About 97.5 per cent of the earth's water resources are saline while freshwater accounts for only 2.5 per cent of the balance (Shiklomanov, 2000:11).

Desalinisation does provide a technical path to transform part of the seawater into freshwater and, thus, enhance the amount of available freshwater. However, at the moment this is an expensive process, which restricts the scope for its wide application. Thus, for all practical considerations, the percentage of freshwater on Earth would remain at this level of about 2.5 per cent of the total water in the hydrosphere. Of this total stock of freshwater, about 68.7 per cent is locked up in the form of ice and permanent snow cover. Groundwater accounts for 29.9 per cent of the total global freshwater, while sources of surface water like rivers, lakes and reservoirs account for only 0.26 per cent of the total stock of freshwater on earth (Shiklomanov, 2000:11). These sources of water, in rivers, lakes and reservoirs, have historically constituted the main supplies for the direct consumptive human needs. This quantitative picture of the hydrological cycle will remain almost the same, unless the global climatic conditions undergo a substantial change.

Water reaches the earth's surface in the very minute forms of raindrops, ice-rain and snowflakes. As shown in Fig. 4.1, it is estimated that in a year approximately 568,000 cu km of water

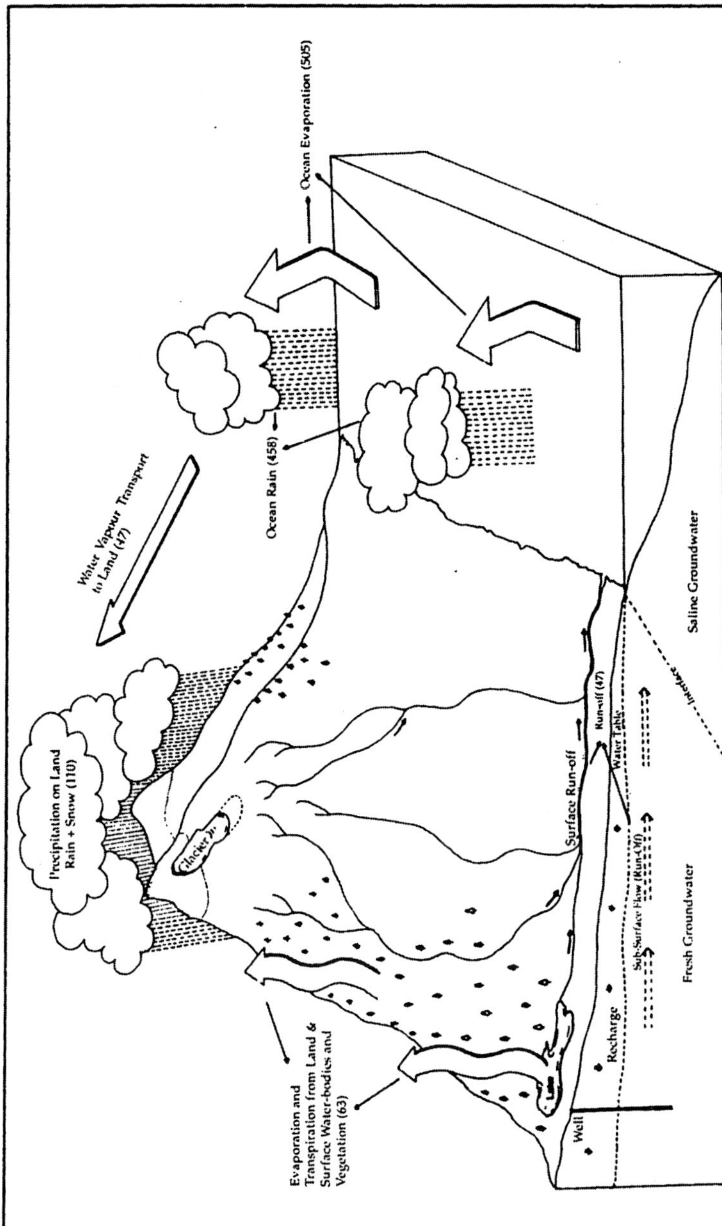

Fig. 4.1: Water Reaches the Earth's Surface

is evaporated and precipitated back on the earth's surface, both on the oceans and on the land (Ayibotele, 1992:1.11). Of this, about 42,750 cu km of water is estimated as the volume of the annual freshwater supplies for the terrestrial surface of the earth. This is the difference in the estimated volumes of total annual precipitation on the land surface and the evapo-transpiration from the same. The share of the Asian continent in this water resource is about 13,500 cu km, followed by South America at 12,000 cu km (Shiklomanov, 2000:18). However, in spite of receiving the largest aggregate annual supplies, the actual year round supply of freshwater in Asia is not very adequate. Raskin *et al.* (1997:27) in their assessment of freshwater resources of the world, have warned that in the coming decades, China, Central Asia, South Asia, the Middle East and North Africa would face conditions of water stress.[1]

Water Resources of India

India occupies about 2.45 per cent of the terrestrial surface of the earth and receives about 4 per cent of the global precipitation. It is the second largest country in the world in terms of population with about 17 per cent of the world's population living in it. Thus, on a macro scale, the precipitation it receives is larger than the global average. In a year, the Indian landmass receives a total precipitation of about 4,000 cu km, which provides the country with an estimated total available water resource of 1953 cu km (NCIWRDP 1999:12). With this amount, the country may be seen as being in a comfortable position as far as the macro picture is concerned. On the other hand, it is comparatively in a worse position, when the water situation is seen through the per capita availability (Dyson, 2001:341). In the 50-year growth-period since independence, the annual per capita availability of water in India has steadily dwindled from 6008 cu m in 1947 to 2266 cu m in 1997 (TERI, 1998:44) and the declining trend is expected to continue for a few more centuries. This figure of the annual average availability of the water resources in the country is still not very bad when compared with many other countries. However, the magnitude of the problem becomes clear when the highly uneven spatial and temporal distribution of this precipitation is considered. The scenario from the stage of precipitation to the creation of the actually utilisable water resources passes through several steps, some of which relate to the hydrological cycle, some others to the human induced management practices.

From Precipitation to Annual Water Availability

India receives a mean annual precipitation of 1170 mm, making a total average annual precipitation of about 4000 cu km in the form of both rain and snow (NCIWRDP, 1999:12). The interaction of the rain bearing monsoon winds with the mountain ranges of the Himalaya and the Western Ghats largely determines the distribution of precipitation over the various parts of the country. The spatial distribution of this precipitation is very uneven as is shown in Fig. 4.2. For example, the annual average precipitation varies from a high of about 11,000 mm in the Meghalaya Plateau in the east to a low of about 110 mm in the desert areas of Rajasthan in the west. The interaction of the monsoon with the highlands and mountains causes quite a high level of precipitation in the north-eastern parts of the country and on the western slopes of the Western Ghats. For the rest of the country, the precipitation levels are not very high and in many areas scarcity of water is common.

This large spatial variation is further compounded by an acute inequity in the temporal distribution of precipitation. Under the influence of the south-west monsoon, about 75 per cent of the annual precipitation occur during the 2.5 months, spread over July to September. Quite a large part of the precipitation occurs in bursts, as a result more than 50 per cent of the entire annual precipitation is received over a total time span of 15 days an with an actual precipitation time of about 100 hours (NCIWRDP, 1999:12). Such extreme variability in the time, location and intensity of precipitation is the reason for the regular floods or water scarcity, dividing the geographical area of the country in a matrix of regions with their characteristic water management challenges.

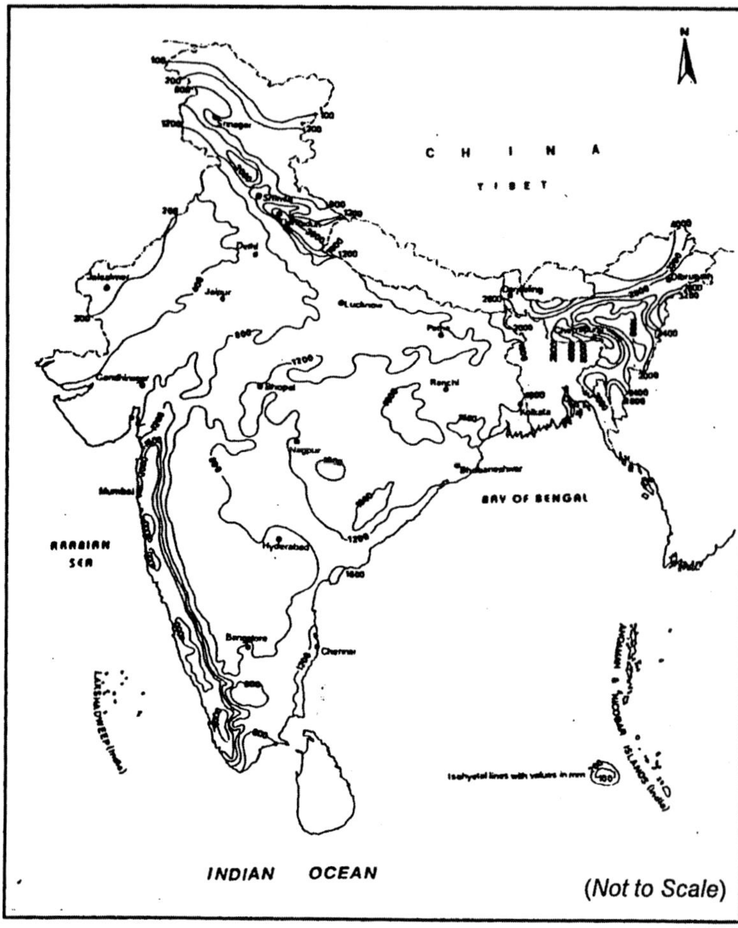

Fig. 4.2: Distribution of Annual Precipitation in India

Of the total annual average precipitation of about 4000 cu km received by India, the quantity of water annually available to the country has been officially estimated at 1953 cu km (NCIWRDP, 1999:12). Another estimate of 1879 cu km has come from Gupta *et al.* (2000). However, the figures of annual aggregate water availability of a large country with drastic variations in rainfall are of little practical use in guiding daily management decisions in the real life situations. In order to understand the challenges in water resources development in a more realistic way, a basin-wise division of this resource will be useful.

Surface Water Availability

For the assessment of the availability of surface water, the country has been divided in 24 convenient areas, mainly on the basis of river basins (Fig. 4.3). The total water availability of 1953 cu km in the country is the sum-total of the water availability in these 24 basin-areas (Table 4.1). In generating the total annual run-off, the monsoon makes the most important contribution. On a very approximate basis, in the rivers of Peninsular India, the monsoon rainfall during June to September accounts for about 90 per cent of the annual run-off, while for the Himalayan rivers, they account for about 80 per cent of the same (Bruijnzeel and Bremmer, 1989:49). The part of the available water flow in a river that can be utilized has largely been estimated on the basis of the capacity of possible storage sites. For example, while the Brahmaputra has an annual average flow of 629 cu km, the net utilisable flow from this river is estimated at only 24 cu km (NCIWRDP, 1999:37). The water flows in both the Himalayan and Peninsular streams reveal a great fluctuation in their flow pattern during the alternate wet and dry seasons. Pre-monsoon dry-season flows of streams are often below sustainable levels, which become distinct in the drier Peninsular parts. The heavier rainfall and the snowmelt make the streams emerging from the Himalaya, the largest contributors of annual run-off in the country. The great temporal and spatial inequity in precipitation poses complex and difficult challenges in ensuring sustainable and equitable supplies of water in all parts of the country. The enormity of the challenge will be clear from the fact, that the amount of water available from both surface and groundwater sources in the Ganga-Brahmaputra river systems is well-nigh two-thirds of the same for the whole country. Even on a smaller scale, in regions lying north and south of the main river in the Ganga sub-basin, sharp inequities in the availability of water

Table 4.1: Surface water endowment in the various river basins of India (in cu km)

River basin	*Area in Sq kms*	*Total Resource*	*Utilisable Resource*
Indus	321289	73.31	46
Ganga-Brahmaputra-Meghna	1101242	1202.43	274
Ganga sub-basin	862769	525.02	250
Brahmaputra sub-basin	197316	629.05	24
Meghna (Barak) sub-basin	41157	48.36	
Subarnarekha	29196	12.37	6.81
Brahmani-Baitarani	51822	28.48	18.3
Mahanadi	141589	66.88	49.99
Godavari	312812	110.54	76.3
Krishna	258948	69.81	58
Pennar	55213	6.32	6.86
Cauvery	87900	21.36	19
Tapi	65145	14.88	14.5
Narmada	98796	45.64	34.5
Mahi	34842	11.02	3.1
Sabarmati	21674	3.81	1.93
West-flowing rivers of Kutchh and Saurashtra including Luni	334390	15.1	14.98
West-flowing rivers South of Tapi	113067	200.94	36.21
East-flowing rivers between Mahanadi and Godavari	49570	17.08	(13.11)
East-flowing rivers between Godavari and Krishna	12289	1.81	
East-flowing rivers between Krishna and Pennar	24649	3.63	
East-flowing rivers between Pennar and Cauvery	64751	9.98	(16.73)
East-flowing rivers South of Cauvery	35026	6.48	
Area North of Ladakh not draining into Indus	28478	0	NA
Rivers draining into Bangladesh	10031	8.57	NA
Rivers draining into Myanmar	26271	22.43	NA
Drainage areas of Andaman, Nicobar & Lakshadweep Islands	8280	0	NA
TOTAL	3,28,7260	1,95,2.87	690.32

Source: NCIWRDP (1999:25,37).

is conspicuous. About 10 to 20 per cent of the total surface area of the Himalayas is covered by glaciers while an additional area ranging from 30 to 40 per cent has seasonal snow cover (Bahadur, 1997). These extensive snowfields form the seat of important glaciers, which are principal flow regulators in the major Himalayan streams in the summer months. High mountain areas in Nepal and Tibet constitute a major source of this snowmelt run-off accounting for roughly 32 per cent (145 cu km) of the total flow of the Ganga (Alford, 1992:20).

The effect of snow and glacier in the period preceding the monsoon cannot be slighted as this greatly augments the summer flows in the northern rivers. Snow and glacier melt-water makes a greater contribution to the annual run-offs of the major rivers originating in the western Himalaya than in the eastern and central areas, where rainfall is very heavy. For example, the dry-season flows (March to June) of Sutlej, a tributary of the Indus, which is about 25 per cent of the annual flow of the river, owes much of its significance from snow and glacier melt (Rao, 1975:16). The seasonal variation in precipitation and the onset of spring snowmelt is to a nicety, the reason for the characteristic flow patterns in the Himalayan streams.

Groundwater Availability

As an integral part of the land-water interactive cycle, groundwater is largely an invisible resource, occurring in a range of water-producing geological structures either as free-flowing artesian water or as confined or unconfined, static or dynamic resource. Apart from forming a major and reliable source for water supply, groundwater also plays an important role in moderating and modulating the surface water regime and surface eco-systems (Newson, 1997:72). Any comprehensive basin-wise approach to water resources development, therefore, needs to address sub-surface water dynamics well in advance, as a prelude to a sustainable and integrated water development plan.

Of the total annual average precipitation of about 4000 cu

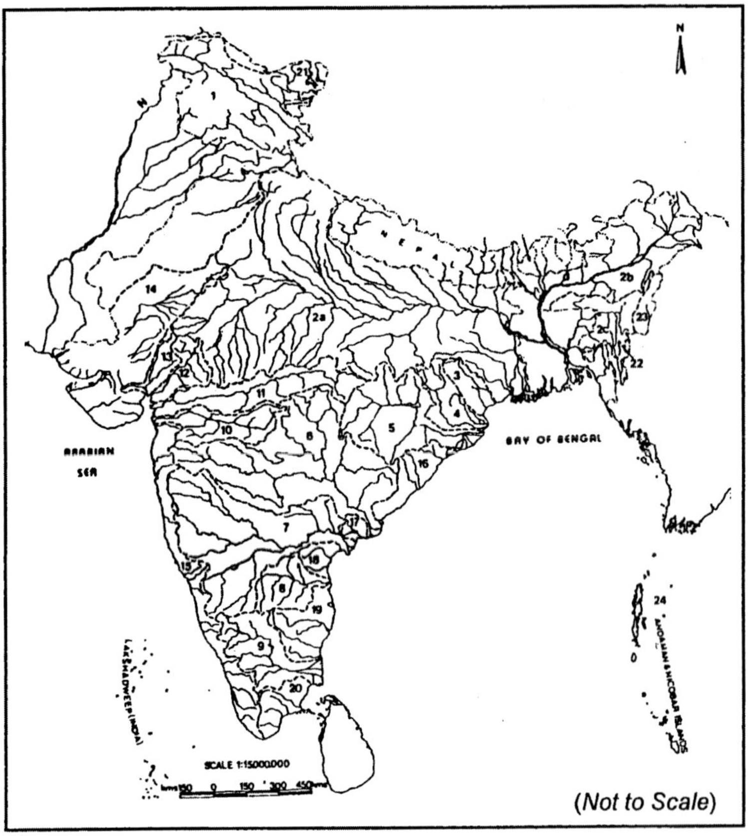

1. Indus
2. Ganga Brahmaputra Meghna (Bearak)
   a) Ganga sub-basin
   b) Barak sub-basin
3. Subarnarekha
4. Brahmani Baitarani
5. Mahanadi
6. Godavari
7. Krishna
8. Pennar
9. Cauvery
10. Tapi
11. Narmada
12. Mahi
13. Sabarmati
14. West-flowing rivers of Kutchh & Saurashtra including Luni
15. West-flowing rivers south of Tapi
16. East-flowing rivers between Mahanadi & Godavari
17. East-flowing rivers between Godavri & Krishna
18. East-flowing rivers between Krishna & Pennar
19. East-flowing rivers between Pennar & Cauvery
20. East-flowing rivers south of Cauvery
21. Area north of Ladakh not draining into Indus
22. Rivers draining into Bangladesh
23. Rivers draining into Myanmar
24. Drainage areas of Andaman, Nicobar & Lakshadweep islands

Fig 4.3: River Basins of India.

Source : NCIWRDP, 1999.

km received by India, roughly 33.3 per cent evaporates and 44.5 per cent flows as surface water. Of the balance 22.2 per cent, 12 per cent remains as a naturally replenishable stock in the form of soil moisture and the remaining 10 per cent as groundwater reserves, in the deeper pores of permeable and weathered soil and fissures in hard rocks (Rao, 1975:92). This part of the precipitation, which percolates into the bedrock, is the available recharge (Holmes, 1992:386). Geologic structure, porosity, permeability and transmissivity of soils and rocks, and localized variations in formation thickness and petrology, influence the renewability and potential for groundwater development.

For example, consolidated formations like clay, silt and sand beds constitute the best repositories of groundwater. Consolidated hard rock formations like trap basalts, slates, granites, etc. have little pore space (secondary porosity)[2] and groundwater can be found only in the weathered mantle or in the joints or fissures of such rock. Large tracts, covering some 1.2 million sq km of the country are underlain by such rock formations (Rao, 1975:102). Semi-consolidated formations with primary porosity resulting from inter-granular pore spaces cover another 0.05 million sq km in states like Rajasthan, Gujarat and Tamil Nadu, where mostly sedimentaries prevail (Rao, 1975:102). Areas consisting of unconsolidated formations (mainly alluvium), reaching a depth of roughly 6000 metres at places and saturated with high quality groundwater extend over about 0.85 million sq km in the Indus and the Ganges-Brahmaputra-Meghna (Barak) basins.

The annual rechargeable dynamic groundwater potential for India is estimated by the NCIWRDP (1999:40) at 432 cu km and by the World Bank (1999b:8) at 452 cu km. About half of this resource belongs to the geo-hydrological set-up of the Ganges-Brahmaputra-Meghna (Barak) basin. Dynamic groundwater aquifers exist in these areas at shallow depths of only 5-15 metres. On account of the significance of groundwater in raising agricultural yields, its role as a catalyst of rural development as well as in poverty alleviation cannot be underplayed. Of paramount importance to most of the rural population, groundwater plays a critical role in maintaining agricultural production during spells of drought (Dhawan, 1988 as cited in World Bank, 1999b:1), it serves 50 per cent of the irrigated area in the country (NCIWRDP, 1999:82), and is responsible for increasing agricultural output by 40 per cent (World Bank, 1999b: 2).

The essentially decentralized, and round-the-year availability of this easily accessible and reliable resource, thus serves as a buffer against drought and normal variations in rainfall, reducing the overall vulnerability of those areas heavily dependent on groundwater (Burke and Moench, 2000: 24). The ubiquity and general high quality of groundwater — as water percolating through the unsaturated zone is improved by filtration and the adsorption of contaminants by clay minerals within the soil or aquifer matrix — makes it an important source of rural domestic water supply. Rural people in India, derive 80 per cent of their domestic requirements from groundwater, whereas about 50 per cent of the urban and industrial supplies are drawn from ground sources (World Bank, 1999b:xv). The annually utilizable dynamic groundwater resource, as estimated by NCIWRDP, is 396 cu km. This amount is arrived at after deducting the estimated requirements of 71 cu km for domestic and industrial needs from 432 cu km and taking a 90 per cent utilization rate in irrigation.

Static and fossil groundwater zones are found below the zone of water level fluctuation. This depth can be more than 450 metres in the alluvium (NCIWRDP, 1999:43). The static and fossil groundwater is taken as a non-renewable resource that does not enjoy the annual recharge, as is the case with the dynamic surface counterpart. The exploitation of static groundwater in India, estimated as 10812 cu km (NCIWRDP, 1999:44), has not advanced very much. The largest potential reserves of the total static groundwater of 10633 cu km are in the alluvium areas. The unconsolidated sediments of the Gangetic trough form the cradle of one of the largest and most prolific reserves of groundwater in the world (Jones, 1989:1) (see Fig. 4.4). The Ganga sub-basin is a major repository of this fossil resource accounting for 7834 cu km of the gross as shown in Table 4.2.

This is a zone of regionally confined, deep freshwater aquifers, which underlie the Ganga plain at depths of 1000 to 2500 metres, and are presently grouped as static resource. As an independent and additional non-fossil resource and likely to be free-flowing with the major recharge zone lying in the Bhabar tract, such rich water-bearing strata are hydraulically discrete from the upper unconfined groundwater aquifer zone, which is currently being exploited (Verghese, 1999:5-6).

The estimated static groundwater potential of hard rock areas in the country, confined mostly to the southern and western parts, stand at a negligible 179 cu km (NCIWRDP 1999:45). These areas are also noted for their poor surface water potential particularly in the non-monsoon period. Thus, both the static and the dynamic groundwater resources of the country are distributed in a very uneven manner. The seemingly great degree of fluctuation involved in the level of groundwater availability and depth from the east to the west as well as from north to south is evinced in Fig. 4.4 and Fig. 4.5. Dynamic groundwater is rechargeable in the country every year, barring certain areas comprising of semi-consolidated formations found in Western Rajasthan, Saurashtra and Kutchh (Rao, 1975:104). The recharge potential gradually diminishes from east to west as from the north towards the south. In Punjab, this potential is roughly 50.75 per cent of the total replenishable groundwater resource, whilst in the middle Ganga plains of Bihar, where the rainfall is ample, this recharge may go up to 84.46 per cent (NCIWRDP, 1999:41). Despite being generously endowed, aquifers of eastern India have experienced a low level of groundwater development. For example, only 9 per cent of the groundwater of Bihar is currently exploited, in spite of a very high groundwater table in the state that varies between 5 and 10 metres from the ground surface. (NCIWRDP, 1999:82).

As in the case of surface water, the development of groundwater resources, has been dominated by the objectives of short term gains made with the help of large-scale, technology-driven, top-down approaches. Uncontrolled and ill-informed extension of the deep tube-well and energized pumping technologies for irrigation in agriculture has, for instance, led to groundwater over-draughts, incurring an abstraction level far exceeding the recharge. The ecological implications of this range from the drying up of dug-wells, dwindling base flows in rivers in the lean season, disappearing wetland areas, land subsidence, a declining groundwater level that falls below the depth of shallow wells, and a mobilization of low-quality water sources into fresh aquifers.

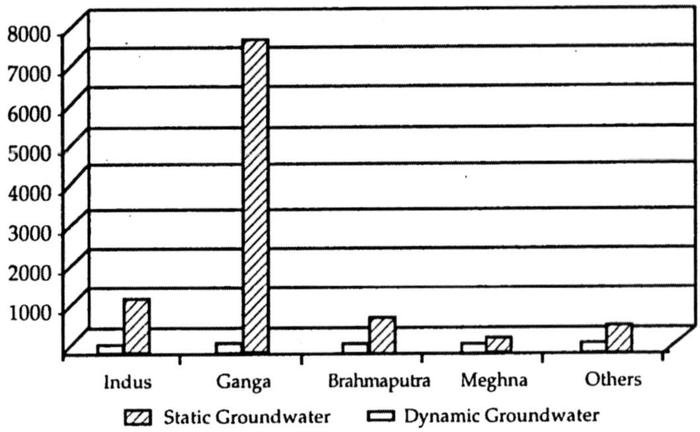

Fig. 4.4: Static Groundwater and Dynamic Groundwater

Source: NCIWRDP, 1999.

Table 4.2: Potential groundwater resources in India (basin/sub-basin-wise)

Basins	*Dynamic Groundwater Resource Potential (cu km)*	*Static Groundwater Resource Potential (cu km)*
Indus	14.29	1338.2
Ganga	136.47	7834.1
Brahmaputra	25.72	917.2
Meghna (Barak)	8.52	101.3
Others (Total)	157.43	621.2

Source: NCIWRDP (1999:42,46).

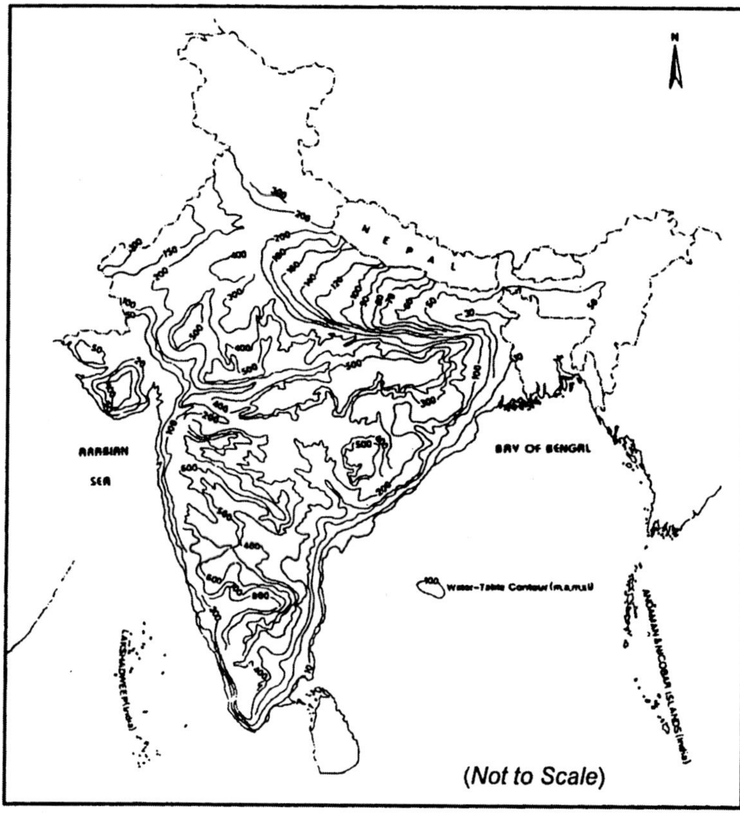

Fig. 4.5: Groundwater Hydrology of India.

Source : National Atlas of India, 1973.

Utilizable Water Resources in India

The utilizable water resources of India, that is the total amount of water that can be utilized annually for the satisfaction of human needs and economic demands, forms a part of the total annually available water resources. For the utilizable surface water sources, several estimates are available. The Irrigation Commission has estimated this as 666 cu km (MIP, 1972), while the National Commission on Agriculture (MAI, 1976) put the figure at 700 cu km. Rao (1975:216) had suggested that the utilizable flow may be half of the total annual flow in a river.

The Ministry of Water Resources (MOWR, 1997) has estimated that the gross utilization of water in India for the year 1990 was 552 cu km, making a per capita consumption of 655 cu m. The total volume of water that can be used for the satisfaction of various human needs and requirements in India has been identified at 1,086 cu km by the NCIWRDP (1999:15). Within this total amount, 690 cu km have been identified as being available from surface sources, while the remaining 396 cu km, are from groundwater sources.

However, even when this picture of the total annual availability at the national level is a comfortable one, this does not act as a guarantee for comfortable access to utilisable water supplies in all parts of the country (Bandyopadhyay, 1995). There have been some unrealistic ideas as for example, Rao's suggestion of the Ganga-Cauvery canal or of the Garland Canal by Dastur (NCIWRDP, 1999:179). Nevertheless, despite the arithmetic attraction of such projections, there are serious ecological and political obstacles in the conversion of this picture of good supplies into easy and countrywide availability of water. In fact, according to Bandyopadhyay and Mallik (2002) the gross estimates are of very little use in the making of the strategies for water resource development in the diversely endowed regions of the country.

Water Requirements in India

Water is utilized with regard to its need in direct consumption and use in human economic activities. Easy access to water is a necessary condition for socio-economic development. It is necessary for vital processes in the field of health, industrial and agricultural production. Water is also needed to sustain the natural eco-systems both on land and in aquatic bodies. The country already faces critical situations in water availability at the basin and local scales with six of the 20 major river basins in the throes of a less than 1000 cu m/year per capita availability with pervasive localized shortages present (World Bank, 1999f: 3). The growing urbanization and concentrated nature of human settlements and industrial activity in the country further exacerbate the inequalities in water resource endowments between the various basins and regions.

Both increasing demand for and active exploitation of water in India in the past few decades has been driven by growth in population, urbanization, irrigation and industry. Out of the total amount of water presently utilized in India, irrigation gets 83 per cent, domestic water supply and sanitation takes 4.5 per cent, and about 3.5 and 3 per cent go to energy development and industry respectively. Patterned on this category of utilization and on the basis of four independent studies on demographic trends and GDP growth rates (Natarajan, 1993; RGI, 1996; UN, 1995; Visaria and Visaria, 1996), an estimate for the future requirements of water in India has been made by the NCIWRDP (1999:49). For the scenarios of high and low growth rates for population, the data generated by Visaria and Visaria (1996) and the UN (1995), respectively have been used. The NCIWRDP (1999:50) has made population projections for the years 2010, 2025 and 2050, as shown in Table 4.3. The trend of urbanization in the country has been steady with a considerably large section of the increasing population concentrated in the major metropolis (World Bank, 1999c: 1).

The figure for the total projected water requirement generated by the NCIWRDP is the sum total of the requirements from the economic sectors of irrigation, domestic supplies, energy and industry, as well as an amount marked for the needs of conservation of the natural eco-systems.

Table 4.3: Urban and rural population estimates

Population Group	*2010 (Low)*	*2010 (High)*	*2025 (Low)*	*2025 (High)*	*2050 (Low)*	*2050 (High)*
Rural	790	759	810	730	700	610
Urban	367	387	476	603	646	971
Total	1157	1146	1286	1333	1346	1581

Source: NCIWRDP, 1999:50.

The nature and scale of the above water requirement scenarios will be discussed in the context of the various sectors as well as the ecological needs within the river basins. However, aggregated indicators of water requirement underestimate local conditions and the seasonal variability status of water availability and therefore, do not provide a satisfactory picture of development of water resources at the micro-level in the country, or an insight into the viable solutions. The growth in water requirements does not have any link with the natural distribution of the water resources in the country. In all the scenarios on sectoral water requirements, the same drawback applies — that the national totals mean very little in the context of the spatial and temporal variations in supply at the local levels.

The total water requirements of India can be estimated within reasonable error limits by summing up the requirements of specific sectors. Priority for allocation goes first to domestic household supplies, and then to irrigation, industries and power generation. However, in terms of the total requirement, irrigation tops the list. In this section, the projections of future requirements for these sectors have mainly been taken from the report of the NCIWRDP. In addition to these sectoral requirements, an important need exists for ensuring adequate in-stream water flow for navigation and eco-systems conservation. In the sections below, the various projections will be presented and commented upon.

Water for Irrigation

According to Rao (1975:109), whenever and wherever in India the average annual rainfall is less than 1150 mm, a need for irrigation becomes evident for sustaining food production. At the time of India's independence, some 22.6 Mha of arable land was irrigated by different sources. Irrigation made possible by large storage reservoirs in particular, has since then assumed a place of great importance, providing roughly 35 per cent of the water needed for irrigation in the country. It is largely through the development of irrigation with major and minor schemes that the country reached the acme of its grain production capacity in the period of Green Revolution. Keeping reasonable parity with population growth, the nation's food production has thus been enhanced four fold in the past four decades from 51 million tonnes in 1950-51 to 208 million tonnes in 1999-2000 (MOWR, 2002). In the official records some 52 per cent of this spectacular increase have been attributed to irrigation alone (see Fig. 4.6). The regional inequities in the natural availability of water are also reflected in the distribution of irrigated areas in the country. The greater share of the total irrigated area in the country exists in the Indus basin and the Ganga sub-basin in the North (Fig. 4.7). These areas are comparatively drier and the demand for dry-season field water needs are significantly high. However, the availability of snowmelt in the Himalayan rivers satisfies much of this requirement. Of the 24 river basins in the country, the Ganga sub-basin, with round about 54 per cent of the gross cropped area under irrigation is more intensively watered than the rest of the country (NCIWRDP, 1999:422). The present share of irrigation in the total water resources annually utilized in India is 83 per cent. According to official projections, even in 2050 irrigation would be the largest user at 79 per cent (NCIWRDP, 1999:101). Such a demand-driven physical expansion of the irrigation potential would be directed mainly towards using up the remaining 24 per cent of the development potential latent in both major and minor irrigation schemes within the next two or three decades. From Table 4.4 it is apparent that such an expansion of irrigated area will reach the estimated ultimate irrigation potential of 113.05 Mha in the country. The additional input of water for such an expansion would also be demanding, especially if, the present low efficiency of use of water in irrigation continues unchanged.

In reality, however, the actual water requirement will not be a linear function of the area irrigated. There are many technological innovations that may reduce the total water needs for irrigation. According to the projections of the Ministry of Water Resources, the gross irrigation water demand in India in 1990 was 460 cu km. It is expected to reach 628 cu km for the low growth and 807 cu km for the high growth scenario by the year 2050 (Table 4.5). About 41.06 per cent of the proposed water requirement for irrigation in the country come from the Ganga basin alone. With one-third of the ultimate irrigation potential available in this area, and the high water availability on agricultural land, the region will see a major expansion in irrigation potential in the future decades. However, the total water use for irrigation alone in the basin as projected for the year 2050 is 258 cu km, an amount larger than the entire utilizable surface water resources of 250 cu km in the basin (NCIWRDP, 1999:422). This means that the additional demand will have to be satisfied mainly by the exploitation of groundwater resources.

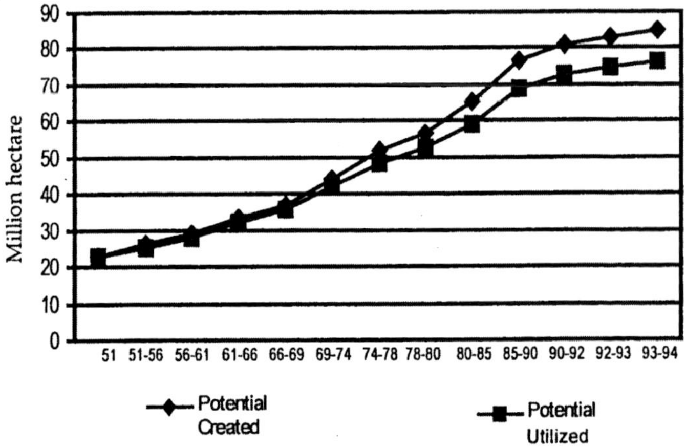

Fig. 4.6: Irrigation Potential Created and Utilized over Plan Periods 1951-1994.

Source: Central Water Commission (1995) as cited in World Bank (1999e: 111).

Over the last three decades, the rapid expansion in the use of groundwater primarily for irrigation has contributed significantly to agricultural and overall economic development in India. Such rapid development in groundwater, however, has a heavy ecological cost. The objective of enhancing irrigated area for increasing food production has major externalities, like an inefficient pattern of water use, reduction in groundwater availability during the critical dry season, dwindling base flows of rivers, declining water levels below the level of shallow wells, mobilization of poor-quality water sources and aquifer

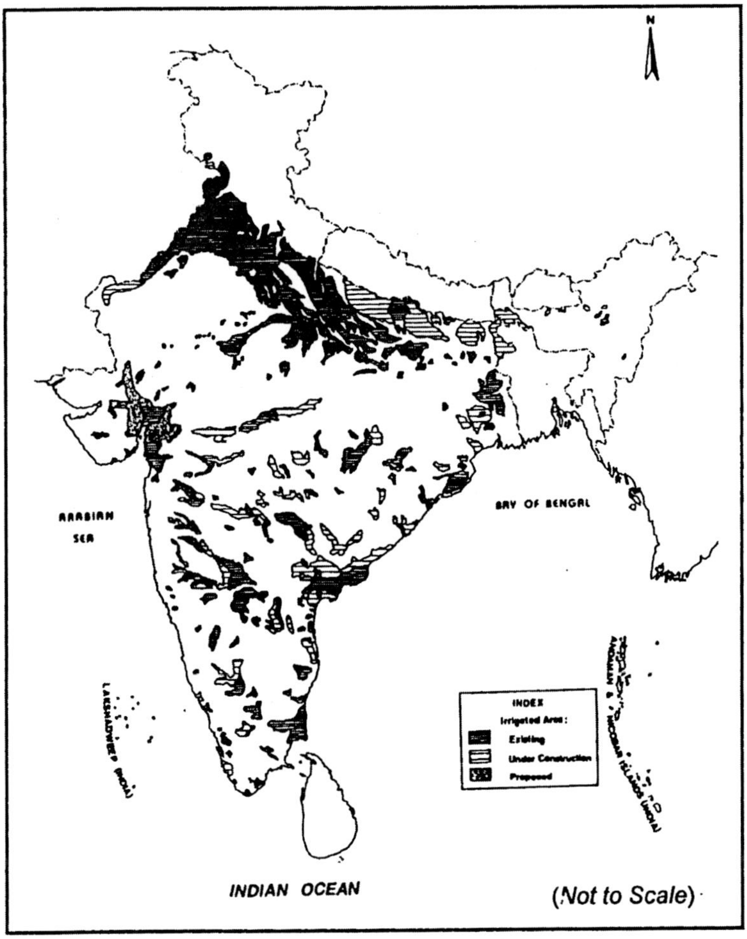

Fig. 4.7: Irrigated Area of India

pollution (Burke and Moench, 2000: 47). Quality declines associated with groundwater over-extraction and its impact on the sustainability of uses and of the environment, is a matter of serious concern. High lateral permeability in unconsolidated sediments, often leads to high transmission of pollutants and the migration of low quality or polluted water into aquifers. This can cause irreversible damage to groundwater resources. Large areas in Haryana, covering some 65 per cent of the agricultural land, are affected by saline groundwater as a consequence of over-pumping and the inflow of native saline water into adjacent high quality freshwater aquifers (Gangwar and Panghal, 1989 as cited in World Bank, 1999b:14).

Table 4.4: Development of irrigation potential from the fifth plan onwards (M ha)

Plan	*Total Irrigation (Million ha)*			*Gross Irrigated Area as Per Land Use Statistics*
	Potential Created	*Potential Utilized*	*Ultimate Potential*	
Fifth (1974-78)	52.02	48.46	113.5**	46.08
Annual (1978-80)	56.61	52.64	Do	49.21
Sixth (1980-85)	65.22	58.82	Do	54.53
Seventh (1985-90)	76.53	68.59	Do	61.85
Annual (1990-92)	81.09	72.86	Do	65.68
Eighth (1992-97)	89.56	80.76	Do	70.64
Ninth (1997-2002)	106.61	93.39	Do	—
2050 LOW*	—	—	Do	113.05
2050 HIGH*	PROJECTED		Do	146

Sources: * NCIWRDP, 1999:79,422. ** World Bank, 1999e: 111.

Table 4.5: Projected water requirement for irrigation

Sl. No.	*Details*	*Year 2010*		*Year 2025*		*Year 2050*	
		Low	*High*	*Low*	*High*	*Low*	*High*
1.	Surface Water required for Irrigation	330	339	325	366	375	463
2.	Groundwater required for Irrigation	213	218	236	245	253	344
3.	Total Water required for Irrigation	543	557	561	611	628	807

Source: NCIWRDP, 1999:59.

Rapid growth in the utilization of groundwater with little concern for the prevailing hydro-geological, ecological and climatic factors, may convert a temporary meteorological process into a long-term ecological one of prolonged water scarcity (Bandyopadhyay, 1989:287). Such couched scarcity is often aggravated by the "high volume" but "low quality" needs of agriculture and "low volume" but "high quality" needs of the domestic sector, increasing the scale and extent of competition amongst sectors. Due to the expansion of groundwater-based irrigation in the arid and hard rock areas of the Peninsula, increases in overdraft areas and associated problems of water-quality are fast emerging. Table 4.6 gives some numerical information on the limitations in the recharge. Falling groundwater levels adversely affect the sustainability of human-economic uses that are dependent on it, without immediately threatening the physical exhaustion of the resource base (Burke and Moench, 2000:62).

If the present trends in the extraction and rapid development of groundwater are to continue, the optimum irrigation potential economically available from the same will be attained shortly, by 2007 (World Bank, 1999b:9). Thus, the number of over-exploited and critical blocks, at present covering only a small fraction of the total groundwater irrigated area, would double every 12.5 years, indicating that, 36 per cent of the 4248 blocks would be defined as dark or critical within a brief span of two decades (World Bank, 1999b:10).

Table 4.6: Comparative groundwater recharge rates in various rock formations

Sl. No.	*Rock Type/Formation*	*Natural recharge due to infiltration in various rock types/formations (%)*
1.	Hard Rocks (Crystalline)	10
2.	Consolidated sedimentaries	15-20
3.	Gangetic Alluvium	20
4.	Coastal Alluvium	10-15
5.	Western Rajasthan Dune Sand	2

Source: Bandyopadhyay, 1989:287.

As irrigation is associated with an inefficient water distribution system which incurrs heavy transmission and field distribution losses, this has not only affected crop productivity at times due to supply system failures of the canal irrigation network, but has also been the harbinger of several complex ecological problems. The result has been, increasing damage to the ecological foundations of agriculture, declining per capita availability of land and water, and non-availability of technologies that can actually increase yield in major food crops on a sustainable basis. Salinization and alkalinization for instance, claims as much as two-fifths of the nation's irrigated area and have direct implications for the environment (Repetto, 1994 as cited in World Bank, 1999b: 14). Waterlogging, in the command areas of both major and minor surface irrigation schemes, which covered 2.46 million hectares of productive land in 1991, (NCIWRDP, 1999: 93) has already reached chronic proportions. All this has made the future food security appear bleak for the burgeoning population (Swaminathan, 1999:10). If the current level of water use efficiency and mismanagement remains unchanged, the required water for irrigated land would be much higher.

With an awareness of the fact that the surplus availability is going to dwindle rapidly in the near future after the attainment of the ultimate irrigation potential (see Fig. 4.8), there has to be a well concerted effort to increase the efficiency of water-use in irrigation by a minimum of 50 per cent (see Table 4.7). Considering the low productivity status of the rain-fed farms, which currently stagnates at 1 tonne/ha as against 2.33 tonne/ha of irrigated land (NCIWRDP, 1999: 57), there is ample scope for an increase in their productivity through conservation-based water management.

Table 4.7: Irrigation water demand with 40 and 60 per cent use efficiency in cu km

Irrigation Efficiency	*2000*	*2007*	*2025*
40 per cent	460	630	770
40 per cent *	460	822	1090
60 per cent	460	610	800

* According to Navalavala.
Source: WAPCOS as cited in Shah (2000).

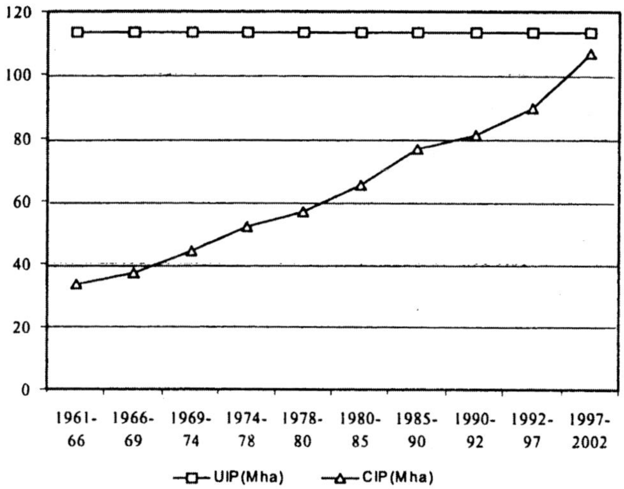

Fig. 4.8: Irrigation Expansion Vs Ultimate Irrigation Potential
CIP= Created Irrigation Potential; UIP= Ultimate Irrigation Potential.

Source: NCIWRDP, 1999; World Bank, 1999e: 7.

Water for Domestic Supply and Sanitation

Water for meeting domestic needs is not only required in the right quantity, but needs also to be of the right quality. It should be available at the required time and place. Being densely populated, India has been facing scarcity situations with domestic water supplies in the arid and semi-arid areas for a long time. The recommended rate of minimum daily per capita domestic water supply is 40 litres in rural areas and 125 litres in urban areas (NCIWRDP, 1999:60). The differential growth rates of the urban and rural population have also contributed to the steeply rising demands. By 2025 this high growth rate will lead to the doubling of domestic water requirements from a present urban and rural use of 25 cu km to 62 cu km (NCIWRDP, 1999:61). Such national aggregates again, are not reflections of the seasonal difficulties and do not provide a very clear estimate of the specific regional requirements.

With India's First Five-Year Plan, the privilege of piped and treated drinking water supply was conferred only to a few major cities which served just less than half the urban population (TERI, 1998:53). For much of the rural areas, fetching water from wells or springs formed the main daily activity especially for women. It cannot be denied that significant progress has been made in India since then to provide access to safe water supplies. The official statistics proclaims that 90 and 97 per cent of the urban and rural population, respectively has access to reliable water supplies (NCIWRDP, 1999:146-47). The Economic Survey conducted in the year 1996-97 reported that over a period of almost ten years the percentage of population with access to safe drinking water has actually increased, with spectacular improvement in the rural areas (see Table 4.8). However, the bad news is mainly about sanitation. So far, sanitation facilities in the rural areas have significantly failed to show any signs of substantial improvement during the same period .The UN estimated that "80 per cent of India's 800 million still live in unsanitary conditions" (BBC World News, 13 March 2000). Such figures become credible, if we consider that the percentage of urban and rural populations that currently covered by sanitation facilities are, 50 and a mere 5, respectively (NCIWRDP, 1999:146).

It is important to note however, that when government statistics claim that a very high proportion of villages are covered by safe drinking water supplies, it more often implies that hand-pumps or stand-posts have been installed in these villages. It does not necessarily signify that water is being provided through these facilities or that the village community is utilizing this water. Thus, merely introducing new technologies/hand-pumps in the villages is not enough to change usage practices or to reduce the time and effort spent by women in water collection, as they continue to use traditional dug-well sources. In a similar vein the supply-focused engineering schemes have largely failed to take into consideration the cultural context and the level of knowledge of local communities, especially women, while planning new technologies.

Table 4.8: Percentage of population covered by urban and rural water supply (drinking and sanitation)

Type	*1985*	*1990*	*1996*
Drinking water supply			
Rural	56.3	73.9	82.0
Urban	72.9	83.8	85.0
Sanitation facilities			
Rural	0.7	2.4	4.6
Urban	28.4	45.9	50.0

Source: TERI 1998:28.

Unavailability of potable water in the desired quantities hold grim implications for the quality of life of the rural poor, in terms of the time spent in collecting water (by women), while consumption of contaminated water has adverse impacts on human health and productivity. According to the 1994 validation survey of the Government of India, about 70 million rural people consume unsafe drinking water laced with varying degree of pollutants (World Bank, 1999d: 4). The government's claim that, 90 per cent of the urban and 85 per cent of the rural population (WAPCOS, 1993 as cited in Shah, 2000:43) enjoy protected water supply facilities, disregards the fact that availability and supply are highly location and season specific. This is highlighted by the wide variation in per capita water supply, which veers from a low of 9 lpcd to a high of 584 lpcd in the urban and from 5 lpcd to 70 lpcd in the rural areas (NCIWRDP, 1999:147). When this is coupled with factors such as quality decimation (as quality is the determinant of effective water availability for actual human use), different levels of economic development and extent of water use it can be seen that fully reliable and safe domestic water supply yet a far-flung dream. Despite the inadequate supplies, large amounts of water are wasted and lost on account of "unaccounted for" losses in the urban water supply system, due to bad maintenance and operation which can lead to a 30 to 60 per cent difference in the net supply (TERI, 1998:54).

To meet the growing challenges in domestic water supply the National Commission for Integrated Water Resources Development (1999:60) has set a supply target of 220 lpcd for

Class I and 150 lpcd for non-Class I cities, respectively and 55 lpcd for rural areas by 2010 (see Table 4.9). Selective urban areas, home to a million and above population, will be withdrawing a large quantity of water in the figure (see Fig. 4.9). This would imply an additional pressure on the available resources in these urban areas. The changing demand for water for the two scenarios of high and low population increase is given in Table 4.10. The rate of projected water supply (see Table 4.10) must be realized if the nation's future water requirements are to be met.

Table 4.9: Recommended standards for domestic water supply in lpcd

Population type	*2010*	*2025*	*2050*
Urban (Class I)	220	220	220
Urban (Non-Class I)	150	165	220
Rural	55	70	150

Source: NCIWRDP, 1999:60.

Table 4.10: Water requirements for domestic and municipal use (in cu km)

Scenario	*2010*	*2025*	*2050*
Low Demand			
Total (Surface + Ground)	42	55	90
High Demand			
Total (Surface + Ground)	43	62	111

Source: NCIWRDP, 1999:61.

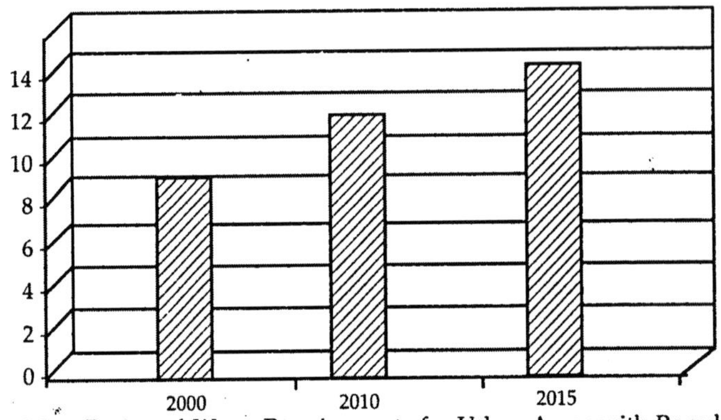

Fig. 4.9: Projected Water Requirements for Urban Areas with Population of a Million and above.

Source: NCIWRDP, 1999:61.

Water for Industry

In terms of priority the manufacturing sector has been enjoying an increasing water supply, next to agriculture. The major water using industries are chemical and petrochemical, iron and steel, paper and pulp industries, etc. The total amount of industrial water use is usually a very small fraction of the total water withdrawal. This figure is usually about 10 per cent but in the case of India it is much less. Almost freely available water, however, makes industries use this resource rather inefficiently. In addition, the use of streams and the wetlands for the disposal of polluted water is also common. The Central and State Pollution Control Boards have identified 1,532 "grossly polluting" industries in 24 states (World Bank, 1999f: 98). Inefficient use of water by Indian industries can be estimated from the value of the output per unit of water used which is only $5 per cu m compared to that of many developing countries for example, Argentina, where it is around $20 per cu m (TERI, 1998:55). The relatively tardy growth process in the industrial sector in the 1960s has been replaced and presently the annual growth rate has touched 11.6 per cent in 1995-96 (World Bank, 1999f: 2). As industrial growth continues in a high gear, water requirements for the growing industries will increase accordingly. A lower estimate of industrial water requirement has been made by Chopra and Goldar (2000:33), on the basis of the average growth rates of 6.24 per cent for business-as-usual and 7.95 per cent per annum for high industrial growth scenarios for the period 1995-2020 (Table 4.11).

Table 4.11: Water requirement for the industry in cu km (low demand scenario)

Details	*1995*	*2020 Business-As-Usual*	*2020 High Growth*
Organized manufacturing	6.99	24.24	36.10
Unorganized manufacturing	0.81	3.67	5.48
Total	7.80	27.91	41.58

Source: Chopra and Goldar, 2000:71.

The latter estimate also suggests a significant increase in the water requirements of the manufacturing industries in the states of Gujarat and Maharashtra between 1995 and 2020. In the high growth scenario, Maharashtra is expected to increase its water requirement to about 11 cu km, while Gujarat would need almost 7 cu km to sustain the water-demand for its industrial activities (Chopra and Goldar, 2000:72). Water demand for industries in India stands at 14 cu km for 1997. This, according to the high demand scenario estimated by the National Commission, is expected to go up to 37 cu km, 67 cu km, and 81 cu km in the years 2010, 2025 and 2050, respectively (Fig. 4.10)[3], at an estimated growth rate of 9 per cent as contemplated in the Ninth Plan Draft (NCIWRDP, 1999:62). The values in Fig. 4.10 do not include water demand from the power generation sector in India.

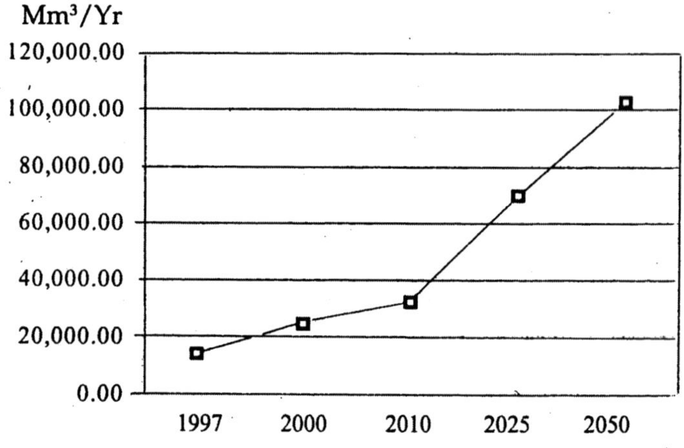

Fig. 4.10: Water Requirements of 17 categories of Indian Industries[4].

Source: NCIWRDP, 1999:455.

Water for Energy Development

Large-scale generation of thermal power needs large quantities of water. For thermal (coal, oil and gas) and nuclear power stations a substantial quantity of water is needed for cooling, while hydropower generation involves reservoirs/dams. The installed total capacity of power in India amounts to 92281 MW, of which the largest contribution is from thermal (73%) followed by hydro (24%) and nuclear (3%) (NCIWRDP, 1999:62). For efficient management and a balance between thermal and hydropower, the contribution from the latter needs to be 40 per cent (NCIWRDP, 1999:133). An annual growth of demand for power at 9 per cent, and a 7 per cent growth in the installed capacity in the country, implies a substantial water requirement. Table 4.12 gives the water requirements for power generation for low and high demand scenarios. About 80 per cent of the total requirements are likely to be provided by surface water sources and the rest from groundwater (NCIWRDP, 1999:63). The water requirement for power generation in Uttar Pradesh, Maharashtra, and Madhya Pradesh will be high.

Table 4.12: Water requirement projections for power generation in cu km

Year →	2010		2025		2050	
Year	*Low*	*High*	*Low*	*High*	*Low*	*High*
Total Water Required for Power Generation	18	19	31	33	63	70

Source: NCIWRDP, 1999:69.

About 65,000 MW or 78 per cent of the mammoth untapped hydro-potential of 84044 MW estimated at 60 per cent load factor, is locked up in the fast-flowing rivers of the Himalayan region of the Ganga, Brahmaputra and Indus basins (NCIWRDP, 1999:135). As water requirement for the generation of hydropower is by far the largest, involving the construction of head storages, *viz.* reservoirs/dams which also lead to great evaporation losses[5], harnessing of this water gains in significance in view of the projected need for water and energy requirements in 2050, even while attaining a high degree of water-use efficiency (Rangachari and Verghese, 2001:124). Evaporation losses from the existing and upcoming reservoirs would amount to 42 cu km, 50 cu km and 76 cu km by 2010, 2025 and 2050 respectively, according to official estimates (NCIWRDP, 1999:68).

Water for Eco-systems Conservation and Navigation

The economic analyses on water, though limited in extent so far, relates mainly to the consumptive utilization of water in irrigation, industry and domestic supplies. The economic utilization and productivity of water resources related to other forms of non-consumptive economic utilization, for example, in navigation, the productivity of downstream water bodies, eco-system conservation, etc. has remained very marginal in the economic literature on India. What is more difficult, such partial approach to the utilization of water resources has almost invariably led to rapid disruption of the essential ecological processes and services associated with water in all parts of river basins.

Navigation makes economic use of water through transportation without withdrawals. There are, however, other damages caused by it. It directly causes major disturbances to the aquatic eco-systems, and carries the risk of major pollution, in case of accidental spill. As such the ecological integrity of the river basin get constantly damaged by withdrawals of high volumes of water and discharge of polluted water. This is manifest in the increasing decline in water quality, hydrological changes, stress in the ecosystems and loss of biodiversity. If the principle that, "science of ecology pertaining to rivers is of primary importance", as proclaimed by Rao (1975:187), is to be upheld, then there is a need to ensure that withdrawals of water from the natural eco-systems are made with full recognition of the in stream and *in situ* needs of water for conservation of the eco-systems and the biodiversity and protection of other downstream economic uses of water.

The NCIWRDP (1999:68) has accepted that it is not possible to estimate the volume of water needed for managing ecological standards. The knowledge base for making any approximate calculation of this requirement is very limited. A provisional projection of the water needed in maintaining ecological stability has been given as 5 cu km, 10 cu km and 20 cu km in the years 2010, 2025 and 2050, respectively. The water requirement for "environment and ecology" has been marked at 2 per cent of the net national water requirements, as ascertained for the high growth scenario of 2050 (NCIWRDP, 1999:69). This is an ad hoc estimate, since there appears to be no specifications of the requirements not only for rivers but also for wetlands, surface water bodies, and groundwater, as well as for aquatic life. The filling of this gap is crucial and requires that a serious assessment of the environmental and ecological water needs be made not only at the national level, but also for each eco-region in the country. It is in this assessment that the application of ecological economics is very much in demand.

Role for Ecology and Economics in the Emerging Water Scenario in India

The projections of gross national water requirements made by the NCIWRDP for meeting the likely high and low demand scenarios till 2050 are given in Table 4.13. While making these projections, the NCIWRDP has considered possible improvements in the end-use efficiencies and technological changes. Even though the comparison of gross requirement and utilizable stock is not very realistic at the macro-level, it should still be pointed out that, according to the high demand scenario, the figure for the projected total water requirement of India in the year 2050 is 1180 cu km. This amount is larger than 1086 cu km, the officially estimated total utilizable water resource of India (Table 4.13). In the National Water Policy, this figure of total availability has been reiterated (MOWR, 2002). There are also some other estimates of water requirement made by Chopra and Goldar (2000:49) in 2020. These are 900.92 cu km and 1044.77 cu km, respectively for the BAU and high growth scenarios. While such an arithmetic comparison can hardly be taken as realistic, the total national requirement approaching or crossing the estimated total availability figure of 1086 cu km should be recognized as a signal for serious concern.

Water is utilized at the micro-levels. The gross national figures are not of much use in addressing the reality of water management in a country as large and as diverse as India. In this respect, unlike cash, water is not that liquid, as a result, income at any point of space and time cannot be freely transferred to another point and used. Thus, any comparison of the aggregate supply and demand at the national level should only be taken as a rough indicator of the comparative status of supply and requirement of water. The acute temporal and spatial distribution of precipitation and also the availability of storage sites only in some areas of the country generate a mosaic with diverse levels of water endowments. Each of these small areas has a separate water audit.

Table 4.13: Total water requirement projections for India in cu km

Year	2010		2025		2050	
Total	Low	High	Low	High	Low	High
Requirements	694	710	784	843	973	1180

Source: NCIWRDP, 1999:69.

For developing any practical and realistic understanding, the water balance estimates are to be made in such smaller spatial units, within which the precipitations are reasonably uniform or the mobility of water within that spatial unit is high. Once the mosaic conditions are recognized, there will not be any scope for surprises over simultaneous conditions of drought and floods in parts of the country. In fact, such small scale observations can establish that in many areas distributed all over the country, the water utilization has already outstripped the availability. This is more starkly established in the areas poor in surface water and dependent almost exclusively on groundwater resources. In particular, reference may be made to the areas listed by the Central Groundwater Board as those with no further use of groundwater (NCIWRDP, 1999: 43). In many other areas, where human needs are satisfied more from surface water sources, the situation of scarcity regularly arises in the pre-monsoon periods when these sources start to run dry. The scarcity hits the domestic water supplies most critically.

Shortcomings of the Present Strategy for Water Resource Development

Meeting the projected demands for water from a large and growing population in India is going to be a challenging task. The main official strategy as outlined in the Report of the NCIWRDP, is mainly based on enhancing the availability of large storage for water. For a long time, Indian planners have contemplated building large storage dams in the water rich basins and transfering this to the water-deficit ones (NCIWRDP, 1999:181). In this way, the strategy has been to provide greater mobility to the water resources stored behind engineering structures. Such a long-standing preoccupation with technologically impressive hydrological interventions draws upon ideas implicit in Rao's proposal of a 2640 km long Ganga-Cauvery Link Canal, and the concept of a Garland Canal along the Himalayan foothills, proposed by Dastur (NCIWRDP, 1999:179).

The past fifty years of water development in India has been following this overall trajectory, led by supply-side concerns and based on large-scale centralized interventions in the river basins. The official projections for the enhancement of available supply have, thus, been made on the assumption of the successful implementation and functioning of a number of large engineering interventions and storage projects, many of which are still in the stage of design. In view of the availability of large amounts of run-off in the Ganges-Brahmaputra-Meghna (Barak) river system, most of these large storage projects are planned to be built within this basin. However, the issue of large dams is at the centre of serious debate in terms of their social, economic and ecological impacts. In addition, the major human problem resulting from a bad record of resettlement and rehabilitation of people displaced by dam projects in India, the existence of critical gaps in knowledge, specially in the case of Himalayan dams make scientific decision making on these dams more difficult (Bandyopadhyay, 2002:133). The narrow engineering approach to the conceptualization and design of the dams has systematically underplayed the broader societal and ecological aspects of such projects. The unexpected or ignored consequences of large hydrological projects range from the human issue of inadequate rehabilitation following involuntary displacement to the level of complex economic risks and uncertainty about the performance of the water projects. As a result, there is a big question mark on how and when, if at all, the storage dams would be constructed and the storage volumes available.

This, to a great extent undermines the approach taken by the NCIWRDP. Storage projects, no doubt, would be needed to satisfy the growing water requirements as well as to store the large monsoon run-offs in the surplus river basins. Further, the idea of the storage dam could mean tapping the immense hydro-potential latent in the waters of many Himalayan rivers and some peninsular rivers. It follows that a good number of storage options would be explored.[6]

The recently published National Water Policy (MOWR, 2002) recognizes the need for an integrated water science in a limited way. Right from the stage of initiation, governmental thinking on water resources has been dominated by concepts of engineering. The non-governmental thinking and actions in India have contributed greatly in bringing the much-needed disciplinary balance and also towards sensitising people on the crucial importance of small and micro-level conservation of water (Agarwal *et al.*, 2001). Water security of the people of India in the future years substantially depends on how the broadening of the official vision takes place. This will require an integration of water management more completely with the concepts of social acceptability, economic viability and ecological sustainability. The shrinking per capita water resources and the need for its most efficient and sustainable use needs both economics and ecology to get their due place in the making of water policy. The next section contains some indicative suggestions regarding the possible role ecology and economics could play in advancing sustainable water resource development so that a new pathway may have to be charted for a secure water future of the Indian people.

Towards Sustainable Water Resource Development – Role of Ecology and Economics

An important starting point for any role that ecology and economics could play is in economists taking a close look at the global hydrological cycle (Fig. 4.1). As has been mentioned earlier, economists have traditionally viewed water through its consumptive uses. However, prior to the stage where a volume of water comes to be used in obvious consumptive economic activities as irrigation, industry or domestic supplies, it does also perform several eco-system services in the upstream that are not visible by itself, and hence are mostly ignored. The same is the case with eco-system services in the downstream, that may have been damaged due to the engineering interventions. A clearer knowledge of the hydrological cycle will help the ecological-economists to put a more realistic value on the available water resources, and thus, its allocation for further use may be made with greater responsibility. There is a new role that ecological economics should take upon itself in redefining sustainable water resource development.

For example, in the case of integrated watershed development, a knowledge of the eco-system services provided by water would be crucial for designing an appropriate landuse plan that would be the most suitable for optimizing integrated water resource development at the level of the whole river basin. In the case of design of storage structures, economists informed of the hydrological cycle can develop a more comprehensive benefit-cost assessment of the proposed structures, by assessing and internalizing the economic costs of downstream ecological transformations to be caused by the structure under design. This will provide a more objective basis for an ecological-economic benefit-cost analysis of the projects and valuation of water resources stored and provided by the concerned structure. In turn, this approach will help in the development of an appropriate mechanism for the pricing of the water used. This will go a long way in realizing the much needed increase in the end-use efficiency for water.

A strengthening of the knowledge base at the interface

between ecology and economics could facilitate an understanding of the costs and benefit of natural hydrological processes not seen from an ecological economic viewpoint. The annual inundation of the floodplains of river basins may be taken as an example. Such an assessment can provide important inputs for a better informed flood management strategy. In a similar manner, an ecologically informed estimation of the environmental costs of the discharging of pollutants to water bodies can help fix a more realistic compensation that the polluter must pay. This may encourage the polluter to take measures for the prevention of the emission of pollutants.

In facing the task of providing water security to the people of India, ecology and economics can jointly play a crucial role in the ushering in of the paradigm shift in the policy maker's perception of water resources development, both at the macro, as well as the micro-levels. By assigning values to the diverse eco-systems existing along a river basin, it can generate new ways of looking at the economics of water use in a river basin. On the other hand, by valuing the economic contributions from micro-level conservation and management of water, it can help in the reversal of the growing vulnerability of communities living in the arid and semi-arid regions, or regions dependent on meagre groundwater resources. As and when the concerns about the present strategies for water resource management as expressed by the ecologists can be examined with the help of ecological economics, the paradigm shift will be greatly facilitated. India has got eminent professionals in both ecology and economics. They have been mainly on a confrontational path so far. This cognitive challenge of promoting sustainable water resource management provides an immense opportunity for collaborative interdisciplinary research. This is a responsibility the professional community of ecologists and economists may take up as a challenge. The initiative taken by the INSEE is a very important step in this direction.

Notes

1. Water stress or the Falkenmark Indicator denotes reaching the limits of water quantity as well as quality. This stress begins when there is less than 1,700 cubic metres of water per person a year for domestic, industrial, agricultural, and natural eco-systems functions and becomes severe when there is less than 1,000 cubic metres per capita (World Water Vision, 2000:26).
2. The ratio of the pore volume to the total volume of a rock is its porosity. Rocks with primary porosity result from the presence of inter-granular pore spaces and those with secondary porosity due to structural features like joints and fault fissures (Holmes, 1992:387-388).
3. The projection graph does not include changes in consumption brought in by technology. The growth rate of the 17 industrial categories shown in the graph is at the present level of 9 per cent.
4. Iron & Steel, Smelters, Petrochemicals & Refinery, Chemicals — Caustic Soda, Textile & Jute, Cement, Fertilizer, Leather Products, Rubber, Food Processing, Inorganic Chemicals, Sugar, Pharmaceuticals, Distillery, Pesticides, Paper & Pulp and General Engineering.
5. Evaporation losses account for some 200 km^3 of water each year from reservoirs, which is 7 per cent of the water consumed by human activities (Shiklomanov, 1998).
6. A considerable part of the storage options are situated in the Ganges, with a gross storage capacity of 53.96km^3 including projects completed and those under construction (NCIWRDP, 1999:36).

References

Agarwal, A., S. Narain and I. Khurana (eds) (2001). *Making Water Everybody's Business: Practice and Policy of Water Harvesting*, Centre for Science and Environment, New Delhi.

Ahmed, S. (1990). 'Cleaning the River Ganga: Rhetoric and Reality', *AMBIO* 19(1): 42-5.

Alford, D. (1992). 'Hydrological Aspects of the Himalayan Region,' Occasional Paper No.18, ICIMOD, Kathmandu.

Anon (2000). *Dams and Development: A New Framework for Decision-Making*, Earthscan, London.

Ayibotele, N. B. (1992). *The World's Water: Assessing the Resource*, Keynote paper for International Conference on Water and the Environment, Dublin.

Bahadur, J. (1997). 'Himalayan Ecohydrology — An Emerging Subject for Restoration of the Mountain System', *Vigyan Prasar*, New Delhi. Website: www.mtnforum.org/resources/library/bahaj97a.htm

Bandyopadhyay, J. (1989). 'Riskful Confusion of Drought and Man Induced Water Scarcity', *AMBIO* 18(5):284-92.

Bandyopadhyay, J. (1995). 'Water Management in the Ganges-Brahmaputra Basin: Challenges for the 21st century', *International Journal of Water Resource Development* 11(4):411-41, December.

——— (2002). 'A critical look at the World Commission on Dams Report in the context of the debate on large dams on the Himalayan rivers', *International Journal of Water Resource Development*, 18(1):127-45, March.

Bandyopadhyay, J. and B. Mallik (2002). 'Population and Water Resources in India: Crucial Gaps in Knowledge for Sustainable Use in Future', Proceedings of the Symposium on *Population, Life Support and Human Development in India* held at Centre for Development and Environment Policy, IIM, Calcutta, February 2001, forthcoming in 2002.

Briscoe, J. (1996). *Water as an Economic Good: The Idea and What it Means in Practice*, Paper presented at the World Congress of ICID, The World Bank, Washington.

Bruijnzeel, L. A. and C. N. Bremmer (1989). *Highland-Lowland Interactions in the Ganges Brahmaputra River Basin: A Review of Published Literature*, Occasional Paper No. 11, ICIMOD, Kathmandu.

Burke, Jacob J. and Moench Marcus H. (2000). *Groundwater and Society: Resources, Tensions and Opportunities*, New York.

Centre for Study of Man and Environment (CSME) (1984) *Basin Sub-Basin Inventory of Water Pollution — The Ganga Basin (Part-II)*, S. P. Das Gupta (ed.), New Delhi.

Chopra, K. and B. Goldar (2000). *Sustainable Development Framework for India: The Case of Water Resources*, Institute of Economic Growth, Delhi.

Cosgrove, W. and F.R. Rijsberman (2000). *World Water Vision: Making Water Everybody's Business*, Earthscan, London.

Dyson, T. (2002). 'India's Population — the Future', *Workshop paper on India's Future: Population, Environment and Human Development*, New Delhi.

Dyson, T. (2001). 'The Preliminary Demography of the 2001 Census of India', *Population and Development Review* 27 (2):341-356, June.

Falkenmark, M. and C., Folke (2002). 'The ethics of socio-ecohydrological catchment management', *Hydrology and Earth System Sciences* 6(1).

Ghosh, D. (2000). Wastewater Conservation: Lessons from Wastewater Treatment and Resource Recovery Projects in West Bengal, Calcutta.

Gibbons, D. C. (1986). *The Economic Value of Water* Resources for the Future, Washington.

Gleick, P.H. (1993). *Water in Crisis: A Guide to the World's Freshwater Resources*, Oxford University Press, Oxford.

Gupta, S.K., P.S. Minhas, S.K. Sondhi, N.K. Tyagi and J.S.P. Yadav (2000). as cited in Chopra and Goldar (2000:53).

Holmes, A. (1992), Ed by P. McL. D. Duff, *'Principles of Physical Geology'* London.

Jones, P.H. (1989). 'Potential for Large-Scale Groundwater Recharge in the Bhabar Belt in India and Nepal to mitigate floods and recover floodwater for irrigation use in the Ganges Plain', February, Louisiana.

Krishna Murti *et al.* (ed) (1991). *'The Ganga: A Scientific Study'*, New Delhi.

Ministry of Irrigation and Power (1972). *Report of the Irrigation Commission*, New Delhi.

Ministry of Agriculture and Irrigation (1976). *Reports of the National Commission on Agriculture*, New Delhi.

Ministry of Water Resources (1997). *Annual Report 1996-97*, New Delhi.

——— (2002). *National Water Policy*, April, New Delhi.

National Commission for Integrated Water Resource Development Plan (NCIWRDP) (1999). *Integrated Water Resource Development: A Plan for Action*, September, New Delhi: Ministry of Water Resources.

Newson, M. (1997). *Land, Water and Development: Sustainable Management of River Basin Systems*, Routledge, London.

Rangachari, R. and B.G. Verghese (2001). "Making Water Work to Translate Poverty into Prosperity: The Ganga-Brahmaputra-Barak Region", in Ahmad, *et al*, (Ed) *Ganges-Brahmaputra-Meghna Region: A Framework for Sustainable Development*, Dhaka.

Rao, K.L. (1975). *India's Water Wealth* (New Delhi: Orient Longman).

Raskin, P. *et al.* (1997). *Water Futures: Assessment of Long Range Patterns and Problems*, Background report for Comprehensive Assessment of the Freshwater Resources of the World Stockholm Environment Institute, Stockholm.

Shah, T. (2000). 'Water Against Poverty: Livelihood-oriented Water Resource Management' in P. P. Mollinga (Ed) *Water for Food and Rural Development*, Sage, New Delhi.

Shiklomanov (2000). 'Appraisal and Assessment of World Water Resources,' *Water International* 25(1) March.

Sengupta, N. (2000). 'Large Dams: India's Experience', *WCD Thematic Review.*

Sengupta, R. (2000). *Ecology and Economics: An Indian Perspective*, Oxford University Press, New Delhi.

Swaminathan, M.S. (1999). *I predict: A Century of Hope — Harmony with Nature and Freedom from Hunger*, MSSRF, Chennai.

Tata Energy Research Institute (TERI) (1998). Pachauri, R.K. and P.V. Sridharan (Eds) *Looking Back to Think Ahead*, New Delhi.

Verghese, B.G. (1999). *Waters of Hope: From Vision to Reality in Himalaya-Ganga Development Cooperation*, Dhaka.

World Bank (1999a). *South Asia Rural Development Series — India Water Resources Management — Inter-sectoral Water Allocation, Planning and Management*, New Delhi.

World Bank (1999b). *South Asia Rural Development Series – India Water Resources Management — Groundwater Regulation and Management*, New Delhi.

——— (1999c). *South Asia Rural Development Series –India Water Resources Management — Urban Water Supply and Sanitation*, New Delhi.

——— (1999d). *South Asia Rural Development Series –India Water Resources Management — Rural Water Supply and Sanitation*, New Delhi.

——— (1999e). *South Asia Rural Development Series –India Water Resources Management — The Irrigation Sector*, New Delhi.

——— (1999f). *South Asia Rural Development Series – India Water Resources Management — Initiating and Sustaining Water Sector Reforms: A Synthesis*, New Delhi.

5

Demand Management in the Face of Growing Water Scarcity and Conflicts in India

Institutional and Policy Alternatives for Future

M. DINESH KUMAR

Introduction

Economic growth is essential for poverty alleviation (Pearce and Warford, 1993). Since independence, India has made substantial progress on the economic front with the per capita net national product recording a compounded growth rate of 1.7 per cent. This is evident from a decline in the poverty rate from 45 per cent in 1951 to 35 per cent in 1993-4 (Datt, 1997). The contribution of agricultural production to this progress in GDP growth during this period has been quite phenomenal, as its value grew 3.2 times in real terms (TERI, 1998). Irrigation has been the key to enhancing agricultural production, and ensuring food security at the national level.

Providing safe water for drinking and sanitation are social goals and water is also needed for industrial production purposes, pollution assimilation and a variety of environmental services (Frederick, 1993; Glieck, 1997; Seckler *et al.*, 1998). India has also made substantive achievements in the provision of protected water supplies for drinking and domestic uses, and water supply for industrial uses (GoI, 1997).

The social, economic and environmental value of water is being felt more strongly than ever before as demographic and socio-economic processes are putting an enormous pressure on the limited freshwater reserves. According to Engelman and LeRoy (1993), the per capita freshwater availability in India, which stood at 6000 M^3 per annum in 1947, would decline to 2266 M^3 in 1997 and further to 1500 M^3 by 2027. Six of the major river basins in India are already water stressed and local scarcity and environmental problems are endemic in all basins (World Bank — GoI, 1998).

The approach to planning, development and management of water resources has been, by and large, centralized, sectoral and segmented (World Bank — GoI, 1998; Kumar and Ballabh, 2000). This approach, which is mainly intended to maximize the benefits of water to individual sectors, not only led to unsustainable development, but also led to several negative social, economic and environmental consequences. The government responds to such problems with an increasing number of environmental legislations and regulations which are, by and large, ineffective (Kumar and Ballabh, 2000).

While a fast growing urban population, an increase in per capita income, along with industrialization continue to raise the demand for water, technological and institutional interventions to manage the same are not forthcoming (Kumar and Ballabh, 2000). As water becomes scarce, the financial and environmental cost of its use increases enormously. Demand management is the key to minimizing the environmental stresses caused by water scarcity (Frederick, 1993; Postel, 1993).

The Future Water Requirements in India

The national-level aggregate water requirements are estimated for 2025, means of projected future demographic and socio-economic trends using 1990 as the base year. The water requirement for various purposes such as drinking and domestic needs, livestock uses, industry and agricultural production are analysed separately. The basic assumptions involved in the analysis are: (*i*) population growth will increase the domestic water requirement; (*ii*) population growth will also increase foodgrain requirement, which implies that a larger area will come under irrigation leading to increased water requirements for the same; and (*iii*) industrialization will increase the demand for water for production processes and indirect demand for pollution assimilation.

Visaria and Visaria (1995) have estimated India's future population till the year 2100. According to their estimates, the population would grow to 134 crores by the year 2025. A salient feature of the projection is that the urban population would grow from 25 per cent of the gross population (21 crores) in 1991 to 45 per cent (60.3 crores) by the year 2025.[1]

The domestic water demand will grow proportional to the population. The actual growth rate in domestic water demand would, however, depend on where the population actually grows, *i.e.* whether in the rural areas or urban areas, and how? The rise in domestic water demand induced by a growth in urban population would be higher than that created by the same growth in the rural population, as the water requirement for the basic survival needs are higher in urban areas. A faster growth in urban population will impact demand for water positively (WRI, 1995). Also, the per capita water demand will rise with an increase in income levels.[2] The available evidences indicate that water demand is highly elastic at low income and low water use levels and that elasticities for domestic water will decline gradually as income and water use rise (Rosegrant *et al.*, 1999).

We have begun our analysis with an assumption that the rural areas would require a bare minimum of 50 litres per capita per day of freshwater to maintain basic health and hygiene, and urban areas would require 150 litres per capita per day in the year 1990. For an estimation of domestic water requirements, a compounded growth rate of 2.0 per cent is assumed for GDP, with an income elasticity of water demand of one. These estimates suggest that the domestic water requirement would grow from 2.29 Mha m in 1990 to 9.30 Mha m in 2025 with 71 per cent of the growth coming from urban areas.

The water requirement for livestock has been estimated for the years 1990, 2000, 2010, 2020 and 2025 by interpolating the figures (estimates) given by CMIE for the years 1985 and 2025 (CMIE, 1991). As per the estimates, water requirement for livestock would be doubled during this period. But overall, it will continue to be a very small proportion of the total water requirement.

The projections of industrial water requirement for the future years were made on the basis of the following: (*i*) actual industrial production figures in 17 identified sub-sectors of industries for the year 1997; (*ii*) water requirement for unit production for each of the 17 sub-sectors; and (*iii*) the industrial growth rate for each of the sub-sectors of industry envisaged for the Ninth Plan period (data sources and further assumptions are given in end note)[3]. The projected industrial production figures and the water requirements for all the 17 sub-sectors are given in Table 5.1.

Increasing population will increase the requirements for foodgrains that include fine cereals such as wheat and rice, coarse cereals and pulses. The total foodgrain production in the country which was 176.6 million tonnes in 1990, went up to 198 million tonnes in 1996. In order to estimate the future foodgrain production requirement, it is also important to know the historical trends in foodgrain consumption. In fact, the per capita foodgrain consumption has declined over the last one decade in both rural and urban areas, in spite of an increase in the per capita foodgrain availability.[4] We have assumed a per capita foodgrain requirement of 500 gram per day, with the fine cereals such as wheat and rice contributing 80 per cent.[5]

Using the above assumptions, it is projected that by the year 2025, the foodgrain requirement of the country would grow to 282 million tonnes from 220 million tonnes in 2000 through 248 million tonnes, and 270 million tonnes in 2010 and 2020 respectively. A substantial proportion has to be produced within the country to ensure food security. As stated above, wheat and rice constitute 80 per cent of our diet so that ensuring a sufficient production of wheat and rice would be the key to food security. Sugar cane is another crop which requires a significantly higher proportion of the water used in agriculture. The consumption of sugar, which was around 13 kg per capita

Table 5.1: Industrial production figures for selected industrial sectors (1996-97) and the water use rates for unit production

Sl. No.	*Category of Industry*	*Annual Compounded Growth Rate (per cent)*	*Water Requirement (M^3)/Tonnes/ Year*	*Year 1997*	
				Production ('000) Tonnes	*Total Water Requirement MCM/Year*
1.	Integrated Iron & Steel	6.5	22	119480	2626.58
2.	Smelters	7.8	82.5	746.28	61.5681
3.	Petrochemicals and Refinery	11.7	17.0	3839.8	65.2766
4.	Chemicals — Caustic Soda	3.2	5.5	3,035.90	16.69745
5.	Textile & Jute	8.07	200	23,908.80	4781.76
6.	Cement	8.25	4.5	76,000	342
7.	Fertilizer	6.21	16.7	11,155.30	186.29351
8.	Leather Products	8	30	912.5	27.375
9.	Rubber	5.8	6.6	403.24	2.661384
10.	Food Processing	5.75	6.8	96,450	655.86
11.	Inorganic Chemicals	13.9	200	2,200	440
12.	Sugar	4.97	2.2	14,500	31.9
13.	Pharmaceuticals	4.3	25	4,190.00	104.75
14.	Distillery (Req. Per 1,000 litres)	12.2	22	1,790.80	39.3976
15.	Pesticides	11.9	6.5	86	0.559
16.	Paper & Pulp	9.11	200	3,100	620
17.	General Engineering	4.2	2.2	5,779	12.7138
	Total	7.75			10,015.39

Sources:

(*i*) Report of the National Commission on Integrated Water Resources Development titled, *Integrated Water Resources Development: A Plan for Action* (GoI, 1999).

(*ii*) Ninth Five-Year Plan, Government of India (1999).

per annum in 1994, was expected to become 25 kg per capita per annum by the year 2000 (Rangaswami, 1994). To meet this level of consumption, India needs to produce around 400 million tonnes, assuming that nearly 40 per cent of the cane would be used for jaggery making. The area under sugar cane stands at 3.4 million hectares. The production of sugar cane stood at nearly 240 million tonnes with an average national level yield of 65 tonnes/hectare (GoI, 1997 as quoted in Hindu Survey of Indian Industries, 1997: 480). The increase in production has to come from an expansion in irrigated sugar cane, though a 25 per cent increase in sugar cane yield is expected at the national level by the year 2000.[6] A lion's share of the irrigation water goes into the production of wheat, rice and sugar cane.

The water requirements for these crops are estimated on the following assumptions given in endnote[7]. The estimates of water requirement for agriculture for the period from 1990 to 2025 are presented in Table 5.2 for 5-year intervals. As per the estimates, the total water requirement for producing 180.32 million tonnes of cereal (wheat and rice) and 536 million tonnes of sugar cane from the irrigated areas in 2025 is 73.73 M ha m. Adding the water requirement for other agricultural produce, the total water requirement in 2025 is estimated to be 84.79 M ha m. Water requirement for rice production alone accounts for 42 per cent of the total required for agricultural production. The water requirements for wheat and sugar cane production were estimated to be 27.12 per cent, and 17.20 per cent respectively of the total water requirement.

The total water requirement for human and animal use industrial production and agricultural production is estimated to be 104.50 M ha m in the year 2025. Figure 5.1 shows by graphic representation, the growth in water requirement in each sector during the period of projection. As per the estimates, agriculture would continue to be the major user of water in the year 2025 at 81.13 per cent.

A salient feature of the estimates is the expected growth in domestic water requirement from 4.70 per cent in 1990 to 8.9 per cent in 2025 and industrial water requirement from a mere 2 per cent in 2000 to 8.83 per cent in 2025. These estimates, however, do not take into account the indirect demand for water to clean the natural freshwater reserves, which are polluted and contaminated by industrial effluents and municipal waste. The water required for pollution assimilation and other environmental services could be much more than the consumptive water use by industries.

Table 5.2: Projected water requirements in agriculture in M ha m (1990 to 2025)

Sl. No	*Year*	*Rice*	*Wheat*	*Sugarcane*	*Total Food*	*Oil seeds*	*Grand Total*
1.	1990	19.32	10.56	8.08	37.97	5.69	43.66
2.	2000	23.1	16.25	10.9	50.25	7.54	57.79
3.	2010	29.06	19.38	12.43	60.86	9.13	69.99
4.	2020	34.5	22.01	13.84	70.35	10.55	80.90
5.	2025	36.1	23.03	14.61	73.73	11.06	84.79

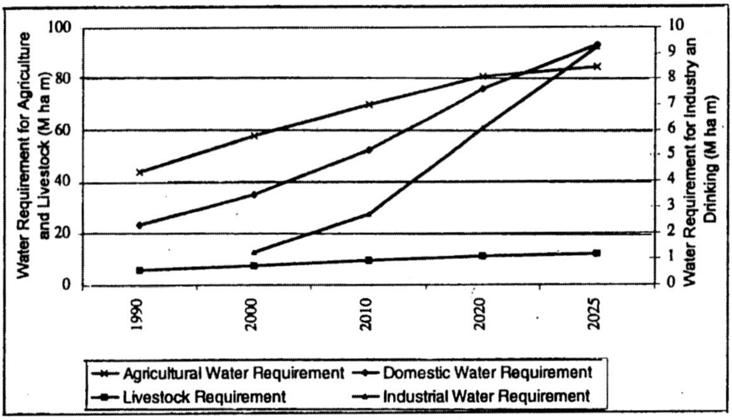

Fig. 5.1: Projected Future Water Requirement in Different Sectors

Researchers have, in the recent past, come out with estimates of future water requirements in India using various methodologies (*see* Rosegrant *et al.*, 1999; Seckler *et al.*, 1998), the most recent being the estimates by David Seckler and others from the International Water Management Institute (IWMI), Colombo. The IWMI's estimates were part of the study carried out to assess global water scarcity using a demand-based approach. The estimates were for three major sectors, namely, agriculture, domestic use and industry for the period from 1997 to 2025. The estimate of future agricultural water requirement was based on the assumption that the net irrigated area per capita would remain the same.

With this assumption, the irrigation water requirement was estimated for two scenarios. The first *business as usual* scenario, assumes the existing irrigation efficiencies would continue till 2025. The second scenario assumes that the country would achieve an irrigation effectiveness of 70 per cent of the gross irrigated area by the 2025. For domestic and industrial purposes, they have assumed a basic per capita requirement of 40 M^3 per annum throughout the projection period (Seckler *et al.*, 1998). Therefore, the only driver of change that they have assumed in their demand projections is the population growth. Whereas in our projections, the drivers of change are population growth, growth in per capita income levels, growth in industrial production and the change in food consumption levels.

Future Water Supplies

Our estimate of the future water supply potential is based on three important considerations: (*i*) the ultimate water resource utilization potential of our country; (*ii*) the status of development of water resources as of the year 1990; and (*iii*) the past growth trends in water development. The growing concerns about the social and environmental costs of water development projects and the emerging problems associated with water resource exploitation were taken into account while extrapolating the past growth trends to arrive at the future levels of exploitation.

The Central Water Commission estimates the total renewable freshwater falling in the country's landmass to be 400 M ha m, contributing a total annual runoff of 180 M ha m. According to them, nearly 38 per cent of this runoff (68.4 M ha m) could be harnessed by conventional technologies. The average annual groundwater recharge is estimated to be 45 M ha m. Considering the physical sustainability goals, utilizable groundwater is therefore 45 M ha m. Hence, the ultimate utilization potential of surface and groundwater together is 113.4 M ha m.

A total of 55.4 M ha m of water resources was developed in the country through large and small projects by 1990. Of this, 40 M ha m consists of surface water. The total water resource potential created was 60.6 M ha m in 1997 (Kumar, A., 1997). But, the growth in irrigation potential created through major and medium irrigation has slowed down during the last decade. This is evident from the fact that during the Seventh Five-Year Plan, the growth in irrigation potential created from major and medium irrigation schemes (surface) was only 2.01 million hectares, against a target of 4.30 million hectares (CMIE, 1991). One reason being that most of the surface irrigation projects that are under investigation or execution are bogged down by controversies to do with environmental and other issues. The other reason being a slackening of government investment in irrigation (Prasad, 1998) during recent years. In view of the growing resistance by environmental groups against large dam projects in India, the actual growth rate that might be achieved in future could be still lower. We have assumed that the utilizable surface water potential would grow steadily to become 75 per cent of the utilization potential by the year 2025. This keeps the annual growth rate pitched at 0.34 M ha m.

Now, the total utilizable potential of groundwater for irrigation is 38.50 M ha m. As of 1990, the net groundwater draft in the country was estimated to be 11.58 M ha m, with 30 per cent aggregate utilization (Kittu, 1995; Raju, 1995). However, this seems to be an underestimation, considering the gross irrigated area of 35.6 M ha from groundwater. Therefore, a total utilization level of 15 M ha was assumed for analysis. Three factors are considered for assessing the future growth trends in groundwater utilization potential.[8] Considering these factors, it is assumed that the utilization potential could grow to 60 per cent of the ultimate potential by the year 2025, with the annual growth rate pitched at 0.4 M ha m. In sum, putting groundwater and surface water together, an annual growth rate of 0.74 M ha m and a total water utilization potential of 78.30 M ha m is expected.

Growing Water Scarcity and Increasing Competition

As per the estimates, by the year 2025, the gap between the total water requirement and the total utilizable supplies will be 26.20 M ha m, indicating the magnitude of the impending water scarcity. This will have two major fall-outs.

First, there will be greater competition between various sectors for the scarce water. The greatest challenge will be to integrate the variety of concerns, namely social, economic, fundamental rights, environmental, in inter-sectoral water allocation. Failure to integrate such concerns will lead to conflicts between users and user groups. In the absence of proper legal and institutional regimes under which water can be allocated among the competing uses, rights will be politically contested. Second, as per the estimates, urban domestic water requirement constitutes 8.9 per cent of the total water requirement. Unlike rural domestic and irrigation demands that are dispersed, urban demands are concentrated. As urban areas have limited local sources, they are already dependent on rural water resources. Further as urban areas in India are economically and politically more powerful than rural areas, it is very likely that they manage the additional supplies required from the rural areas, depriving the socially and economically weaker groups of their share of water. Already, there are conflicts in the water scarce regions of Gujarat and Rajasthan over the diversion of water from irrigation reservoirs to urban areas for drinking during scarcity.

Even greater will be the environmental impact of the quality of wastewater generated in urban areas. Even today, only 10 per cent of the urban wastewater in the country is treated before disposing this off into the natural sinks like rivers and streams and lakes.

Industrial water use can conflict with rural drinking, irrigation and urban drinking in future (Ballabh and Singh, 1997; Ballabh *et al.*, 2000; Kumar *et al.*, 1999). As per the estimates, industrial water demand would increase rapidly in volumetric terms. With an enormous increase in the financial and environmental costs of water use, in future there will be greater pressure to reallocate the available water for economically most efficient uses such as industry. Agriculture is far less economically efficient than industry in terms of water use (Xie *et al.*, 1993). Also, the current political economy of growth based on industrialization and urbanization encourages re-allocation of water meant for other uses in favour of industries and urban areas leading to conflicts (Ballabh *et al.*, 2000). This has already started happening.

As the direct demand of water for industrial production purposes increases, the indirect demand of water for assimilating the pollution generated by effluents also increases. As untreated effluents find their way in to rivers and watercourses, pollution can deprive the communities living in areas around these water bodies of their share of water, on which they are directly dependant for their survival. Pollution of river water due to the disposal of industrial effluents has already led to several conflicts in the country.

Managing Demand

The crisis perpetuated by the growing water scarcity can be mitigated only through interventions that are aimed at achieving higher efficiencies in the existing uses and transferring water to high valued uses (Frederick, 1993; Postel, 1993). While doing so, we also need to ensure that adequate quantities of water are available for low value uses like crop production. Also, the instream flows will have to be protected and natural water quality will need to be maintained.

In a normal year, the available water supplies in a region may be sufficient to meet all the requirements while employing all the conservation technologies and pollution control measures. But, there could be several years, wherein the available water is insufficient to meet all the requirements. In such situations, a transfer of water from "low valued uses" to "high valued uses" (here it refers to uses that give high economic return per unit volume of water used) need to be encouraged, provided those uses which suffer due to such transfers are adequately compensated. Thus, demand management solutions should fulfil two important criteria: efficiency and cost effectiveness. Therefore, demand management in water will have three distinct stages: (*i*) allocating water among alternative uses; (*ii*) encouraging conservation; and (*iii*) maintaining in-stream flows and protection of water quality. The third component can be treated as integral parts of the first two because water transfers across sectors can ensure maintaining of in-stream flows, and conservation of water used for pollution assimilation through treatment measures would ensure water quality protection (Frederick, 1993).

Institutional Changes for Demand Management in Water

Technologies and institutions, both are required for demand management. But, the behavioural changes required to facilitate the adoption of technologies can be affected only through institutional changes. Therefore, institutional mechanisms and market-based (economic) instruments will be the fountainhead of the water management strategy for future.

Institutional Interventions for Transfer of Water from Low to High Valued Uses

The growing competition and concomitant conflicts between different sectors are major issues that need to be addressed in water allocation. Allocation of water for high valued uses holds the key to addressing them. The fundamental issue to be tackled is the diversion of water for uses that have low economic efficiency. Inequity in access to water from canals and groundwater is another major concern.

Markets and regulations can be sought as instruments for water allocation (Frederick, 1993; Howe *et al.*, 1986). Howe *et al.* (1986) suggest six criteria for comparing alternative institutional arrangements to allocate water: *(i)* flexibility in allocating supplies in response to both short-term and long-term changes; *(ii)* security of tenure to encourage investment in and maintenance of water-using system while allowing for users to respond voluntarily to incentives to reallocate supplies; *(iii)* whether the user is confronted with the real opportunity costs of the resource; *(iv)* predictability of the outcome of the transfer; *(v)* equity impacts; and *(vi)* whether public values are adequately reflected in the process. Frederick (1993) adds low-transaction costs of moving water from one use to another to this list.

Both markets and regulatory approaches are likely to fall short of satisfying all these criteria. Let us take the case of maintaining in-stream flows in natural watercourses and rivers, which forms an important demand management objective to be achieved through water transfers. As this is a public good, the users of in-stream flows have no incentive to pay for the services it provides, so that markets are likely to fail in this sector. The enormous geographic and temporal diversity in water supply and demand situations suggest that no single institutional arrangement is likely to be preferred in all instances (Frederick, 1993). Howe *et al.*,(1986) argue that markets meet their six criteria better than any likely alternative in many situations.

This argument, by and large, is valid in the Indian context. The spatial and temporal variation in water availability is very high in India, caused by heterogeneity in hydrology and geo-hydrology and high inter-annual variability in rainfall. For instance, 62 per cent of India's water resources are concentrated in the Indo-Gangetic basin (GoI, 1999: 21). So is the variability in demand situation. In socially and economically backward regions such as Bihar and Eastern Uttar Pradesh, irrigation demand is very low, though water resources are abundant, and therefore, problems of water logging due to rising groundwater levels caused by flooding and excessive irrigation from canals are encountered (Shah, 2000). The demand for water for industrial and urban uses also remains very low in these regions. The demand management challenge here is to create increased demand for groundwater through market institutions, as investment in public tube wells has not been very effective.

On the other hand, demand for water is extremely high in

the arid and semi-arid regions of Gujarat, Rajasthan, Tamil Nadu and Maharashtra where water resources are very scarce, and groundwater is the major source of water for all purposes. Pumping regulations in areas facing over-development problems through groundwater legislation, control of institutional financing for well development and restrictions on power connections for pumps has been by and large ineffective in these regions (Kumar, 1995; 1999/2000; Janakarajan, 2002). Further, the large distances involved in conveyance of water between regions of abundance and shortage reduces the ability of the government to invest in public water systems for the supply of water in bulk as it has serious financial and environmental imperatives.

But, the absence of well-defined private property rights regimes can be a major source of uncertainty about the negative environmental impacts of resource use, leading to sustainability use (Pearce and Warford, 1993; Kay *et al.*, 1997). The other issue is one of efficiency. Since there are no restrictions on the amount of water that can be pumped by a well owner, even in regions like north Gujarat, where water markets are extensive, the well owners are not confronted with the real opportunity costs of using water. This is because the price at which water is sold to irrigators is still lower than the net economic return from irrigating the crops. In north Gujarat, which is one of the most water scarce regions in India, the price that tube well owners charge for water is in the range of Rs. 1-1.5/m^3, while the net return from irrigated crops is close to Rs. 5/m^3 (Kumar and Singh, 2001). Also, the lack of opportunity to transfer water to more demanding uses such as industry and urban drinking is an issue in many regions. So the well owners would prefer to pump extra water and sell this rather than using water his/her fields more efficiently and use the saved water for selling.

Lack of well defined private property rights regimes also lead to the resource rich and the powerful over-using the resource and depleting it. This can significantly reduce the ability of the poor to access groundwater, perpetuating negative impacts on access equity. Therefore, establishing privately owned property rights is critical to establishing conditions under which, markets and prices provide individuals with opportunities and incentives to develop, transfer and use the resource (Frederick, 1993). This will address the issue of equity.

In the Indian context, many researchers in the recent past have suggested the establishment of property rights as a means to build institutional capability to ensure equity in allocation and efficiency in use of water across sectors (Singh, 1995; Chaudhari, 1996; Kumar, 1997a; Saleth, 1996; Kumar *et al.*, 1999, Kumar, 1999/2000). But, again if the rights are allocated only to use water, it can create incentives to use it even when there is no good use for it (Frederick, 1993). Therefore, water rights have to be tradable (IRMA/UNICEF, 2001; Kumar and Singh, 2001). The volumetric use right of different use sectors and individual users can be defined and established by the agency concerned by using a variety of physical, social and economic parameters. River basin organizations (RBOs) can be a viable institutional arrangement to evolve norms regarding water rights of different use sectors and individual users, and allocate the rights and enforce them.

A user who needs more water than the actual entitlement can purchase the water rights from another user, by paying prices that are determined by the supply-demand interactions. When property rights are tradable, the price of water will reflect the opportunity cost for using water. The markets and market determined prices could induce incentives among users to shift to alternative uses that provide higher or same economic returns as the price of water, or continue the existing uses with more efficient practices or resort to selling (Frederick, 1993). Such transfers can promote access equity and efficiency in use (Kumar, 1997a; Kumar *et al.*, 1999, Kumar, 1999/2000).[9] Empirical evidences collected on the functioning of groundwater irrigation institutions in north Gujarat show that under a system of fixed volumetric water use rights, farmers prefer to grow mustard, which is less water intensive, in larger area as compared to wheat, though the earlier one has much lower land use productivity than wheat, but getting same water use productivity (*see* Kumar 2000 for details).

Tradable private property rights are suggested for

groundwater and water supplied from public reservoirs for irrigation and domestic uses. In the case of groundwater, as individuals enjoy access to the resource, private property rights for individual users are envisioned. Also, in the case of canal water supplied for irrigation, private individual property rights are suggested. But, in the case of municipal water supplies, the rights of the water utilities over the water from public systems can be recognized.

Frederick (1993) suggests that for markets to function efficiently, the full benefits and costs of transfer should be borne by the seller and buyer. Generally, he states this is not possible due to the third party effects of water transfer.

> Water resources may also produce public goods that are affected either positively or negatively by a water transfer. Non-paying individuals cannot be excluded from enjoying the benefits of public goods. As a result, the users of such goods have an incentive to free ride, and producers have an incentive to under-invest in these goods from a social perspective.

Allowing the user to transfer only the consumptive portion of the water he/she uses can reduce the third party effects of water markets. Government will have to play a great role in reducing the third party effects of water transfers, apart from defining and establishing the individual property rights. Similarly, government has to invest in protecting or retaining the public goods that are affected by water transfers.

Efficient water markets are more likely to come up in groundwater irrigated areas due to the following reasons: extensive use of conduits for water conveyance increases the transferability of water and reduces the third party effects; ability to measure the rights purchased; and, relatively low distance of water transfer, which reduces the transaction cost.[10]

In order to meet the ecological and environmental needs such as maintaining in stream flows in rivers, that are public goods, the river basin organization can allocate sufficient water from the basin in volumetric terms. Thus, if the water market operates under well-defined tradable private property rights for competitive uses enforced by institutions such as RBOs, with some regulations to provide water for environmental needs and reduce third party effects of water transfers, it can bring about demand management.

Water Saving

Water conservation has three distinct components: *(i)* conservation by preventing the loss of stored water; *(ii)* conservation by preventing the loss of water from the system during conveyance from supply source to the point of usage; and, *(iii)* conservation of water at the user level by adopting efficient water use technologies, so as to enable the user maintain the same services. First of all, we will analyse the scope of each of these options in increasing the effective water availability by examining the current efficiency levels in use.

In irrigation, the storage losses are very high for surface reservoirs. The potential for saving this water is very high.[11] Again, the conveyance or the network losses are very high for the surface irrigation systems in the country. It is believed to be in the region of 45-55 per cent for many of the large surface irrigation systems with extensive distribution network consisting of several hundred kilometres of unlined channels.

In India, the farm-level efficiencies in surface irrigation systems are very poor on account of the very high field evaporation, evapo-transpiration, percolation, and runoff losses due to flood irrigation and the poor on-farm water management practices being adopted by farmers. There is enough empirical evidence to substantiate this. For instance, Vamadevan and Destane (1968) found that a 70 per cent water loss in rice fields was due to percolation through the sandy loam soils of Delhi and only 480 mm out of 1200 mm of water was actually used consumptively by rice; but, the use of irrigation technologies such as drips (for row crops) and sprinklers (for field crops) that can result in significant saving in water used for irrigation is in just one per cent of the total irrigated area (cited in El-hanbali, 1992).

There are no storage losses in groundwater irrigation. The network losses are also very low due to the generally short conveyance systems used for water conveyance from the source to the fields (Xie *et al.*, 1993). The field efficiencies are also generally higher due to greater control over water and manageable discharges. But, ideally, if the quality of power supply is good, under flat rate pricing, farmers use irrigation water excessively, resulting in economic efficiencies that are sub-optimal as the marginal cost of pumping water is zero. However, the quality of power supply in rural areas is very poor. Uncertainties about the time and duration of power supply create an incentive among farmers to use excess water when the supply is available. This not only leads to inefficient use, but also low levels of return (Kumar and Singh, 2001).

In urban water supplies, distribution losses comprising technical losses due to leakage and breakage and non-technical losses due to pilferage are very high. The percentage of unaccounted for water to the total water supplied in the three metropolitan cities, namely Mumbai, Delhi and Culcutta are 25 per cent, 30 per cent and 35 per cent respectively (ADB, 1993). In Ahmedabad, the loss in the distribution system of municipal water supply is as high as 45 per cent (AUDA, 1994). The technical losses are huge, where the distribution systems are very old. Pilferage is huge where water is transported over long distances, unprotected. In the rural water supply sector, the potential for water saving exists only in regional water supply schemes, as distribution losses are high. In village level water supply schemes, the distribution losses are potentially low.

The levels of production technologies in use in Indian industries are that of other third world countries and therefore, the potentials for efficient use of water in production systems in water-intensive industries such as cement, ceramics, petrochemical and chemical industries, paper industries, thermal power stations, etc. are great. The potential for conserving the water used for pollution assimilation through reductions in the pollution load (biological, chemical) of industrial waste is tremendous. The following section will examine what kind of measures will work for bringing about conservation at various levels.

Pricing of Water: The fact that water is a scarce economic input should be a major decisive factor in determining the price of water (Kay *et al.*, 1997). As a general principle, the price of water for competitive use such as in the irrigation and water intensive sectors, means that the pricing of water should be fixed in such a way that it discourages uses that are economically inefficient (Kay *et al.*, 1997; Pearce and Warford, 1993).

Pricing of Irrigation Water from Canals: In India, irrigation water is heavily subsidized. The annual irrigation subsidies are estimated to be around 5400 crore rupees (Wolf and Hubener, 1999). After independence, the Indian governments saw irrigation as a means of welfare, and were reluctant to raise the irrigation fee charged to poor farmers. As irrigation services declined and the agencies weakened, farmers became reluctant to pay the water charges (Brewer *et al.*, 1999). Also, the charges are paid on an acreage basis and are not reflective of the volume of water used. It is believed that the lack of linkages between volumetric water use and water charges, and lack of agency capability to recover water charges and penalize free riders create an incentive for overuse or wasteful practices.

In spite of the recommendations of the second irrigation commission, the state irrigation bureaucracies have failed to hike water charges that make economic sense due to the potential social and political ramifications such measures can create. The failure has its roots in the absence of an institutional capability to improve the quality of irrigation services and correctly monitor water use, the lack of institutional arrangements at the lowest level to recover water charges from individual farmers, and to enforce penalties on free riders. A few successes have been seen in areas where the PIM programme is implemented, where farmers have shown the willingness to pay more for irrigation services to the Water User Associations (Ballabh *et al.*, 1999).

A recent analysis for Gujarat shows that the negative impact of a rise in the cost of irrigation water can be offset by the differential yield due to increased reliability of irrigation water delivery (Kumar and Singh, 2001). Thus leads us to the conclusion that the rates for canal water can be increased to substantially higher levels, provided the quality of irrigation water is enhanced. But, water pricing for irrigation can impact poor farmers adversely, if pitched at higher levels (Frederick, 1993). One of the ways to reduce the negative impacts on access equity is to introduce a progressive pricing system. An appropriate pricing structure for water followed by clearly recognized private property rights and a good quality irrigation service could help achieve the desired effect of pricing changes on demand management.

Pricing of Urban Water Supplies: As urban areas are likely to emerge as the second largest users of water by the end of the first quarter of this century, promoting efficient use in the urban water sector will create a strong leverage in managing the overall demand of water to match the supplies. For a critical appraisal of this, we would have to examine the institutional factors that influence domestic water use in Indian cities.

The first institutional factor is subsidy. Domestic water supply is highly subsidized in many Indian cities (*see* Table 5.3). In three out of the four metropolitan cities of India, the operation subsidy ration is in the range of 1.25 to 1.50. The average water tariff is ridiculously low in these cities when compared to many large cities in the Asia-Pacific region. Against a rate of 0.325 USD per cubic metre of water (for use up to 20 M^3 per month) in Singapore, and 0.15 USD in Bangkok, a unit rate of 0.01 USD is levied in Delhi, Chennai and Mumbai.

Table 5.3: Cost of Production, Price and Unaccounted Losses in Water Supplies

Name of City	*Cost of Water Production (USD/m^3)*	*Price of Water (USD/m^3)*	*Percentage of Unaccounted Losses*
Kolkata	0.04		36.0
Chennai	0.095	0.01	N.A
Mumbai	0.04	0.01	24.0
Delhi	0.02	0.01/0.025*	30.0

Note: * Above 20 M^3 of water per month.
Source: ADB, 1993.

Secondly, domestic water supply is not fully metered even in the largest Indian cities. In Kolkata, Delhi and Chennai, public taps account for nearly 25 per cent of the domestic water supplied by the urban water utilities. But, standpipes are not metered in these cities and no payments are levied for the use of water from the public taps. Consumer metering is again partial in these cities.

The third institutional factor is the practice of fixing the water charge. In Mumbai, Chennai and Delhi, a combination of metered use and flat rate is followed for payment of water charges for house connections. In Kolkata, the payment of the water charged is attached to property tax. Together, these factors do not leave any incentive for an efficient use of water (ADB, 1993).

Incentives for Pollution Control: Although the existing pollution control norms are stringent, the legal powers to enforce them are not given to the pollution control agencies. Industries and municipal authorities often get away with flouting the pollution control norms (Ballabh *et al.*, 1999; Kumar and Ballabh, 2000). The fact that the agency that monitors "pollution" is the same agency that enforces the "norms", which is against the institutional design principle for sound water management (Frederiksen, 1998), further weakens the agency. Over and above, the polluting industries and municipal authorities are not confronted with the opportunity costs of pollution as the penalty paid for pollution is much lower than the investment in effluent treatment facilities to avoid or mitigate pollution.

To reverse the trend, pollution taxes that reflect the volume of effluent discharged and the level of toxicity of the effluents need to be introduced. Incremental block rates can be introduced along with the creation of two separate institutions, one for monitoring pollution, and the other for levying pollution taxes and carrying out corrective measures against pollution. In doing so, the industries and municipalities will have strong economic incentives to minimize the use by investing in low water consuming technologies, water treatment and recycling.

The effectiveness of economic instruments in affecting pollution control, to a great extent, would depend on the water quality monitoring (WQM) systems established by official agencies. Water quality monitoring in India is very poor and there are several technical issues associated with water quality monitoring (Kumar and Ballabh, 2000 on the basis of Moench and Metzger, 1992; Moench, 1995a; Biswas, 1996). These issues need to be adequately addressed.

Pricing of Electricity: Many researchers have suggested the rational pricing of electricity as a potential fiscal tool for sustainable groundwater use in India (Moench, 1995b; Saleth, 1997). Many argue that the flat rate-based pricing structure in the farm sector creates incentive for farmers to over-extract it as the marginal cost of extraction is zero. However, empirical evidence does not seem to suggest any impact of the cost of extraction on the use of groundwater for irrigation in water scarce areas (Kumar and Patel, 1995; Kumar, 1995b), and in areas where water charges reflect the scarce value of the commodity (Mohanty and Ebrahim, 1995). Diametrically opposite views about the impact of electricity pricing on groundwater use also exist. For instance, Shah argues that flat rate pricing will induce a positive impact on access equity in groundwater (Shah, 1993).

Nevertheless, conventional wisdom suggests that if the pricing structure is designed in such a way that it reflects the actual volume of water and electricity consumed, with a progressive increase in unit rates, it can induce incentives among farmers to use water more efficiently. But, such fiscal measures for indirect management have not been attempted in any of the states where groundwater depletion problems actually exist, due to their social and political ramifications. In Gujarat, unit pricing had been followed till 1987. But, due to a widespread pilferage of electricity through tampering of meters and illegal connections, the State Electricity Department was compelled to opt for a flat rate system of pricing.

The pricing of electricity in the farm sector has strong social and political ramifications. The influential farm lobby is likely to protest against any move by the government to either change the tariff structure from flat rate to unit horsepower and/or to increase the tariff. Therefore, policies with regard to water and electricity pricing are guided by strong political and economic considerations (Moench, 1995a).

Studies carried out in the Mehsana district of north Gujarat and coastal Saurashtra on diesel and electric well commands show that control over watering will have a greater bearing on the net returns from irrigation than the cost of irrigation. In spite of the higher direct cost of irrigation using diesel pump as compared to the implicit cost of irrigation using electric motors, the command areas with diesel pumps were found to receive more irrigation and give much higher yield levels as compared to those with electric motors (Kumar and Patel, 1995; Kumar, 1997b). This means that the desired impacts of changes in the pricing structure of electricity on the economic efficiency of irrigated crops can be realized only if the quality of power supply is ensured (IRMA/UNICEF, 2001). Further, the effectiveness in implementing pricing policies would depend heavily on the ability to supply high quality in rural areas and meter its use, levy the charges without default, prevent thefts, and penalize free riders. Involvement of private sector partnership in power distribution in rural areas is the major reform suggested (Kumar and Singh, 2001).

Role of Institutions: Technological and institutional changes alone will not be sufficient to bring about desired changes in the demand situation. New institutions such as RBOs will also have to be created by the agencies for enforcing rules and norms relating to volumetric water rights, water prices etc. Apart from allocating and enforcing tradable water rights among different sectors and users, RBOs can provide a coordination structure for the agencies looking after water-related services such as irrigation, drinking water supply in rural and urban areas and industrial water supply, and water resource management functions such as soil and water conservation, catchment protection and pollution control for sustainable basin management.

In the case of canal water, the volume of water delivered is metered only at the outlet level. The farmers do not pay for the water lost in the large distribution systems and hence do not have incentives to conserve it. Unauthorized withdrawal of water from canals is also very common. A technological approach alone will not be sufficient in this case. If water is delivered to the farmers' organizations at the distributory level, instead of minor outlets, and charged on a volumetric basis, it will create incentives among them to conserve it and protect it from theft. The farmers' organizations at the distributory and outlet levels can, in turn, charge for water on the basis of volumetric delivery to the outlets and individual fields, respectively. Such institutional arrangements can ensure efficient distribution, proper measurement of water delivery, increase recovery and prevent unaccounted losses in transmission. The major institutional innovation required to implement electricity-pricing change effectively is to set up electricity users' co-operatives at the village level. The private agency involved in power distribution can sell electricity in bulk to the cooperative, which in turn can sell it to individual farmers.

Similarly, in urban water supplies, the technical losses in the distribution system can be prevented by an investment in regular maintenance and repairs. But, the non-technical losses can be prevented only through putting the proper institutions in place. Also, in regional water supply schemes that supply water to a large number of villages, proper institutional arrangements have to be made at the system and village level for protecting the water systems and for maintenance and repair, ensuring equitable water distribution at the village level and a recovery of water charges.

Overall Impact of Economic Instruments on Agricultural Water Demand: The overall impact of economic instruments such as pricing of water and energy on demand for water in agriculture is analysed on the basis of the assumption that introduction of these economic instruments would force farmers to adopt efficient water use technologies in irrigation. For the analysis, we assume that with the introduction of a rational pricing of electricity, well irrigators growing wheat would use sprinkler irrigation and those growing sugar cane would use drip irrigation leading to greater farm efficiency and an increased yield. We also assume that with institutional innovations in the management of canal irrigation, the farmer organizations would introduce a more efficient conveyance technology.

The following assumptions are made with regard to the extent of water saving per unit area due to an adoption of water-saving technologies. *First*: adoption of sprinkler irrigation will help to reduce the water requirement at the farm level by 25 per cent and increase the yield by 20 per cent. *Second*: adoption of drip irrigation would reduce the water requirement at the farm level in sugar cane by 50 per cent, and increase the yield by 30 per cent. Further, the conveyance efficiency in the canal system is assumed to be 75 per cent instead of 55 per cent. This way, the supply requirement for wheat would be 0.48 ha m in the case of canal water, and 0.32 metre in the case of groundwater. Similarly, the supply requirement for sugar cane would be 0.8 ha m instead of 2.18 ha m. Here it is assumed that the irrigation of sugar cane is mainly from canal water. However, no water saving technology is considered for paddy.

Due to the decrease in water supply per unit area (on account of improvements both in field as well as conveyance efficiency) and increase in yield per unit area, the water use efficiency would increase for wheat and sugar cane significantly for both well and canal irrigation. The new water use efficiency figures would be 7.5 t/ha m for wheat grown with canal water and 11.0 t/ha m for wheat grown with well water. The water use efficiency would be 130 t/ha m for sugar cane (*see* Table 5.4).

The total water requirement for growing wheat and sugar cane using water-saving technologies would be only 16.12 M ha m (*see* Table 5.4) instead of 37.64 M ha m (23.03 M ha m for wheat and 14.61 M ha m for sugar cane as provided in Table 5.2). The total reduction in water demand for growing cereals and sugar cane would, therefore, be 21.5 M ha m. Here, we have not included the reduction in demand for water in growing oil seeds due to the adoption of water saving technologies for these crops. As a matter of fact, this will be quite significant. For instance, crops such as mustard, castor, cotton and groundnut are highly adaptable to water saving technologies such as drips and sprinklers. While sprinklers are suitable for mustard and groundnut, drip system can be used for castor and cotton.

Table 5.4: Overall impact of conservation measures on demand for water in agriculture

Type of Irrigation Source	*Improved Water Use Efficiency for Wheat (t/ha m)*	*Improved Water Use Efficiency for Sugarcane (t/ha m)*	*Irrigated Wheat Production in 2025 (Million Tonne)*	*Irrigated Sugarcane Production in 2025 (Million Tonne)*	*Water Requirement for Wheat in 2025 (M ha m)*	*Water Requirement for Sugarcane in 2025 (M ha m)*
Canal water	7.5	130.00	55.27	536.00	7.40	4.12
Groundwater	11.0		55.27		4.60	

Legal and Policy Frameworks

To achieve sustainable water use, the approach to water development should be based on river basins. This would help to realistically assess the total water availability and the sustainable levels of extraction and enable the analysis of various socio-economic systems affecting demand and use and their inter-dependencies (Mitchell, 1990). Taking the river basin as the unit for planning helps to identify various supply and demand side interventions to manage the basin's water resources. It also helps integrate the variety of concerns in water development such as economic growth, social advancement, environment, and ecosystem management and minimize the conflicts between different sectors of water use.

Policy change should facilitate a paradigm shift from designing strategies for increasing supplies to altering the socio-economic systems affecting water use to manage the demand of water. Allied sectors for policy change would include irrigation, agriculture, rural water supplies, urban water supplies and industry. Some of the highlights of the policy changes suggested are: (*i*) involvement of farmer organisations and their representatives in irrigation management beyond the outlet level should be made mandatory for farmers in the command to claim irrigation water; (*ii*) involvement of village water users' groups to be made mandatory in operation and maintenance of water supply schemes and institutions of higher order in the management of large water supply schemes, especially their distribution systems; and (*iii*) electricity board to involve private agencies and rural electricity co-operatives in the distribution of electricity in the farm sector.

Establishment of tradable private property rights is the major institutional change suggested in the water sector. For proper enforcement of the property right regimes, it is essential to have a proper legal framework. There should be clear-cut laws on water and such laws should define the rights of the state and every potential user over water resources in all its natural forms. The laws should also explicitly state the different types of rights such as rights for basic human survival, rights to drinking water, rights to instream uses, riparian rights, etc. It should also clearly indicate the circumstances under which the rights can be said to be "violated", the offences, and how the violation of such rights should be compensated. The legal framework should also provide for communities to establish their own norms, rules and regulations with regard to water use and management, to protect and manage their water resources and water supply infrastructure.

Summary and Conclusion

The solution to the growing water crisis lies in institutional changes that alter the social systems to manage the demand for water. The institutional changes suggested for this are: (*i*) establishing tradable private property rights in groundwater and canal water to enable and promote economically efficient uses; (*ii*) appropriate pricing of water to encourage efficient use of water from public water supply systems including canals and urban water supply systems including industries; (*iii*) unit pricing of electricity used for groundwater pumping to promote efficient use of water in irrigation; and (*iv*) taxing industrial and urban wastewater on the basis of volumetric effluents and pollution levels to encourage conservation of water. Drawing upon the analysis, the role of economic instruments in tackling water scarcity and mitigating conflicts over water use will be quite prominent. Multiple hierarchies of organizations are suggested in surface irrigation for increasing system efficiency as well water use efficiency at the farm level. In the power sector, it is important to strengthen institutional capabilities through power sector reforms such as private sector involvement in distribution across villages, and the formation of co-operatives in power supply among users. Major changes are required in the existing legal framework to provide for creation of RBOs, enable enforcement of tradable private property rights in water.

Notes

1. These projections, by and large, match with those made by the World Resource Institute, World Bank and United Nations Development Programme.
2. This argument is supported by the fact that only 4.6 per cent of the rural households and 50 per cent of the urban households enjoyed adequate access to sanitation in 1996 (*Source*: Economic Survey 1996-97 as quoted in TERI, 1998). Improved economic conditions will encourage people to have better sanitation facilities, which would means increased water requirement.
3. The actual industrial production figures and the expected growth rates during the Ninth Plan period (1997-2002) were obtained from the Report of the National Commission on Integrated Water Resources Development titled *Integrated Water Resources Development: a Plan for Action* (GoI, 1999). They were later on verified and corrected on the basis of the data provided in the Ninth Plan document (GoI: 661). The per unit water consumption figures—as taken from the Report of the National Commission on Integrated Water Resources Development (Annex 5.7: 455), were based on information available from Central Pollution Control Board (CPCB) and the Industrial Association of the particular sector of industry. For demand projections, the growth rates envisaged for the Ninth plan period was continued till 2025.
4. Results of National Sample Survey data show that during the period 1987 to 1997, the per capita aggregate cereal consumption (including course cereals) declined from 173 kg to 154 kg in the rural sector and 134 to 124 kg in the urban sector. The weighted average for the whole country is 148 kg per capita per annum (405 gram per day). The per capita availability of pulses in the recent years is on an average of 35 grams per day (Thamarajakshi, 1999). However, it is important to note that the per capita foodgrain availability per day steadily increased from 456.1 grams during 1981-85 to 484.2 gram per day during 1991-95 (Dev, 1997).
5. This is based on the assumption that the real demand for cereals will not decline due to the consumption pattern shifting towards meat and eggs, the production of which require foodgrains for animal and poultry feed.
6. The sugar cane yield has shown rising trends in most of the states. According to R.S. Paroda, Director General, Indian Council of Agricultural Research, the sugar cane yield can be increased by 15-20 per cent by increasing the efficiency of use of fertilizers and irrigation water (Paroda, 1998).
7. The assumptions are: *(i)* The requirement of rice and wheat in the

food consumption basket would become equal by the year 2000 and would continue till 2025. *(ii)* The contribution of rain-fed paddy to the total paddy production would reduce from 55 per cent to 40 per cent and that of rain-fed wheat from 8 per cent to 2 per cent during the period from 1990 to 2025. *(iii)* The percentage contribution of groundwater and surface water to production of rice is assumed to become equal by 2020 (30 per cent each) from 15 per cent: 30 per cent in 1990, 20 per cent: 30 per cent in 2000 and 25 per cent: 30 per cent in 2010. *(iv)* The yield levels of wheat and rice for canal and groundwater irrigated areas were assumed as 3.0 t/ha and 3.2 t/ha and 2.70 t/ha and 3.0 t/ha respectively. The yield level for sugar cane was taken as 65 t/ha in the base year and 80 t/ha in 2000. *(v)* The total water supply requirement for irrigating one hectare of wheat is taken as 0.48 hectare-metre (*ha m*) and 0.80 ha m for groundwater and canal irrigated areas. For rice, values taken are 1.2 ha m and 1.8 ha m, respectively and for sugar cane, the value is 2.18 ha m. Here, the physical efficiency of surface irrigation (including the farm efficiency) was taken as 55 per cent including the field efficiency for the entire time frame and that for groundwater irrigation was assumed as 100 per cent. *(vi)* Water requirement for producing oil seeds, vegetables and other crops was assumed to be 15 per cent of the water required for producing cereals and sugar cane and treated as constant over time.

8. First, there are great spatial variations in groundwater potential across the country. In areas where groundwater resources are abundant such as the Indo-Gangetic plains and eastern India, the growth would continue to be low. Second, the arid and semi-arid regions such as Western India and the North-west, which are facing problems of over-development, are unlikely to contribute to future growth. Third, the fact that increasing number of legislations and regulations on the use of groundwater will have some impact on the withdrawal of groundwater.
9. For instance, let us assume that the rights to use water from a reservoir are allocated between irrigators and the municipality. In a drought year, the supplies from the reservoir can fall far short of the total demand from both the sectors, leading to major cuts in the supplies to each sector. But the potential impact of the cuts will be severe on urban water supplies. Drinking being a high priority demand, the urban area can be willing to pay heavy prices for the water, pushing the market prices upwards. This may result in irrigators selling their water rights to the municipality.
10. Well-developed groundwater markets exist in many parts of India and are more widespread in the groundwater-irrigated areas of Tamil Nadu, Gujarat, Haryana, Bihar, Eastern Uttar Pradesh and Coastal Orissa.

11. During the droughts of 1987-88, 30 per cent of the water—which otherwise have been lost in evaporation from reservoirs—was saved by spreading chemicals and plastics on water surface at the cost of Rs. 3 per cubic metres, which is far less than the cost of transporting water from elsewhere (Chitale, 1991).

References

Asian Development Bank (1993). *Water Utilities Data Book Asian and Pacific Region*. Philippines: Asian Development Bank, November.

AUDA (1994). "Synopsis of Technical Feasibility Report of Water Supply for West Ahmedabad Urban Complex Area," report submitted to Ahmedabad Urban Development Authority (AUDA), Multimedia Consultants Private Ltd., November.

Ballabh, V. and Katar Singh (1997). "Competing Demands for Water in Sabarmati Basin: Potential and Future Conflicts," paper presented at the IDPAD Seminar on Managing Water Scarcity: Experiences and Future Prospects, Amersfort, The Netherlands, October 11-17.

Ballabh, V., R. Prabhakar, M. Dinesh Kumar, Vina Vani, Arun Mudgerikar and Bishwas Joshi (2000). "Western Regional Consultation on Freshwater Issues," report of the Workshop held at Institute of Rural Management Anand, October 11-12.

Ballabh, V., M. Dinesh Kumar and Jayesh Talati (1999). "Food Security and Water Management: Institutional Challenges for Early 21st Century," paper presented at the National Symposium on Building and Managing Organisations for Rural Development held at Institute of Rural Management, Anand, December 13-14.

Biswas, A.K. (1996). "River Water Quality Monitoring in Developing Countries," in Aly M. Shady (ed.) *Management and Development of Major Rivers. Water Resources Management Series-3*. Oxford University Press: Culcutta.

Brewer, J., Shashi Kolavalli, A.H. Kalro, G. Naik, S. Ramnarayan, K.V. Raju and R. Shakthivadivel (1999) *Irrigation Management Transfer in India: Policies, Processes and Performance*. Oxford & IBH: New Delhi and Calcutta.

Chaudhari, S. (1996). "To Sell a Resource," *Down to Earth*, February 15.

Chitale, M.A. (1991). "Comprehensive Management of Water Resources: India's Achievements and Perspectives," unpublished paper for the 1991 June International Conference on Water Resources Management, Washington D.C.

Centre for Monitoring Indian Economy (1991). *Basic Statistics Relating to the Indian Economy, Land, Water and Minerals, Economic Intelligence Service*, Vol. 1, August 1991, Table 3.3.

Datt, G. (1997). *Poverty in India and Indian States*. Washington D.C.: International Food Policy Research Institute.

Dev, S.M. (1997). *"Growth, Employment, Poverty and Human Development," Review of Development and Change, Vol. 2 (2), July-December, 1997.*

El-Hanbali, U. (1992). "Sprinkler and Drip Irrigation Project Identification Mission," Back to Office Report, ASTAG, World Bank, March.

Engelman, R. and P. LeRoy (1993). *Sustaining Water: Population and Future of Renewable Water Supplies*. Washington, D.C.: Population and Environment Programme, Population Action International.

Frederick, K.D. (1993). "Balancing Water Demand with Supplies: The Role of Management in a World of Increasing Scarcity," Technical Paper 189, Washington D.C: World Bank.

Frederiksen, H.D. (1998). "Institutional Principles for Sound Management of Water and Related Environmental Resources," A.K. Biswas (ed.) *Water Resources: Environmental Planning, Management, and Development*. Tata McGraw-Hill: New Delhi.

Glieck, P.H. (1997). "Human Population and Water: Meeting Basic Needs in the 21st Century," in R.K. Pachauri and Lubina F. Qureshi (eds.) *Population, Environment and Development*. Tata Energy Research Institute: New Delhi, pp. 105-121.

Government of India (1997). *Economic Survey (1996-7)*. New Delhi: Government of India.

——— (1999). *Integrated Water Resource Development: A Plan for Action*. Report of the National Commission on Integrated Water Resources Development, Volume I, New Delhi: Government of India.

——— (1999). *Ninth Five Year Plan*. New Delhi: Government of India.

Hindu Survey of Indian Industry (1997). *The Hindu Survey of Indian Industry*. Kasturi & Sons: Madras.

Howe, Charles W., Dennis, R. Schurmeier and W. Douglas Shaw, Jr. (1986). "Innovative Approaches to Water Allocation: The Potential for Water Markets," *Water Resources Research*, 22 (4), pp. 439-445, April.

Institute of Rural Management, Anand/UNICEF (2001). *White Paper on Water in Gujarat*, report submitted to the Government of Gujarat, Gandhinagar, April.

Janakarajan, (2002). "Wells and Illfare: An Overview of Groundwater Use and Abuse in Tamil Nadu, South India," paper presented at the annual partners' meet of the IWMI-Tata Water Policy Research Programme, Anand, February 19-20.

Kay, Melvyn *et al.* (1997). *Water: Economics, Management and Demand*. E & FN Spon: London.

Kittu, N. (1995). "Status of Groundwater Development and its Impact on Groundwater Quality," in M. Moench (ed.) *Groundwater Availability and Pollution, The Growing Debate over Resource Condition in India*. Monograph, Ahmedabad: VIKSAT-Natural Heritage Institute.

Kumar, A. (1997). "Water Management: Its Future Challenges," *Agro India*, January 25-April 25.

——— (1995a). "Drinking Water Scarcity in Rural Gujarat: Physical, Social and Institutional Dimensions," paper presented at the Workshop on Drinking Water Problems in Rural Gujarat: Issues and Perspectives, GIDR, Ahmedabad, January 25.

——— (1995b). "Over-exploitation and Approaches to Management: An Overview of Groundwater in Gujarat, Western India," paper presented at the 3rd South Asian NGO Summit, Kathmandu, Nepal, February 21-23.

——— (1997a). *Managing Common Pool Groundwater Resources: Identifying Management Regimes*. Monograph, Ahmedabad: VIKSAT.

——— (1997b). Water Use Study, Report submitted to the Water Resources Development and Management Project, Rajkot.

——— (1999/2000). "Institutional" Framework for Managing Groundwater: A Case Study of Community Organizations in Gujarat, India," *Water Policy*, 2 (6).

——— (2000). "Institutions for Efficient and Equitable Use of Groundwater: Irrigation Management Institutions and Water Markets in Gujarat, Western India." *Asia-Pacific Journal of Rural Development*, Vol. X (1), July.

Kumar, M.D. and Praful J. Patel (1995). "Depleting Buffer and Farmers Response: Study of Villages in Kheralu, Mehsana, Gujarat," in M. Moench (ed.) *Electricity Prices: A Tool for Groundwater Management in India?* Monograph, Ahmedabad: VIKSAT-Natural Heritage Institute.

Kumar, M.D. and O.P. Singh (2001). "Market Instruments for Demand Management in the Face of Scarcity and Overuse of Water in Gujarat," *Water Policy*, 3 (5).

Kumar, M.D., Shashikant Chopde, Srinivas Mudrakartha and Anjal Prakash (1999). "Addressing Water Scarcity: Local Strategies for Water Supply and Conservation Management in Sabarmati Basin, Gujarat," in M. Moench, E. Caspari and A. Dixit (eds.) *Rethinking the Mosaic: Investigations into Local Water Management*. Kathmandu and Boulder: Nepal Water Conservation Foundation and Institute for Social and Environmental Transition.

Kumar, M.D. and Vishwa Ballabh (2000). "Water Management Problems and Challenges in India: An Analytical Review," Working Paper 140, Anand: Institute of Rural Management Anand.

Kumar, M.D., Shashikant Chopde, Srinivas Mudrakartha and Anjal Prakash (1999). "Addressing Water Scarcity: Local Strategies for Water Supply and Conservation Management in Sabarmati Basin, Gujarat," in M. Moench, E. Caspari and A. Dixit (eds.) *Rethinking the Mosaic: Investigations into Local Water Management*. Kathmandu and Boulder: Nepal Water Conservation Foundation and Institute for Social and Environmental Transition.

Kumar, M.D. and Vishwa Ballabh (2000). "Water Management Problems and Challenges in India: An Analytical Review," Working Paper 140, Anand: Institute of Rural Management Anand.

London and New York: Belhaven Press. Mitchell, B. (1990). *Integrated Water Management*: pp. 1-2.

Moench, M. and H. Metzger (1992). *Groundwater Availability for Drinking in Gujarat: Quantity, Quality and Health Dimensions*. Monograph, Ahmedabad: VIKSAT.

Moench, M. (1995a). "When Good Water Becomes Scarce: Objectives and Criteria for Assessing Over-development in Groundwater Resources," in M. Moench (ed.) *Groundwater Availability and Pollution: The Growing Debate Over Resource Condition in India*. Monograph, Ahmedabad: VIKSAT-Natural Heritage Institute.

——— (1995b). *Electricity Pricing: A Tool for Groundwater Management in India?* Monograph, Ahmedabad: VIKSAT-Natural Heritage Institute.

Mohanty, S. and Ebrahim, A. (1995). "Groundwater Management Perspectives for Junagadh District of Saurashtra," in M. Moench (ed.) *Electricity Prices: A Tool for Groundwater Management in India?* Monograph, Ahmedabad: VIKSAT-Natural Heritage Institute.

Paroda, R.S. (1998). "Strategies for Doubling Food Production in India in Next Decade," Agricultural Situation in India, Vol. 55(5). August. pp. 259-65.

Pearce, D.W. and Warford, J. (1993). *World Without End: Economics, Environment, and Sustainable Development*. (published for the World Bank) Oxford University Press: London and New York.

Postel, Sandra (1993). *Water Tight*. Washington D.C.: World Watch Institute.

Prasad, K. (1998). "Accelerating Growth of Food Production—Some Basic Issues and Options," *Agricultural Situation in India*, August, pp. 279-82.

Raju, T.S. (1995). "Emerging Problems, Options and Strategies in the Development and Management of Ground Water Resources," in M. Moench (ed.) *Groundwater Availability and Pollution, The Growing Debate over Resource Condition in India*. Monograph, Ahmedabad: VIKSAT-Natural Heritage Institute.

Rangaswamy, G. (1994). "Potentials for Increasing Agricultural Productivity in India," *Indian Food Industry*, 13(I), January-February 1994.

Rosegrant, M.W.C. Ringler and R.V. Gerpacio (1999). "Water and Land Resources and Global Food Supply," in G.H. Peters and Joachim Von Braun (eds.) *Food Security, Diversification and Resource Management: Refocusing The Role of Agriculture?* Ashgate Publishers.

Saleth, R. Maria (1996). *Water Institutions in India: Economics, Law and Policy*. Sage Publications: New Delhi.

——— (1997). "Power Tariff Policy for Groundwater Regulation:

Efficiency, Equity and Sustainability", *Artha Vijnana*, Vol. XXXIX. No. 3. September, pp. 312-22.

Seckler, David, Upari Amarasinghe, David Molden, Radhika de Silva, and Randolph Barker (1998). "World Water Demand and Supply 1990 to 2025: Scenarios and Issues," Research Report 19, Colombo, Sri Lanka: International Water Management Institute.

Shah, Tushaar (1993). *Water Markets and Irrigation Development: Political Economy and Practical Policy.* Bombay: Oxford University Press.

——— (2000). Wells and Welfare in the Ganga Basin: Public Policy and Private Initiative in Eastern Uttar Pradesh, India, Research Report 54, International Water Management Institute, Colombo, Sri Lanka.

Singh, K. (1995). "Cooperative Property Rights as Instruments for Managing Groundwater," in M. Moench (ed.) *Groundwater Law: The Growing Debate.* Monograph, Ahmedabad: VIKSAT- Natural Heritage Institute.

TERI (1998). *Looking Back to Think Ahead Growth with Resource Enhancement of Environment and Nature.* New Delhi: Tata Energy Research Institute, March.

Thamarakshi, R. (1999). "Food Production and Consumption in the New Millennium," *Banking Finance*, October, p. 19.

Vamadevan, V.K. and N.G. Destane (1968). *Suitability of Soils for Irrigated Rice.* IL RISO, Anno. XVII (3): 243-50.

Visaria, L. and Pravin Visaria (1995). "Prospective Population Growth and Policies for India," Working Paper 75, Ahmedabad: Gujarat Institute of Development Research.

World Bank-Government of India (1998). "Inter-Sectoral Water Allocation Planning and Management," World Bank-Government of India, India Water Resource Management Sector Review, Vol. I (Main report), New Delhi.

Wolf, Peter and Rolf Hubener (1999). "Irrigation in the World — The Future will not be like the Past," *Natural Resources and Development*, Vol. 51, Institute of Scientific Co-operation, Tubingen.

World Resources Institute (1995). World Resources 1994-95: People and the Environment, Report prepared by World Resource Institute in collaboration with United Nations Environment Programme and United Nations Development Programme. Oxford University Press: Bombay.

Xie, Mei, Ulrich Kuffner and Guy Le Moigne (1993). "Using Water Efficiently: Technological Options," Technical Paper 205. Washington D.C.: World Bank.

SECTION-II

SPECIFIC PROBLEM-RELATED OR REGIONAL ISSUES

6

Salinization and the Importance of Rainwater Harvesting*

NIRMAL SENGUPTA

Deserts are formed not only by natural processes but also by human action. In the more recent period land degradation by human action creating a desert-like barrenness, has become an alarming phenomenon. This led to the convening of the United Nations Conference on Desertification (1977). The Conference defined desertification as the diminution or destruction of the biological potential of the land, which could lead ultimately to the formation of desert-like conditions. In 1991, the United Nations Environment Programme (UNEP) redefined desertification as land degradation in arid, semi-arid, and dry subhumid areas resulting mainly from an adverse human impact. Dregne and Nan-Ting Chou (1992) describe desertification as a human-induced process of land degradation that can range in severity from slight to very severe and in cause from erosion or salinisation or toxic chemical accumulation to vegetation degradation, irrespective of climate. They also identify three principal desertification processes. Rangeland desertification is primarily a matter of degradation of the vegetative cover through overgrazing and the cutting of woody

* Keynote address, Second Biennial Conference of the Indian Society for Ecological Economics, Bhopal, 19-21 December 2001. An earlier version of this paper was presented to the Seventh International Conference of the International Rainwater Catchment Systems Association, Beijing, 19-25 June 1995.

vegetation. Rainfed cropland desertification is commonly expressed as increased water and wind erosion. Salinization and waterlogging are the principal degradation processes on irrigated land. From various available data it is estimated that the world is losing at least three hectares of arable land every minute because of soil salinity.

By their chemical characteristics the salt-affected soils can be grouped under two broad classes: saline and alkaline. *Saline soils* are those which contain neutral salts, *e.g.* chlorides and sulphates of Na, Mg and Ca. *Alkali soils* contain salts capable of alkaline hydrolysis, *i.e.* sodium carbonate, bicarbonate and silicate.[1] For reasons of brevity the title of the paper refers only to salinity. But we will discuss both the types. The term *'salinity'* is often used as a generic term to cover both salinity and alkalinity.

Salinity in India

Table 6.1 shows the global distribution of salt-affected lands. The data is about all land, not just arable or irrigated land.

Table 6.1: Regional distribution of salt-affected soils

(in million ha)

Regions	*Total area*	*Saline soils*	*Per cent*	*Sodic soils*	*Per cent*
Asia, Australia and the Pacific	3107.2	195.1	6.3	248.6	8.0
Europe	2010.8	6.7	0.3	72.7	3.6
Latin America	2038.6	60.5	3.0	50.9	2.5
Africa	1899.1	38.7	2.0	33.5	1.8
Near East	1801.9	91.5	5.1	14.1	0.8
North America	1923.7	4.6	0.2	14.5	0.8
TOTAL	12781.3	397.1	3.1	434.3	3.4
India	297.3	6.2	2.1	2.4	0.8

Source: FAO data based on the FAO/UNESCO Soil map of the world.

We have also included in the above table the data for India taken from the same source. For reasons such as lack of standards, seasonal variability of the phenomenon and unreliable estimation methods, findings differ widely. Other estimates of the Indian situation are different. But all sources confirm what is very evident from Table 6.1, that the condition of India is definitely better than several other countries. Nevertheless, it is a serious problem even in India, particularly for the irrigated land.

Salinization is a serious problem of irrigated land. Of the 230 M ha of irrigated land in the world, 45 M ha (19.5 %) are salt-affected soils to varying degrees by human-induced processes (Oldeman *et al.*, 1991). In contrast, of almost 1,500 M ha of dryland agriculture, only 32 M ha (2.1 %) are salt-affected soils. In 1991 a Working Group was appointed by the Ministry of Water Resources of the Government of India, to identify the problems of waterlogging, salinity and alkalinity in irrigated areas. As per the Report of the Working Group, waterlogging problems had developed in 2.46 M.ha areas, salinity in 3.06 M ha and alkalinity in 0.24 M ha. Assuming that 70 per cent of this was in major and medium projects, Mohile *et al.* (in NWDA, 1994: 5) estimate that the total affected area under major and medium irrigation would be around 15 per cent.

However, the Ministry of Water Resources Working Group (1991) had adopted the following norms for identification:

- Waterlogged area – water table within 2 m of land surface (due to rise in water table).
- Potential areas of waterlogging – water table between 2 to 3 m of land surface.
- Safe area – water table between 3 m of land surface.

The estimates of the Working Group was about "waterlogged area" as per this definition. Understandably, the extent of affected area is far more. Some idea is obtained from the land degradation data collected by the Ministry of Agriculture in 1984. It shows that an overall 17.91 M.ha. land was fully or partially affected by waterlogging, salinity and alkalinity (Table 6.2).

Partially affected area, or even all of the fully affected area, is not completely out of production. But productivity reduces in such land. Abrol (1992: 67) suggested that the productivity, in nearly 50 per cent of the area under canal irrigation has been adversely affected in varying degrees due to problems of waterlogging and salinity. Dregne and Nan-Ting Chou (1992) use a 4-point scale namely, slightly desertified irrigated land has had yields depressed by less than 10 per cent, moderately desertified by 10 to 25 per cent, severely desertified by 25 to 50 per cent, and very severely desertified by more than 50 per cent. Fig. 6.1 shows the estimates made in this scale:

Table 6.2: Estimates of salt-affected land in irrigated areas, India

Problem	*Area affected (m. ha.)*
Waterlogging	8.53
Saline soils	3.88
Alkali soils	5.50
Total	17.91

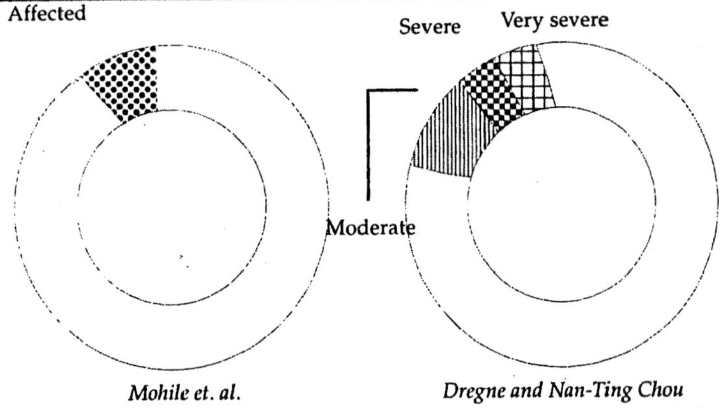

Fig. 6.1: Desertified Irrigated Land in India. Total Irrigated Area Represented by the Ring.

Those affected moderately or worse, *i.e.* a total of 8.1 M ha of irrigated area that has suffered at least a 10 per cent reduction in productivity, may be considered as desertified to some extent. This was about 35 per cent of the total irrigated area at that point of time.

All these can only give some idea of the magnitude of the problem. Exact estimates cannot be obtained. The watertable fluctuates over a wide area, from year to year, from season to season. Fig. 6.2 shows the yearwise data of area with depth of the watertable at 0-3 m in Haryana between October 1988 to October 1991 (*data source*: Tanwar and Rana in NWDA, 1994: 320).

The Working Group defined the area where the watertable is between 3 m of land surface as a "safe area". But how does one infer the extent of safe area from Fig. 6.2?

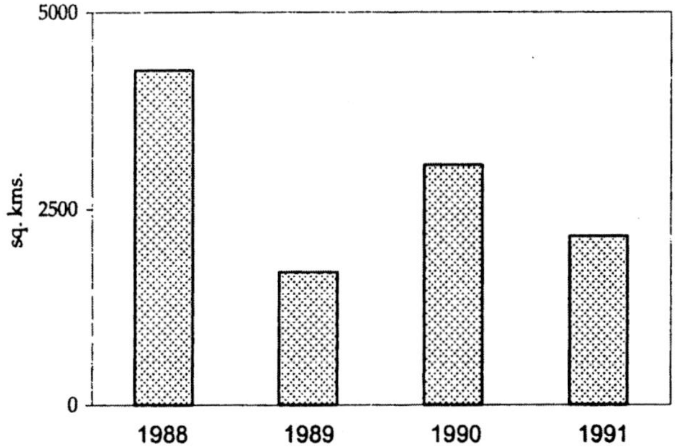

Fig. 6.2: Area with Depth of Watertable 0-3 m (Haryana, October, 1988 to October, 1991).

Based on the climatic and meteorological conditions in different parts of India, salinity-prone soils have been identified as of three broad types (Mamoria, 1982: 148-9).

1. The western part of India is a semi-arid to arid region with hot climate and a dry winter. In this climate evaporation is always greater than precipitation. Hence the soil profile development by eluviation is greatly retarded. The soil is marked by a concentration of salts—saline and alkaline.
2. The marked seasonality of rainfall effects salt release over a very large part of the country. In south India and the Gangetic valley a warm rainy climate is

followed by a dry winter. During the rainy season precipitation is greater than evaporation which induces leaching of soluble salts down the profile. If the water table remains high, the soluble salts will remain in the profile. In northern India, in particular, the salts are transported in solution by the Himalayan rivers which later percolate in the sub-soils of the plains and goes on accumulating in the area of inefficient surface drainage. Thereafter, during the dry season these salts are drawn upwards through the capillary spaces by evaporation from the surface. In many cases salts are deposited on the soil showing white or black patches of efflorescence on the soil surface. Impervious sub-soil or hard pan also contributes to the phenomenon.

3. A very large part of the deltas and estuaries of rivers are affected by the sea tide carrying salt laden deposits. Large parts of the sea coast are subjected to periodic inundation by tidal water.

A rough idea of the extent of the problem in different areas can be obtained from Table 6.3.

Table 6.3: Estimates of salt-affected areas, by regions

Regions	*States*	*Area (m. ha.)*
Indo-Gangetic plains	Haryana, Punjab, U.P., Rajasthan, M.P., Bihar	2.500
Arid and semi-arid regions	Gujarat, Rajasthan, U.P., Haryana, Punjab	1.000
Coastal areas:		
(a) arid zone	Gujarat	0.714
(b) deltaic	W.Bengal, Orissa, A.P., Tamil Nadu	1.394
(c) sea water ingress	Kerala and other western coast	0.016
Total		5.624

Source: Abrol and Dhruva Narayana ed., 1990: 319

Merits of Rainwater

The quality of river waters in India, unless contaminated otherwise, are of a very high order (Gupta, 1990: 19-26). Those of the northern rivers, like the Ganga and its tributaries, the Indus and the Brahmaputra are the best. The Yamuna water is not as good, but is still of a sufficiently high quality for irrigation. Waters of south Indian rivers like that of the Tungabhadra and Cauvery, as well as those which drain into the Rann of Kutchh, have relatively high EC and RSC values. Canal waters originating from the rivers or their reservoirs as the main source, represent the parent rivers in quality, unless contaminated otherwise. The proportion of salts are safe, usually less than 500 $\mu S\ cm^{-1}$ with a appreciable concentration of calcium ions. The latter leads to a decrease in actual salinity though it increases alkalinity.

But even this small quantity of salt may play havoc if due care is not taken. Minute quantities of salts are added to the soil with each irrigation. Crops remove much of the applied water from the soil to meet the evapo-transpiration demand but leave most of the salt behind. With each successive irrigation more and more salt is added. Hence a portion of the added salt must be leached from the root zone before the concentration affects crop yield. Leaching is done by applying sufficient water so that a portion percolates through and below the entire root zone carrying with it a portion of the accumulated salts. After many successive irrigations, the salt accumulation in the soil will approach some equilibrium concentration. A successful water management consists in keeping the equilibrium level within a certain limit which is best for crop growth. This equilibrium level is decided by three factors (FAO, 1985):

(i) The salinity of the applied water;
(ii) Depth of water leached below the root zone; and
(iii) Depth of water applied at the surface

A good salinity management strategy should take care of all the three factors. Canal water in India may be very good. Still it is necessary to provide drainage for leaching below the root zone, and a judicious use of water applied on the surface. If these conditions are met the use of canal waters can be practised for indefinitely long periods for sustaining stable agricultural production in areas where drainage is not impeded (Gupta, 1990: 24). Yet, no matter however good canal water may be as a leaching agent, rainwater is certainly the best. Rainwater has the lowest salt content of all types of water available for irrigation. It may contain some dissolved gases and some dissolved salts originating from terrestrial and marine sources. The composition varies as one moves away from the sea, depending on local pollution sources and from shower to shower. But overall, the rainwater has an EC range between 5 to 50 $\mu S\ cm^{-1}$ (Gupta, 1990:16). Thus the equilibrium concentration of salt accumulation in the soil attained after many successive irrigations, is lower if the leaching agent is rainwater. This is one great merit of rainwater harvesting irrigation systems in comparison to the use of canal water. Also, for partial leaching to attain tolerable limits for crop growth, a lesser amount of water is necessary if the leaching agent is rainwater. This, in turn, contributes to the reduction in problems of waterlogging in areas having drainage problems. Summarizing, the merits of rainwater over canal water are:

- It has a modifying effect on the number of irrigations required to meet the evaporative demand as part of it is met by rainfall, and
- It helps in leaching of accumulated salts, because it is the best quality water available for leaching.

The maximum utilization of rainwater is, therefore, a very desirable strategy for salinity control. Gupta (1990: 264-73) made some rough estimates of salt balance and leaching requirements. He found that in monsoon seasons, when the rainfall is short and concentrated, rainwater may be sufficient to leach out salts from the root zone. Without going into the details let me draw attention to the rule of thumb that he developed. "The depth of soil from which the soluble salts (> 80 per cent of it) may be leached during the monsoon season, corresponds roughly with total rainfall during the monsoon season" (Gupta, 1990: 270). Thus, if the root zone depth of the soil is 40 cm. from which accumulated salts are intended to be leached, a minimum amount of rainfall equal to 40 cm. is necessary. The actual requirement will vary depending on the texture of the soil and other percolation considerations. Rainwater is available in plenty for effecting a high proportion of leaching almost everywhere in India.

Traditionally the water management system was based on this principle. It survived for hundreds of years without any noticeable land degradation. Modern extension programmes failed to grasp these principles and have done enormous damage to the agricultural land.

Indo-Gangetic Plains

As noted earlier, the quality of irrigation water in the Indo-Gangetic alluvial plains, both from rain and the canal, are of excellent quality. But this is still one of the worst salinity affected areas. Due to the problems of drainage the soluble salts remain in the profile and increase in concentration. During the season of drought these are drawn upwards through the capillary spaces due to evaporation from the surface. The problem is so acute that saline soils also occur in unirrigated regions. The process is accelerated with canal irrigation. Towards the east, in the northern flood plain of the Ganga, in Bihar, a serious salinity problem does not arise, thanks to the very high rainfall (100-140 cm annual average). Nevertheless, crops suffer because of moderate salinity and/or excessive wet conditions and prolonged submergence. The soils of the Indo-Gangetic plain are very fertile and well supplied with potassium, phosphorous, calcium, iron and manganese. But under waterlogged condition the alkali soils have a zero water infiltration and diffusion rate. In consequence, the trace minerals, though rich, are not available to the plants. Thus, everywhere in this area, the threat of a rising groundwater table must be the first consideration in water management.

If water influx in the substratum is greater than water expenditure [I > Q] then the groundwater level increases and waterlogging occurs [Abrol and Dhruva Narayana (ed.) 1990: 308]. The water influx in this area is very high because of the heavy annual rainfall and an enormous amount of snowfall carried by the Himalayan rivers flowing through this region.

The first priority, therefore, is to drain out the excess water. The traditional method facilitated drainage—both by horizontal and vertical methods. The natural drainage lines were never interfered with, leave alone directing them inland through canals. Besides, the major mode of irrigation consisted of wells, which provided vertical drainage. Fortunately, most parts of this region has sweet groundwater. Around 1860, even after some extension of the canal area, about 15 per cent of cultivated land in the province was irrigated by wells (Whitcombe, 1971: 31). Indeed, in recent years this method has been employed as conjunctive use in the canal irrigated area thereby successfuly reducing the problems of salinity in Punjab and Haryana.

Natural precipitation used to be retained (Whitcombe, 1971:6) in water harvesting structures like tanks and ponds (called *jhils* — surface depressions filled by summer rains). The *jhils* were reported to supply some amount of irrigation water and produced forage (Stone, 1984: 140). Indirectly too, they must have contributed significantly by enriching subsoil moisture. Recharging of groundwater was an important factor; although rainfall is high in this region it is highly variable and irrigation is essential during summer and drought. Rainwater collected in tanks and was also worked as a leaching agent. Canal construction techniques were not unknown but were rarely used. The Western Yamuna canal was constructed in the sixteenth century by the Delhi Sultanate. But it was abandoned, probably because of the experiences of salinity and waterlogging as of today. Yet this was the canal which had inspired the British engineers to undertake the canal construction programme in India.

The same pattern was also found in the lower reaches of the Ganga valley. The eastern states of Bihar, Bengal and Assam depended largely on facilitating drainage along with rainwater harvesting. Here, quite often, additional drainage channels were dug while rainwater was retained in numerous *jhils*, beels and ponds (Sengupta, 1993: 72-3). This region too had faced severe problems due to unthoughtful extension. However, we will not consider this aspect of rainwater harvesting since salinity was never a major problem here.

The nineteenth century construction of canals and distributaries in U.P. did not follow the watersheds and interfered considerably with natural drainage lines. The railways and roadways too did not take the drainage systems into consideration (Stone, 1984: 134-5). Very soon, the areas brought under canal irrigation were ravaged by severe flood, waterlogging, salinity and malaria. As the problems of salinity aggravated, investigations were begun. By the late 1860s, the causes for this were known and a couple of experimental leaching and drainage stations were established. However, the results were not encouraging. A Committee (Reh Committee, 1878) was appointed whose report reflects the unassailable contradictions faced by the government (Whitcombe, 1971: 76-9; 285-9). The Committee could not admit what was clearly evident from all available information—that the introduction of canal irrigation itself was the basic mistake. For heavy water influx the Committee laid the blame on the farmers' practice of over-irrigation, suggested the raising of the water rates to achieve a greater economy of water and recommended a continuation of the reclamation experiments. The Reh Committee had also recommended, albeit reluctantly, that lift irrigation should be substituted for flush irrigation. The Canal Department had earlier offered flush irrigation at lower rates than the existing lift irrigation in order to open up the market (Whitcombe, 1971: 9). Gravity irrigation from canal was being touted as a replacement for the dreary old water lifting (Whitcombe, 1971: 78). So it was a reluctant admission that lift irrigation is preferable to flush irrigation from canals. However, the government could not reverse its policies by squarely admitting the blunders. This had set the course of salinity control efforts. The leaching and drainage experiments, though unsatisfactory, became the only implementable policy. Fundamental reconsideration of canal policies was not on the agenda.

The traditional water management methods were already facing the adverse effects of the introduction of canals in the neighbourhood. As a consequence of the raising of the watertable by the canal system, salinity increased in the command areas of the traditional systems as well. This necessitated a greater application of water for leaching, which increased the workload but aggravated waterlogging. Ultimately, the farmers had no option but to rely on canal water. Drainage created a further problem for rainwater harvesting structures. Large scale implementation of the drainage programme in U.P. began only since the 1880s, when a series of wet seasons created serious waterlogging problems and prompted an outcry against canals (Stone, 1984: 139). By 1900 the aggregate surface drainage channels constructed were more than a third of the total length of canals and distributaries (Stone, 1984: 134-45). Now the problem was of water shortage. Stone (1984: 135) describes a case with details which gives an inkling of the process. The traditional system in that area consisted of the draining of flood water into a rivulet after flowing through two *jhils*. After the construction of a canal, within a few years the water levels of the *jhils* increased inundating the surrounding land, throwing them out of the cultivable category. Then, after the drainage channels were dug, numerous complaints were made by farmers that the drainage system was drying out the country excessively. Often their canal supplies were diminished. Besides, the *jhils*, etc. which stored rainwater were rendered dry for about nine months a year (Stone, 1984: 140). Indeed, it was so excessive that the watertable of places thirty miles away, fell. Some twenty years later, following serious drought, when tubewell irrigation was being considered, some officers felt that the policy of surface drainage had been carried too far. The subsoil water level was lower than it was twenty years ago and had a permanent tendency to fall (Stone, 1984: 141). Following the destruction of rainwater harvesting structures canal irrigation remained the only source of recharging groundwater. Wherever canals went, well-digging was often made impracticable. So the farmers had little choice but to adopt canal water (Voelcker, 1893: 69; Whitcombe, 1971: 80-81).

The old irrigated areas settled to new equilibrium levels after several years. Later, after the introduction of tubewells and power operated pumpsets vertical drainage too has been considered as a possibility. These have solved a part of the problem, but only a part. One would have expected, as did the

Royal Commission on Agriculture in India, in 1928 (India, 1928: 353) that, "the lesson has been learnt and that... a careful drainage survey which should include estimates for drainage construction will, in future, form an integral part of new irrigation projects." Drainage considerations remained an eternal starter. As late as in 1972, the Irrigation Commission noted that

> "The importance of drainage is so well recognised in advanced countries that it is always considered an integral part of any irrigation system.... The Commission is glad to observe that the States are aware of the importance of drainage and have started—taking steps to improve in their irrigated areas.... We (recommend) that adequate provisions for drainage be made all new works.... funds for the maintenance of drains be allotted separately, so that drains are not neglected." (India, 1972: Vol. 1: 312-19).

The National Commission on Agriculture, 1976 (India, 1976: Vol. V, p. 56) was more critical:

> In recent years there has been increasing awareness of this problem and it has been laid down that in all new irrigation schemes, there must be adequate provision for drainage. It is, however, noticed that these provisions are not always adequate and that the drainage is planned comprehensively. Several project estimates have contained only a token or a lump sum provision, which later on has been found to be completely inadequate.

And so, little has as yet been done to provide proper drainage in canal irrigated areas. The problem continues and the threat of waterlogging and salinity becomes more and more acute. In forty years, till 1989-90 the state of Uttar Pradesh was able to reclaim 157,036 ha. of land which accounts for only 12 per cent of the total saline area in the state (Arora, 1994). But a part of this area might have reverted back after the chemical treatments were withdrawn. In Punjab, waterlogging problems have been greatly reduced by the conjunctive use of surface and groundwater. A new equilibrium has been reached with irrigation facilities extended over a much wider area than in the past. One hopes that the equilibrium will not be disturbed again following the removal of subsidies for power.

Central, Western and Southern Region

Salinity is also found in large parts of Gujarat, Madhya Pradesh, Maharashtra, Karnataka, Rajasthan and Andhra Pradesh. The soil characteristics here differ from the Indo-Gangetic pattern which need a different set of considerations. Both saline and alkali soils are found depending on the aridity of the region. In leaching of salt content one has to consider that the soils of this region in general, have slow water infiltration and percolation rates and water diffusivity. A substantial part of this region consists of medium and deep black soil (vertisols) which tend to have a positive water balance fluctuating between shallow depths. Any additional salt carried in by canal water tend to accumulate. Since digging, laying and maintaining is difficult in this soil, the scope of drainage is limited. Groundwater is sodic or saline and therefore, vertical drainage possibilities are also ruled out. Because of slow water diffusivity, the soil degradation process in this region is slow. Only as time passes and in complete absence of leaching, the salt content increases sizably in the surface horizon. But once it appears it is difficult to get rid of. Table 6.4 depicts the extent of waterlogging in some of the projects:

Table 6.4: Waterlogged areas in some projects

Name of the project	*Waterlogged area as p.c. of C.C.A.*
Nagarjuna Sagar	6.70
Sri Rama Sagar	6.04
Ghataprabha	1.11
Ghod	1.14
Krishna Canal	4.70
Mula	5.78
Sirhind Canal	4.35

Source: Keskar *et al.* (in NWDA,1994: 375).

The traditional water harvesting system in this area comprised the use of tanks—both above-surface and submergence tanks. Tanks survive over an extensive area and were there for more than thousand years. Their commands have not suffered from salinity problems though the quality of river water in this part of the country is more saline than that of the Indo-Gangetic plain. In contrast, canals constructed by replacing tanks, like the Nagarjuna Sagar, have developed waterlogging problem. Fortunately, the salinity aspect has been thoroughly studied by the scientists of Central Arid Zone Area Research Institute (CAZARI) in one of the traditional water harvesting structures — the *Khadins*. We introduce these findings below in some detail.

The *khadins* of Rajasthan, which provide surface irrigation in years of good rainfall, are utilized primarily for subsoil storage properties. *Khadins* are found even in the heart of Thar desert. The old *khadins* have been found to be site-specific systems. The soils of *khadins* are medium to fine in texture, with good water-holding characteristics as against the loose, sandy, droughty soils of the rest of the desert. These soils remain wet or moist for a long period, and are favourable for the chemical weathering of soil minerals and for microbial activities. The temporary watertable that is formed in *khadins* after collecting precipitation over the catchment area, slowly recedes into the subsoil and substrata during winter, thus continuously supplying moisture to the deep growing roots of crops throughout its period of growth. Depending on the good or bad rainfall in a year either gravity irrigation or bed cultivation is practised in a *khadin*. It is reported that *khadins* succeed in producing at least one crop even if the rainfall in a year is one-third that of normal.

The region where *khadins* are found is a particularly saline tract. The amount of run-off collected in a *khadin* depends on the extent of the catchment area, which is about 12 to 15 times that of the service area (Kolarkar and Singh, 1990). The rainwater harvest of *khadins* contain quite some amount of salt by leaching out their wider catchments. One would expect that in course of time the salt would accumulate in the profile and turn the land unusable. *Khadins*, however, have remained functional for several centuries. Intrigued by this fact, the soil scientists of CAZARI conducted on this system one of the rare studies of salinity control in traditional systems. Singh and Kolarkar (1983) show that there was not much salt accumulation inside the *khadin*, but heavy deposits were found just outside where water from the seepage passes out. Obviously, the incoming salts were continuously expelled. Their explanation is that the sediments accumulating inside the *khadin* gradually raise the bed level. It reaches in old *khadins* between 0.25 to 1.00 metre. In consequence, the hydrostatic pressure of water body of *khadins* push the salt content out, which deposits on the outer side of the embankment.

The *Bundhies* or submergence tanks (Sengupta, 1993) were extensive in the black soil region. *Bundhies* were basically *in situ* water harvesting structures. They retained water practically throughout the rainy season. After the rainy season water used to be drained out and the bed was used for cultivation. They were not used for gravity irrigation. Submergence techniques were basically conservation of moisture by contour bunding in the watershed itself. Storage of fresh rainwater for months over the whole watersheds could effect some amount of leaching. This was the perfect answer to extremely slow water infiltration and percolation properties of the soil. At the same time the admirable water-retention capacity of the soil was used to devise a more efficient method of water utilization than gravity irrigation. Submerging also helped in soil conservation. The concerned area is prone to severe soil erosion. Until a couple of decades back, the submergence technique was extensive all over the *haveli* tract of M.P. A century ago it was found all over the western region. Reference to the *sailabi* system of Punjab is found in old official records. It is still a recognized system of irrigation in Baluchistan where it is known as the *sailaba* system (Oesterbaan, 1983). Modern canals constructed in these areas face some of the worst problems of waterlogging and salinity. The canal systems of Pakistan are amongst the most severely saline affected systems in the world. The Indira Gandhi Canal Programme (IGNP) is very severely affected by waterlogging.

Epidemics of malaria broke out in 1994 on such a scale that the government offices had to be closed. The problem has led to widespread public protests. In the IGNP command area groundwater is saline. Hence conjunctive use cannot be proper remedy for these problems. Most parts of the Tawa dam, the first major dam built on the Narmada Valley, was waterlogged after its construction. In its report in 1979-80 for M.P., the CAG was critical that the productivities of crops in Hoshangabad district had in fact, decreased after the construction of the canal system. However, after construction of the drainage system and a careful monitoring the Tawa irrigation system is now in a much better shape. Upholding the claim that it is the best designed project, the Sardar Sarovar project is probably the only one in the country where waterlogging possibilities have been considered at the design stage. It need to be seen whether these designs and maintenance principles are put into practice and whether, these actually succeed.

There are several other spheres where the traditional systems can add valuable lessons to modern engineering. None of the hundreds of old anicuts[2] have given rise to the waterlogging problem whereas the Srivaikuntham anicut, constructed by modern engineers, has waterlogging problem. However, we will leave its discussion here in order to remain more focused. The success was due to some traditional civil engineering practices[3] — another important area of traditional knowledge which too would need long discussion.

Coastal Areas

Throughout the western coast, marine cycles affect sea water ingression over an extensive coastal area. These include low-lying, marshy lands found along the rivers, creeks and estuaries, locally called *khar, khajan, gazani, pokkal* and *kole* lands, extend from Gujarat to Kerala. Most parts of the Western Ghats have plenty of rainfall. Hence sweet water is available in plenty. Farmers construct dykes and sluice gates (tidal regulators) to stop sea water from entering the land. The rainwater stored behind these structures creates sweet water surfaces and subsurface barriers preventing saline ingression. In this process the inland coastal belt is leached out of salts and remains fit for cultivation (Sengupta, 1993: 75-6). Creeks and lagoons are not so common on the eastern coast. Though some of these do exist, including the famous Chilika lake. Similar steps, construction of dams and levees to stop sea water from entering the land, and storing rainwater behind as barriers, were adopted in these parts too. In the eastern coast only the *bheri* lands of Bengal are extensive.

During high tide, seawater floods the low-lying land, which is therefore saline (or *khar* in Marathi and *khajan* in Konkani). To protect the land against the sea, earthen bunds were constructed along the creek side. These dams contained flap gates which opened only on one side. After a heavy downpour in the adjacent hills the sluices opened, draining out excess water. But during tides the sluices closed in reverse flow thereby acting as shields preventing sea-water from entering the fields. This was a self-adjusting water management system—the wooden doors of the sluice gates opened or closed, unless purposely blocked or sealed, in response to changing water levels within the streams and particularly the pressure of the very high and low tides during the full and new moons. The protected lands are thus leached during the monsoon, so that the salt content of the upper soil is reduced in successive years. The *khar* lands thus became fit for rice cultivation within a couple of years after the completion of the protection works. After the monsoon, the farmers may allow tide water to enter along with prawn seed, fish seed, and other predators. These are retained for five to six months and are periodically harvested (fish capture). The catch comprises of different varieties of prawns and fishes. The farmers regularly leach the fields in the monsoon seasons and flood these with sea water during the off-season. During the monsoon the salinity reduces to near freshwater conditions and it is possible to raise paddy crops. In the public interest litigation in the Aquaculture case[4], the Supreme Court opined that intense aqua farming violated a number of principles of good environmental management methods. The court also confirmed that the traditional and improved traditional aquaculture for prawn breeding was a sustainable practice.

For a variety of reasons these system are threatened. These factors extend from organizational inadequacies to the construction of roads and railways. The colonial government was of the opinion that the landlords were responsible for these works.[5] Following this notion it was found that after the abolition of landlordism there was no agency with the delegated power. Hence the government had to take up such works, and this in turn encouraged the local people to become more dependent on the government. In *khar* lands too, based on the assumption that landlords maintained the structures, the government stepped in after the land reforms in the post-independence period. But the government locked both the expertise and the information for doing the work properly. The embankments need regular vigilance, even against snails making holes, apart from annual repair. As late as in 1979, a Khar Land Development Circle was constituted within the Irrigation Department of Maharashtra for the maintenance of the 720 *khar* land schemes. By now however, it has been found that proper maintenance of these schemes require close peoples' collaboration. Participatory effort is being tried now.

More recently, the rising importance of fish and prawn production has become a major reason for the reversal of the system of rice production and fish farming. In the past fishing activities were secondary; the sluice gates were there to protect the rice fields from inundation. The people were keen to protect the dykes and gates to save the rice fields. However, the rice fields are now secondary, and catching fish has become one of the major economic activities in the village (Rubinoff, 2001). Further there is also a growing lobby of opinion claiming that all these west-flowing rivers are surplus and need to be diverted to the drier east. Kerala and Tamil Nadu governments are engaged in a bitter contest over this issue. Already the Kuttanad region is suffering from a shortage of fresh water for leaching purpose.

In the east coast, salinity due to sea water ingression is less, but saline delta regions abound. Saline soils of delta and coastal regions include the Ganga delta or the Sunderban, Mahanadi, Godavari, Krishna, Cauvery and small river deltas on the coast.

In the west, the estuaries of Narmada and Tapi have similar problems. These areas suffer from waterlogging and salinity problems, which resemble those of the Indo-Gangetic plains. We will not go into the details here for reasons of space. Lately, because of excess groundwater extraction, seawater is being allowed to enter in several places. Cities like Chennai are facing acute problems of groundwater turning brackish. The sweet water tanks which were once abundant in these areas, have mostly been acquired and utilized for urban development.

The southern part of the east coast and the Kutchh peninsula are the only coastal regions not favoured by heavy rainfall. In these rainfall deficient regions the groundwater too is saline. Data shows that in the Kutchh and Saurashtra region there were thousands of small tanks in the past. The evaporation rate in this region is very high and the stored water would have been lost in no time. But as the soil is porous and the stored water percolated in to the subsoil. That might have contributed partly, to the formation of the necessary sweet water belt against saline intrusion.[6] The Salinity Ingress Preventive Circle (SIPC) of the State Government recommend tidal regulators and bandharas among the useful salinity control methods (Marvania, 2001).

Conclusion

For fighting salinization two types of solutions are adopted: *preventive* and *curative*. Available curative measures include surface and subsurface drainage, chemical treatment and vegetative and biological measures. Standard preventive measures include the lining of canals and conjunctive use. Rainwater harvesting, which has been used for hundreds of years for both preventive and curative purposes, is not included in the inventory of remedial measures. The absence is disturbing. By its physical properties rainwater is the best leaching agent. By its historical record rainwater harvesting has performed well in containing salinization in India. Tradition is the test of sustainability. The Convention on Biological Diversity, 1992 recognizes in clear terms, the importance of indigenous and local knowledge in sustainable development. The UN Convention to Combat Desertification includes programmes based on traditional knowledge. The relevance of the indigenous techniques and principles of water management, that had successfully contained waterlogging and salinization problems in India, is not lost even today.

Lining of canals or conjunctive use produces immediate results. Other rehabilitation methods have a gestation period of three to five years. But all these have almost prohibitive costs. Nor are all these methods applicable everywhere. The much-touted conjunctive use is not applicable to areas where the soil and groundwater are endemically saline. Non-availability of electricity constitutes the arresting factor in many other areas. Farmers usually find that in spite of reduced productivities, the returns from cultivation of rice and wheat are higher than the returns from an application of vegetative and biological measures of salinity control (Raina, 2001).

Yet, under the existing system of assessments, the economics of remedial measures favour land rehabilitation by using these costly methods. Two kinds of costs are involved in assessments of the economic impact of land degradation: *(i)* income foregone as the result of prior land degradation, and *(ii)* cost of controlling and repairing land damage. Land rehabilitation costs are those incurred for arresting further degradation and to restore the land to something approaching its original undegraded condition: the costs of the preventive and curative measures. The benefit component is the discounted aggregate income that would be foregone due to prior land degradation. Considering the fact that nearly 35 per cent of irrigated land is partially affected and about 15 per cent is acutely so, income foregone due to salinization is very high, so much so that the expected returns favour even the costliest method of land rehabilitation. However, there is a flaw here. Income lost due to waterlogging and salinization should be included in the benefit estimates of modern irrigation projects, reducing the net benefits. If this is done many modern projects would not be viable. Needless to say that the project proposals hardly ever show any realistic estimate of possible waterlogging and logging after commissioning of the project.

The progress is limited due to the high costs of land rehabilitation measures. Whatever success has been achieved in land reclamation is largely because of heavy subsidies. The success of Punjab in extensive conjunctive use of groundwater may be largely due to the free power available to farmers. Farmers of the Indo-Gangetic belt have shown considerable enthusiasm for the adoption of the chemical treatment. This has much to do with the heavy subsidies the state governments provide. In Haryana, for example, the subsidy on gypsum is around 75 per cent of cost — under the National Oilseed Production Programme as high as 90 per cent (Raina, 2001). Transportation costs too are low since gypsum is available in abundance in the neighbouring state of Rajasthan. It is questionable whether even the present pace of the remedial measures can be sustained after these subsidies are abolished. Rainwater harvesting, which is the most economic, is not included in the inventory of remedial measures.

This article is a plea for the inclusion of rainwater harvesting in the repository of salinity control techniques and the resurrection of the traditional principles of salinity control. It must not be misunderstood as an argument for a revival of the past. We are arguing only for learning from the past. It is neither possible nor desirable to go back to the pre-modern situation by writing off the networks of canals and other structures. But along with drainage, embankments or long duration storage wherever necessary, traditional principles need to be resurrected in order to find a sustainable solution to the ever-widening problems of desertification of even irrigated land. We also reject the position that relegates rainwater harvesting structures to the dry areas where canal irrigation cannot be extended. We find that even in the intensive canal irrigated areas of U.P. and Punjab rainwater harvesting on an extensive scale is a desirable strategy for salinity prevention and cure.

Notes

1. The parameters are—saline soils [*p*H < 8.5, ESP < 15 and ECe > 2 dS/m], alkali soils [*p*H > 8.5, ESP > 15 and ECe – variable].
2. Sometimes tanks are fed by diversion of river water through canals. Hundreds of such diversion structures on rivers, called anicuts, were constructed in the past. Anicuts and feeder channels are almost identical to canals.

3. In his studies, Dr. Gomathinayagam, Centre for Water Resources, Anna University, Chennai has suggested some explanations which deserve attention.
4. S. Jagannathan *v.* Union of India (1997) 2 SCC 87 and Gopi Aqua Farms *v.* Union of India 1997 6 SCC 577.
5. Similar descriptions of a division of responsibilities have been noted by colonial officials in almost all spheres of activities pertaining to common property resources. But intensive investigations reveal that the actual situation was very different. It was the beneficiary farmers who did the works. Since the colonial government had no need to bother about their upkeep, they could indicate anyone as responsible for the maintenance (*see* Sengupta, 1985).
6. Local people recall that only a few decades back parts of the peninsula had lush green vegetation, known as *gher* land. Another reason is suggested for salinity ingress in recent years. The coastal belt of the Peninsula was once rich in limestone. Over the years substantial parts of the deposits have been extracted by cement factories; in fact the availability of limestone explains the concentration of cement factories around this region. Massive limestone deposits along the coast had once acted as a natural protective belt against salinity. Now this protection is not available — personal talks with Dr. P.G. Marvania.

References

Abrol, I.P. (1992). "Towards Sustainable Use of Our Soil Resources", in *Growth of Indian Agriculture: Issues and Policies*, Deptt. of Economics and Sociology, Punjab Agriculture University.

Abrol, I.P. and Dhruva Narayana, V.V. ed. (1990). *Technologies for Wasteland Development*, Indian Council of Agricultural Research, New Delhi.

Arora, Hema (1994). "Some Emerging Issues in User Reclamation", *Wasteland News*, February-April.

Bowonder, B. and C. Ravi (1984). "Waterlogging from Irrigation Projects: An Environmental Management Problem", Centre for Energy and Technology, Administrative Staff College of India, Hyderabad, unpublished.

Dregne, H.E., and N-T. Chou. (1992). Global desertification dimensions and costs. In *Degradation and Restoration of Arid Lands*. Texas Tech. University: Lubbock.

FAO (1985). *Water Quality for Agriculture*, Irrigation and Drainage Paper No. 29.

FAO Materials from Land And Plant Nutrition Management Service: Global Network on Integrated Soil Management for Sustainable Use of Salt-affected Soils.

Gupta, L.C. (1990). *Use of Saline Water in Agriculture: A Study of Arid and Semi-Arid Zones of India*, Oxford and IBH: New Delhi.

India, Govt. of (1972). *Report of the Irrigation Commission*, Vol. 1, New Delhi

—— (1976). *Report of the National Commission on Agriculture*, Vol. V, New Delhi.

Kanwar, J.S. (1977). "Soil Degradation in India", *Soil Bulletin*, No. 34, F.A.O., Rome.

Kolarkar, A.S. and Singh, N. (1990): "Runoff Farming in Arid Tract of India: A Review", *Indian Journal of Soil Conservation*, Vol. 18, No. 3, December.

Mamoria, C.B. (1982). *Geography of India*, Vol. I, Shiva Lal Agarwala: Delhi.

Marvania, P.G. (2001). "Salinity Ingress and Technological and Institutional Solutions — Some results: A Case Study of Coastal Saurashtra", paper presented at Second Biennial Conference of the Indian Society for Ecological Economics, Bhopal, 19-21 December 2001 (unpublished).

NWDA (National Water Development Agency) (1994). *Proceedings of the Fifth National Water Convention*, sponsored by the Ministry of Water Resources, GOI, Govt. of Haryana and Indian Water Resources Society, Faridabad, 25-27 February.

Oesterbaan, R.J. (1983). "Modern Interferences in Traditional Water Resources in Baluchistan", *Water Resources Journal*, December.

Oldeman, L.R., Hakkeling, R.T.A. and Sombroek, W.G. (1991). *World map of the status of human-induced soil degradation. An explanatory note.* International Soil Reference and Information Center (ISRIC), Wageningen.

Raina, Rajeswari, (2001). "Water quality: am institutional analysis of agricultural science and its ecology", paper presented at Second Biennial Conference of the Indian Society for Ecological Economics, Bhopal, 19-21 December 2001 (unpublished).

Rubinoff, Janet Ahner, (2001). Pink Gold: Transformation of Backwater Aquaculture on Goa's Khazan Lands, Review of Agriculture, *Economic & Political Weekly*, March 31.

Sengupta, Nirmal (1985). "Irrigation: Traditional vs. Modern", Special No. on Natural Resource Utilization, *Economic and Political Weekly*, November.

——, (1991). *Managing Common Property: Irrigation in India and Philippines*, Sage Publications, New Delhi.

——, (1993). *User-Friendly Irrigation Designs*, Sage Publications: New Delhi.

Singh, N. and Kolarkar, A.S. (1983). "Some Physico-Chemical Properties of Soils of "Khadins" in Western Rajasthan", *Journal of Soil Conservation*, Vol. 11, Nos. 2 and 3.

Stone, Ian (1984). *Canal Irrigation in British India*, Cambridge Univ. Press: Cambridge.

Voelcker, J.A. (1893). *Report on the Improvement of Indian Agriculture*, London, reprinted Agricole, New Delhi, 1986.

Whitcombe, Elizabeth (1971). *Agrarian Conditions in Northern India, Vol. I: The United Provinces Under British Rule, 1860-1900*, Univ. of California Press, California.

7

Salinity Ingress and Technological and Institutional Solutions—Some Results

A Case of Coastal Saurashtra

P. G. MARVANIA

Introduction

This paper describes a process of degeneration in groundwater resources and the results of efforts made to restore the same in the Saurashtra region of Gujarat state. The paper is divided in three parts. First, it presents a description of the causes of salinity ingress and its impact in the region. The second part highlights the preventive measures taken by the state and the people. Third, it reviews the main issues in groundwater resource use and conservation in this region.

I. Genesis and Magnitude of Problem

Geo-morphological Settings of the Resource Region

Saurashtra is a peninsular region situated in the westernmost part of the state of Gujarat. It has a coastline of 765 km. Geological formations of this region include alluvium, miliolite, limestone Gaj-bed, Deccan-traps and standstones. These formations have made the underground formations rocky. Much of Saurashtra is a basaltic plate, rising from about 100 to 200 m. above sea level. This extensive sheet of lava covers more than a half of the region and is highly eroded by rivers flowing in all directions from the central plain. The whole region is traversed by 71 small rivers. In view of the topography described above all these river-basins are more or less homogeneous in their hydro-metrological characteristics. The rocky underground formations have a reduced percolation capacity for the groundwater recharge in this region.

Besides, as seen from Table 7.1, rainfall is quite scanty and erratic in this region which is also prone to recurrent droughts.

Table 7.1: Rainfall and normal rain

(In millimetres)

No.	*Year*	*Jamna-gar*	*Rajkot*	*Surendra-nagar*	*Bhav-nagar*	*Amreli*	*Junagadh*
1	1981	1369	641	641	765	805	1387
2	1982	884	470	532	418	843	1131
3	1983	885	637	453	654	1088	2538
4	1984	655	539	498	449	294	1215
5	1985	208	278	423	657	311	449
6	1986	189	200	308	364	456	762
7	1987	36	180	81	121	133	194
8	1988	482	1056	443	761	1317	2178
9	1989	286	433	449	573	880	1071
10	1990	204	454	603	736	640	787
Normal		497	621	507	596	546	881

Source: Socio-Economic Review, 1990-91, Government of Gujarat, pp. 22-23.

There is no perennial river in this region. All the rivers in the region are short in length and are rain-fed and flood during heavy rains. The floods are flashy. They usually last for a few hours. The rising limb of the flood hydrograph lasts only 3 to 4 hours. All these rivers dry up in the fair season.

The geo-morphological settings, rainfall pattern and river system of the region indicate that the region has very limited surface and rainwater resources. Hence, this area is heavily dependent for its water requirements on the groundwater resources.

Surge in Groundwater Demand

Water crisis in this region has now become a regular phenomenon, *i.e.* more or less in drought years and always in the summer months. The worst affected by these crises are the rural parts of the region.

There seem to be two causes for the aggravated depletion of groundwater resources in this region. First, the depletion could be ascribed to indiscriminate expansion in the use of modern water extraction devices in accessing groundwater. Second, it is due to the low level recharging of groundwater during the monsoon.

The energized pumping technology accelerated an extraction rate of groundwater that was beyond sustainability. Prior to 1951, only 25 per cent of the total groundwater resources were harnessed in this region. This was because the availability of pumping equipments and the energy needed to operate them was quite limited then. But with the advent of the "Green Revolution" technology packages farmers were given sufficient incentives to go for modernization of groundwater extraction technology. As seen from Table 7.2, during the 1980s this region witnessed a steep growth in the use of electrified pump-sets and construction of tubewells.

During the 1960s through 1980s, groundwater remained the major source of irrigation in this region. Table 7.3 indicates that during this period, the share of groundwater in total irrigation stood at 81.1 per cent. But in the Junagadh district—which is now the worst affected of all the districts of the region—the share was the highest at 95 per cent.

Simultaneously, changes have taken place in the case of extraction technology and types of well-structures. The traditionally used menial and animal power to lift groundwater was replaced by mechanical force. And the dug-wells were replaced by pneumatically drilled bore-wells. Data in Table 7.2 and Table 7.4 give a picture of this transformation. By the end of March 1990, 17,897 villages from a total of 18,114 villages in the state were supplied with electricity. This sort of technological transformation facilitated a rapid extraction of the resource.

Table 7.2: Districtwise numbers of pumpsets/tubewells electrified

Sl. No.	*District*	*1979*	*1991*
1.	Amreli	7333	17232
2.	Bhavnagar	9921	27535
3.	Jamnagar	6091	20195
4.	Junagadh	21918	53538
5.	Rajkot	11372	30232
6.	Surendranagar	5592	14295
Saurashtra Region		62229 (35.0)*	163027 (38.93)* (161.98)**
7	Banaskantha	8307	29023
8	Mahesana	18158	41097
9	Sabarkantha	18906	50167
North Gujarat Region		45371 (25.52)*	120287 (28.73)* (165.12)**
10	Ahmedabad	9845	18395
11	Gandhinagar	2788	4623
12	Kheda	12394	22107
13	Vadodara	9644	16462
Central Gujarat Region		34671 (19.5)*	61587 (14.71)* (77.63)**
14	Bharuch	4142	6533
15	Surat	8454	18209
16	Valsad	9816	20619
South Gujarat Region		22412 (12.61)*	46361 (11.07)* (106.85)**
17	Dangs	13	24
18	Panchmahals	3415	8136
19	Kutch	9687	19300
Tribal and Desert Area		13115 (7.37)*	27460 (6.56)* (109.4)**
All Gujarat		177798 (100)*	418722 (100)* (135.5)**

* Relative Share of the region in the total; ** Growth rate over 1979

Source: Socio-economic Review, 1990-91, Government of Gujarat.

Table 7.3: Area irrigated by source 1960-61 to 1985-86

(km^2) ('00 hectares)

Sl. No.	*Source*	*1960-61*	*1970-71*	*1980-81*	*1985-86*
1.	Govt. Canals	652 (9.5)	2358 (17.2)	3668 (18.3)	3586 (17.6)
2.	Private Canals	6 (0.1)	6 (0.04)	—	—
3.	Wells & Tubewells	5677 (83.1)	10831 (79.0)	15884 (79.3)	16532 (81.1)
4.	Tanks	128 (1.9)	372 (2.7)	409 (2.0)	253 (1.2)
5.	Other Sources	366 (5.4)	414 (1.0)	65 (0.4)	24 (0.1)
6.	Total Net Area Irrigated	6829 (100)	13708 (100)	20026 (100)	20395 (100)
7.	Gross Area Irrigated	7338	14939	23344	23812

Source: Socio-economic Review, 1990-91, Government of Gujarat.

Table 7.4: Statement showing growth of groundwater extraction structures

('000)

Sl. No.	*Structures*	*1980*	*1985*	*Growth Rate (%)*
1	Dug-well	654.0	672.9	3
2	Pvt. Tubewells	2.5	4.5	80
3	Pub. Tubewells	2.2	3.9	77
4	Electric Pump Sets	202.9	294.1	45
5	Diesel Pump Sets	562.5	732.5	30

Source: Government of India, 1986, "Groundwater Development in India," Ministry of Water Resources, Delhi.

Extent of the Problem

The main salinity prone region is the coastline which is formed by creeks and tides up to 6 km. of the plain region. The normal coastal salinity is also accompanied by marine sedimentary formations. Now, the salinity has invaded as far as 7 km inland from the coast into the aquifer system. Normal water flow is from inland to coastal side. But due to overextraction of groundwater in the adjoining coastal plains, the flow has now reversed. Salinity in the shallow depth is found in an area totalling to 34,625 km^2 with adverse effects on the phreatic aquifers. Salinity below 200 m. of depth is observed over an area of 34,000 km^2. This problem is further aggravated by the inherent salinity of thick coastal territory and alluvial formation deposited under marine conditions. Due to this the possibility of getting freshwater from the aquifers within exploitable depths is lost. Only the upper-bed in these formations, which receives constant replenishment monsoon after monsoon, has moderate freshwater storages.

II. Preventive Efforts

Government Initiative

By taking into account the disastrous effect of a rapid salinity ingress, the government appointed the First High Level Committee, popularly known as HLC-I, in 1976 to study the problem of salinity ingress in the Una-Madhavpur area of coastal

Saurashtra. The government also constituted the second High Level Committee, popularly known as HLC-II, in 1978 to study the problem of salinity ingress in the Una-Bhavnagar and Madhavpur-Malia areas of coastal Saurashtra. Both, the HLC-I and the HLC-II recommended four scientific methods to prevent the spread of salinity and to regenerate the damaged resource. They were as follow:

1. *Resource Management System (RMS)*: The RMS included evolving a system which could control the extraction rate and prevent the over-exploitation of the groundwater resource. Under the RMS, a change in favour of less water intensive crops to reduce the extraction rate was suggested.
2. *Salinity Control Method*: To control salinity, two methods, namely Tidal Regulators and Bandharas were recommended. These are engineering structures constructed to control salinity.
3. *Recharge Method*: To regenerate the damaged resource, the following measures of groundwater recharge were recommended. They include: construction of check-dams, recharge-tanks, recharge-wells, spreading-channels, nala-plugs and forestation.
4. Coastal Reclamation Bunds to reclaim saline affected land near coastline.

The state government created a separate department known as Salinity Ingress Prevention Circle (SIPC) for undertaking construction of the recommended structures. Details of the recommended and completed structures under the Salinity Ingress Prevention Circle are given in Table 7.5. Besides, a department of geologist was created to monitor water quality in the areas where the SIPC had completed the salinity ingress preventive structures.

The High Level Committee-I had recommended works of an estimated cost of Rs. 64 crores in the Una-Madhavpur area. The High Level Committee-II had recommended works estimated at a cost of Rs. 168.7 crores and Rs. 370.42 crores in Una-Bhavnagar and Madhavpur-Malia area, respectively.

Of this, the government has spent Rs.237.31 crores on completed works as of this, the shown in Table 7.6, by June 2000.

Table 7.5: Progress of salinity ingress in Saurashtra region 1977 to 2000

Sl. No.	*Gross Salinity Effect*	*1977*	*2000*
1.	Numbers of wells affected	12562	32700
2.	Numbers of villages affected	120	534
3.	Numbers of people affected	286106	1079479
4.	Area affected (hectares)	100009	700129

Sources: High Level Committee-I, 1978, Govt. of Gujarat; and Salinity Ingress Prevention Circle, Rajkot, 2000 Govt. of Gujarat.

Table 7.6: Works completed under recommendations of HLC-I & HLC-II

Type of Works	*HLC-I* *Una-Madhavpur*		*HLC-II* *Una-Bhavnagar*		*HLC-II* *Madhavpur-Malia*		*Total Completed Structures*
	Reco.	*Comp.*	*Reco.*	*Comp.*	*Reco.*	*Comp.*	
1	2	3	4	5	6	7	8
1. Tidal regulator	7	6	8	0	23	3	9
2. Bandhras	10	12	29	4	48	1	17
3. Recharge reservoir	4	2	22	0	21	0	2
4. Check-dams	75	172	300	172	460	185	529
5. Recharge tanks	7	5	30	0	60	2	7
6. Recharge-wells	200	198	450	135	680	66	399
7. Forestation (hectares)	1000	5867	15750	0	25000	0	5867
8. Nalla-plugs	0	4487	20000	0	45400	0	4487
9. Spreading-channels (km.)	60	20.20	60	0	240	1	21.2

Source: *A Note on Problems and Solutions of Salinity Ingress in the Coastal Saurashtra*, Salinity Ingress Prevention Circle, Rajkot, Water Resource and Water Supply Department, Govt. of Gujarat, August 2000, p.3.

The main objective of all these engineering structures was to store fresh water and facilitate the recharge of the groundwater aquifer for improving water quality and to regulate tidal waves to prevent surface and sub-soil intrusion of salinity. Tidal Regulators and Bandharas have been effective in controlling saline intrusion due to tidal waves. This has raised agrarian productivity in the Una-Madhavpur area. Table 7.7 gives details of improvement in productivity of different crops.

Institutional Initiative

The major source of livelihood in this region is income from the agriculture sector. The semi-arid conditions and scanty and erratic rainfall have made the people more dependent on groundwater for their prosperity. Attempts to reap quick gains by sowing cash crops has created a situation like the "tragedy of commons" with regard to the stock and quality of groundwater resource. This tragedy is caused both by the action of nature and of the people. In response to the rapidly deteriorating groundwater resource situation, many voluntary organizations, self-help groups and NGOs came forward for regenerating the resource.

Institutional efforts are mainly directed towards a recharging of the groundwater aquifers and the creation of structures to store rain water. Among these efforts, well-recharge activity is quite popular. The Jayashree Rural Development Foundation is one of the many such organizations working for the well recharge activity in this region. By 1996-7 it had undertaken the recharge of 96,895 wells in this region (Marvania, 1998). This activity helps in raising the groundwater draft for a short period. During the last two successive drought years (1999 & 2000), many check dams were constructed by self-help groups.

III. Main Issues in Groundwater Use and Conservation

Factors like scanty rainfall, rocky underground bed, water intensive cropping pattern, modern groundwater extraction technology, rising population and overextraction by farmers

Table 7.7: Details of increase in productivity of agro-products due to salinity ingress prevention structures

(Productivity in MT/Ha)

Sr.	*Name of Structure*	*Crop Type*	*Prior*	*After-wards*	*% increase*
1.	Meghal Tidal Regulator and Meghal River Basin Taluka: Malia Dist. Junagadh	Groundnut	11.20	24.00	114.28
		Wheat	20.00	38.00	90.00
		Bajari	17.00	26.00	52.94
		Coconut (pieces)	20000	66500	232.05
2.	River Basin of Sarswati & Bedhara Ta. Veraval Dist. Junagadh	Groundnut	17.78	24.00	34.98
		Wheat	20.74	32.00	54.29
		Bajari	22.52	24.50	8.79
		Sugarcane	1245	1350	8.43
3.	Khari River Basin & Tidal Regulator at Vdodara Ta. Sutrapada, Dist. Junagadh	Groundnut	17.78	23.80	33.85
		Wheat	20.74	32.50	56.70
		Bajari	22.53	25.20	11.85
4.	Shardagram Bandhara Ta. Mangrol, Dist. Junagadh	Groundnut	15.00	20.00	33.33
		Wheat	10.00	19.00	90.00
5.	Sheel Bandhara Ta. Mangrol, Dist. Junagadh	Groundnut	10.00	20.00	100.00
		Wheat	10.00	24.00	140.00
6.	Spreading Channel No.1 Ta. Mangrol, Dist. Junagadh	Groundnut	19.00	27.00	42.10
		Wheat	20.00	37.00	85.00
7.	Noli River Basin Dist. Junagadh	Groundnut	15.00	20.00	33.33
		Wheat	10.00	19.00	90.00
8.	Netravati River Basin Ta. Mangrol, Dist. Junagadh	Groundnut	10.00	20.00	100.00
		Wheat	10.00	24.00	140.00
9.	Vrajami Reservoir Ta. Malia Dist. Junagadh	Groundnut	12.50	37.50	200.00
		Wheat	—	25.00	—
		Bajari	9.5	10.00	5.26
		Mis. Crops	—	25.00	—
10.	Medha Creek Tidal Regulator Ta. & Dist. Junagadh	Groundnut	1.00	3.00	200.00
		Mis. Crops	0.20	0.40	100.00

Source: *A Note on Problems and Solutions of Salinity Ingress in the Coastal Saurashtra*, Salinity Ingress Prevention Circle, Rajkot, Water Resource and Water Supply Department, Govt. of Gujarat, August 2000, Appendix-2.

culminated in salinity ingress and its rapid progression in this region. Efforts both, private and public, to reduce the impact are on rise. However, a large numbers of issues need to be closely scrutinized, namely:

1. Almost all the measures taken for the conservation and regeneration of the groundwater resource lay emphasis on an increase in the supply side. That is, these measures aim at increasing the stock of groundwater and preventing its quality deterioration. However, no effort is directed towards limiting the growing demand for groundwater. A region with geomorphologic conditions such as these, needs to pay an equal attention to limiting the demand for groundwater.
2. There does not exist a well defined regime of rules for extracting groundwater. Evolving a suitable regime of groundwater rights is quite essential. The "commons" like characteristic of the groundwater resource could easily lead to its over exploitation.
3. There is an urgent need for growing crops which are less water intensive and yet fetch an income comparable to the one earned from water-intensive crops. This will reduce the farmers' tendency to overexploit this resource.

References

Govt. of Gujarat (1992). *Socio-Economic Review*, 1990-91; Govt. of Gujarat, Gandhinagar.

Govt. of Gujarat (2000). *A Note on Problems and Solutions of Salinity Ingress in the Coastal Saurashtra*, Salinity Ingress Prevention Circle Rajkot, Govt. of Gujarat, Gandhinagar; August 2000.

Govt. of Gujarat (1978). Reports of High Level Committee I & II, 1978; Govt. of Gujarat, Gandhinagar.

Govt. of India, (1986). "Groundwater Development in India", Ministry of Water Resources, Delhi.

Marvania, P.G. (1998). "Role of Institutions in the Groundwater Management in Saurashtra;" U G C Minor Research Project, Dept. of Economics, Saurashtra University, Rajkot (unpublished).

Pandya, B.A. (1984). A Study of Salinity Problem in Saurashtra; ICSSR Research Project, Dept. of Economics, Saurashtra University, Rajkot (mimeo).

8

Efficient Use of Water in Canal Command of Punjab

C.J. SINGH

Water is the premier determinant of agricultural production and also enriches the environmental quality of life. Hence, the rational development and utilization of water resources should undoubtedly be a priority concern of our agricultural policy. The irrigation development in th Punjab region dates back to Indus civilization. The present canal system (Fig. 8.1) came into existence in the middle of the nineteenth century under British rule and it has been improved upon periodically since then. Dhillon (1992) has described the system in detail. In Punjab, water has become an increasingly scarce and costly resource since the initiation of highly intensive agriculture in the late sixties. This situation has arisen despite the substantial generation of irrigation supplies from both its river and groundwater resources since independence. Prihar *et al.* (1993) draws attention to a net water deficit of 1.77 M ha m in the state with a negligible scope for further water resource development and pinpoints the utmost necessity for the most efficient use of existing irrigation supplies and for diversification of agriculture. However, following irrigation-related changes during 1983-6 and 1997-2000 further ring an alarm bell:

- Annual consumption of electricity per capita increased more than three times.

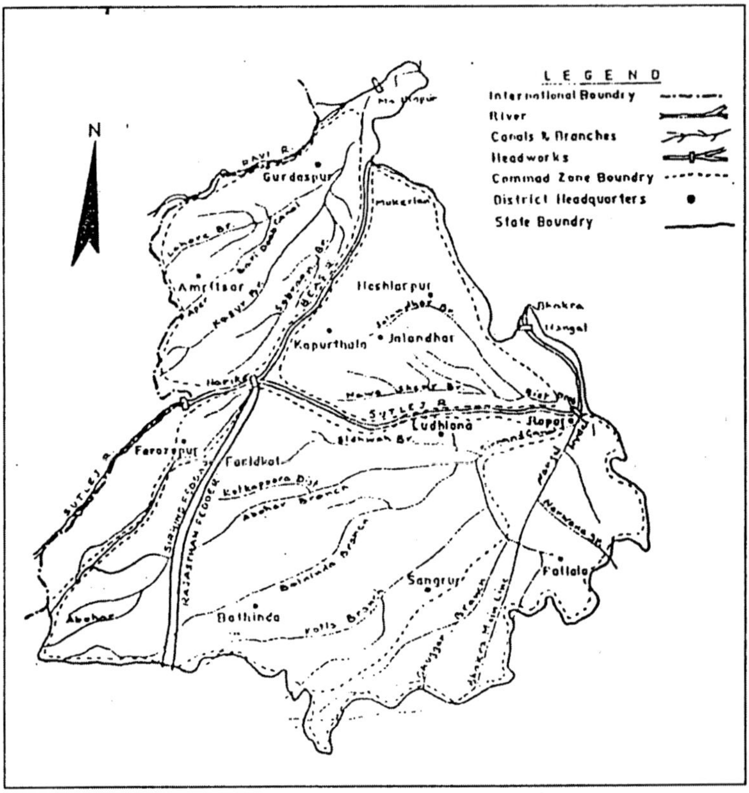

Fig. 8.1: Surface Irrigation System in Punjab.

- Number of tubewells energized increased from 4 lakh to 7.6 lakh.
- Area irrigated by canals decreased by 19.5 per cent.
- Area irrigated by tubewells increased by 29.5 per cent.
- Per cent net irrigated area/per cent net sown area/ increased from 86.3 to 94.7 per cent.
- Per cent gross irrigated area/per cent gross cropped area/increased from 90.5 to 96.1.
- Area under paddy increased by 53 per cent and under wheat by 10.4 per cent.
- Area under wheat decreased by 10.4 per cent.
- Area under cotton decreased by 10.9 per cent.
- Area under fruits decreased by 35.6 per cent.

- Area under vegetables increased by 63.8 per cent.
- The yields of paddy and wheat increased by 6.7 and 30.9 per cent.
- The yield of cotton decreased by 34.7 per cent.

Similar changes were observed in the canal command districts. The increase in area under paddy in south-west Punjab is more harmful because of the high demand of water for the crop which canal supplies cannot meet and the poor quality of the groundwater.

Thus diversification in agriculture has not been successful in the state and irrigation water continues to be used injudiciously especially so in the canal-irrigated south-west Punjab. Punjab Agricultural University, in its research efforts at Ludhiana in the initial years developed land and water management practices (Narang, 1995) and irrigation scheduled for crops (Sandhu and Prihar, 1983). However, there are considerable differences between the natural resources of the central and south western parts of the state. In the central districts and the south western zone, 99.8 and 63 per cent of the area, respectively experienced a falling water table. A lowering of the water table beyond 15 m depth makes the working of traditional centrifugal pumps difficult and further extraction of groundwater dependant on submersible pumps thereby increasing the initial investment as well as maintenance and running costs. The rise of the water table in the development blocks of south west Punjab is shown in Fig. 8.2. The water table in Jalalabad, Guruharsahai, Fazilka, Muktsar, Kotkapura, Nathana, Khuiansarvar was <10 m deep during 1974, which rose to <3 m in the blocks except Nathana and Kotkapura. In 2000, it deepened slightly, still being <5 m deep. From Jhunir to Lambi, the water table was >10 m during 1974, it rose to <5 m in Malout and Kotbhai, and < 10 m in Jhunir and Bathinda; and 10-15 m deep in Talwandi sabo, Abohar, Sangat and Lambi. During 2000, the water table rose to <5 m in Jalalabad, Guruharsahai, Fazilka, Muktsar, Khuiansarvar, Malout, Kotbhai and Lambi, and was 5-10 m deep in Kotkapura, Nathana, Jhunir, Bathinda, Talwandi sabo, Abohar and Sangat.

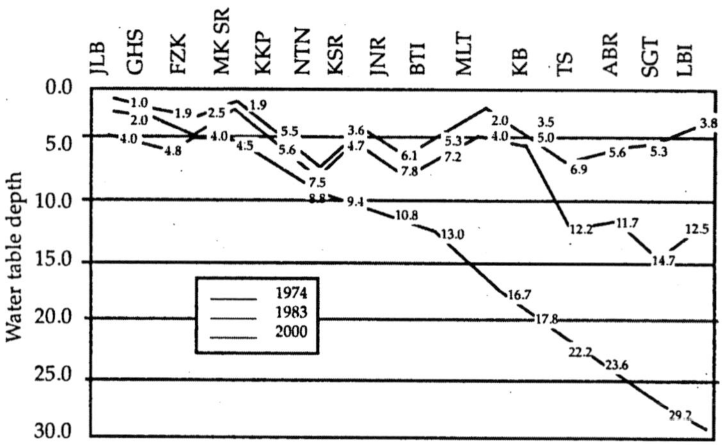

Fig. 8.2: Rise in Water Table Depth in Zone V during 1974-2000.

Source: Groundwater Cell, Directorate of Water Resoruces and Environment, Punjab.

Thus, a judicious use of irrigation water in these canal commands is of utmost importance.

Agro-climate

Punjab is sub-divided into six agro-climatic regions (Fig. 8.3). The reported studies were conducted in Zone V (Fig. 8.4), which is spread over 19.14 per cent of the total geographical area of the state and covers 20.13 per cent of the net sown area. The topography is uneven. The major part of the zone lies at a height of 200 m above the sea level and is part of Indo-gangetic plains. There are three physiographic units *viz.*, alluvial terraces, sand dunes and interdunal areas. The soils are highly calcareous and rich in dispersed lime; sandy to sandy loam in texture with pockets of fine textured soils. Soil profiles are generally well drained, except where highly impermeable layers of clay and/ or calcium carbonate occur in the subsoil. Generally, the soils are poor in nitrogen supply, low to medium in available

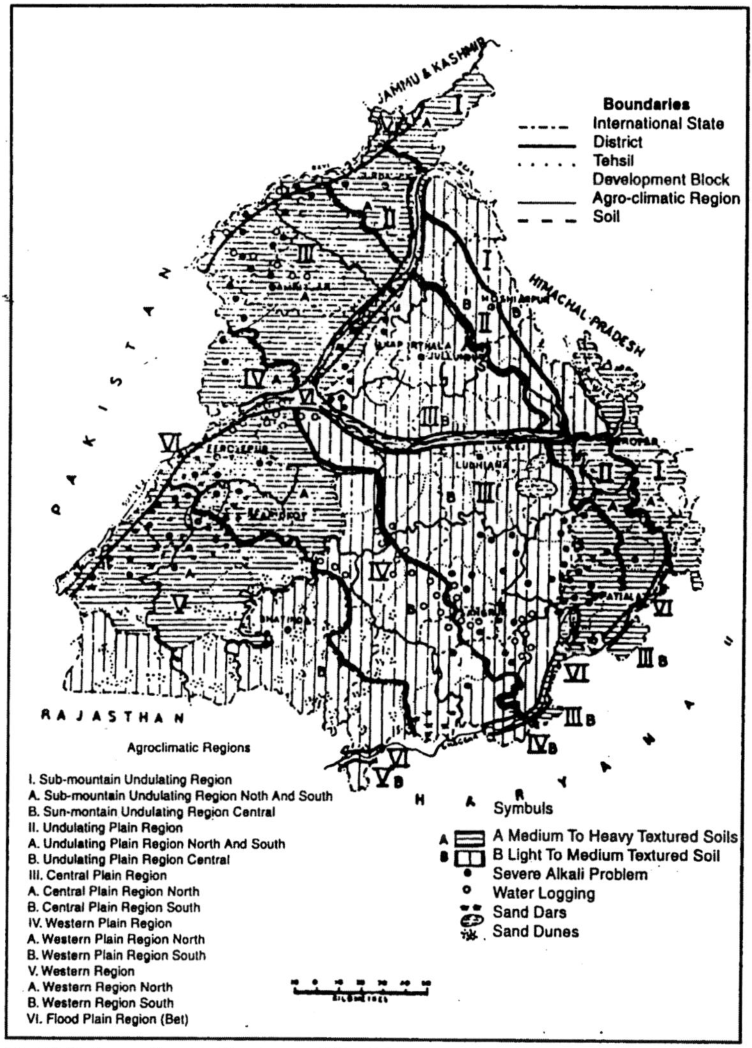

Fig. 8.3: Agroclimatic Regions of Punjab.

phosphorus, and medium to high in available potassium. Deficiencies of zinc, manganese and iron as well as toxicities of boron and fluorine have also been observed in some crops in the zone.

The zone has an arid to dry climate with high temperature

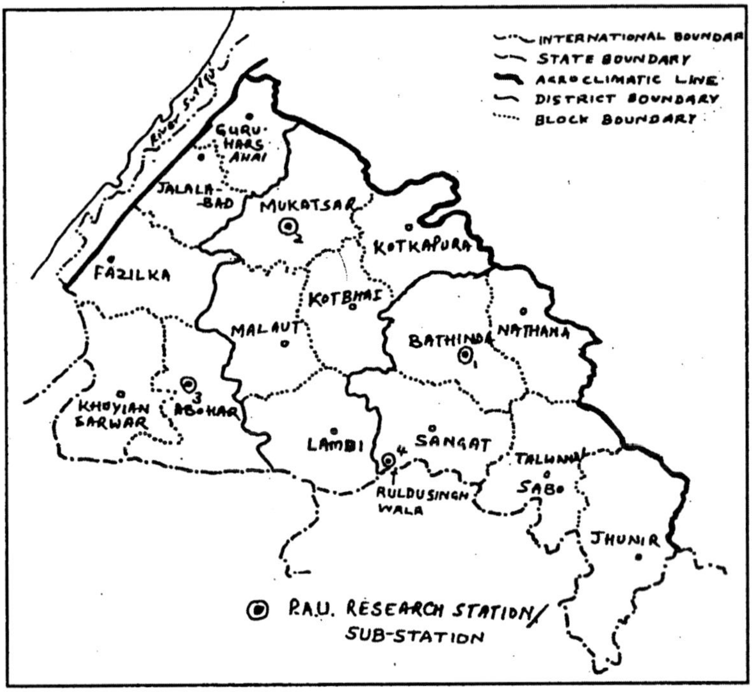

Fig. 8.4: Map of Zone V of Punjab.

and low humidity during May-June and early July followed by a rainy season of 2-3 weeks in late July to early August. During May-June, dry winds and dust storms and the occurrence of frost/chilly winds in December/January is common.

The average annual rainfall over different parts of the zone varies from 200-400 mm. The average monthly rainfall is maximum during July and August, and minimum during October, November and December. During May-June the average monthly rainfall is below 40 mm.

Methodology

The adaptive research-cum-demonstration trials were conducted (1983-4 to 1988-9) in the canal command area of the water course No. 79660 TR of the village Jaisingh wala, at the tailend of the Bathinda distributory, selected in consultation with the officials of the state irrigation and agriculture departments. The trials on wheat, cotton, mustard, berseem and jwar crops were intended to evaluate the soundness of the improved irrigation practices and other farming techniques which directly or indirectly reduce the wasteful expense of scarce irrigation water and increase the water use efficiency. These practices relate to timing of post-sowing irrigations including last irrigation, appropriate depth of pre-sowing irrigation, optimization of width of border strips, balanced use of fertilizers, application of gypsum to offset adverse effect of soil sodicity, etc.

Initially, demonstrations were conducted with cotton and wheat, the major crops of the region, to compare the traditional (farmers') with those of improved practices on fields of individual farmers who volunteered for the purpose. The fields were generally scattered over the command area of the water course. In each growing season 7-8 farmers participated in the programme.

For comparison of these improved package of practices with the existing package, two separate adjoining blocks of 4-6 ha each were identified to establish the validity of improved practices so as to ascertain the scope for a mass adoption of technology assuming this could differ from individual demonstrations in potentialities and limitations. Four to six farmers participated in each block in each season. Various agronomic operations were monitored in both the blocks. The farmers of the improved block were guided to adopt improved practices as far as possible. In the farmers' block the growers followed various agricultural practices of their own accord. We noticed that even in farmer's block, cultivators benefited from our earlier demonstrations and improved upon the traditional practices to some extent.

Besides, a survey of the groundwater in the region was conducted by collecting samples of the working tubewells, analysing these for the chemical composition, and collecting information regarding the use of waters by the farmers to study the extent of conjunctive use prevalent in the region.

A detailed field study on the conjunctive use of canal water and tubewell water was conducted (1993-2000) on the Research Station farm on the Bathinda-Dabwali Road 7 km from Bathinda city. The experimental details have been described by Singh *et al.* (2002).

A randomized and replicated detailed study on the evaluation of drip irrigation was conducted earlier at the Fruit Research Station on Hazi Rattan link road. The experimental details have been described by Singh (1996).

The soils of these sites are representative of the soils of the region and only the indicative results are presented hereunder for reasons of brevity.

Irrigation Practices

Land levelling

An appropriate gradient of land surface is essential for smooth advance of the irrigation stream and high irrigation application/distribution efficiencies. Such a gradient of economic viability depends upon the levelling index of the land, nature of the soil, plot size, irrigation method, and irrigation stream size. Even where the levelling index is small, a micro-levelling of the land could be helpful to check the patchy stand of the crop and salinization of soil and to ensure a uniform application of water. A poor levelling is as bad as no levelling. Levelling increased the yield as well as irrigation efficiency in cotton and wheat (Table 8.1).

Table 8.1: Effect of land levelling on mean irrigation time and yields of cotton and wheat

Condition of land	*Wheat*		*Cotton*	
	Irrigation Time (min/ha)	*Yield Kg/ha*	*Irrigation Time (min/ha)*	*Yield Kg/ha*
Unlevelled	188	2730	187	2050
Levelled	166	2860	156	2320

Border Size

In south-west Punjab, the length of the borders is limited by *kilabandi, i.e.,* acre length and is usually 50-70 m. The border width can, however, be adjusted to the prevailing discharge of the water stream generally available for the pre-sowing and post-sowing irrigation. Such adjustments can improve water distribution and storage efficiencies. These optimum border widths also reduced the time required to irrigate a unit area, thus suggesting irrigation economy. Also these wider borders as compared with the conventional narrow borders reduce the labour cost.

Table 8.2: Effect of border width on the yield and irrigation time in cotton, wheat and raya in 50-70 m long strips

Border width (m)	*Yield (Kg/ha)*		
	Kapas	*Wheat*	*Raya*
5	—	3300	800 (198)
9	1820 (165)	—	—
10	—	3350 (182)	850 (190)
13.5	1860	—	
15	—	3330 (172)	1040 (175)
18	1850 (138)	—	—
20	—	3240 (148)	— —

The optimum width of 50-m long strip border for irrigation with the available stream size of the command area was 15 m for wheat and 18 m for cotton. Thus the present farmer's practice of maintaining a 5-10 m wide borders could be dispensed with to economize labour costs (Table 8.2).

Irrigation Practices and Timing

Cut-off Ratio

An optimum cut-off ratio is one that ensures a uniform distribution of water over the entire border length. It primarily depends upon the stream size, intake rate of soil, slope and width of the plot. It is chosen so that an equal opportunity time for infiltration is provided at both ends of the border. Farmers generally cut off irrigation water from the stream in the plots of cotton and wheat when the waterfront in the plots covers 90 per cent of its length. However, it has been shown that irrigation can be cut off at the 70-80 per cent length thus economizing irrigation water with no adverse effect on the crop yields (Table 8.3).

Table 8.3: Effect of cut-off ratio on yields of cotton and wheat

Cut off ratio	*Mean yield (kg ha)*	
	Kapas	*Wheat*
0.7	2190	2250
0.8	2300	2250
0.9	2240	2180

Pre-sowing Irrigation

Crops are occasionally subjected to water deficits for some periods even in irrigated areas when the supply of canal or tubewell water is either inadequate or delayed. Such a problem in the water supply is often encountered in canal irrigated areas. These water deficits and the associated yield reductions can be averted partly, if not totally, in case the water availability in the root zone of the crops can be enhanced through a repletion of soil profile with heavier pre-sowing irrigation. A heavy pre-sowing irrigation to wheat gave a 100-200 kg/ha more grain yield than the light pre-sowing irrigation. Moreover, an additional input of 4-cm water with the former practice gives an insurance against the water deficits during the growing season (Table 8.4).

Table 8.4: Effect of pre-sowing irrigation on the mean yield of wheat

Pre-sowing irrigation	*Yield (kg/ha)*
Light (7.5 cm)	3600
Heavy (10 cm)	3720

Post-sowing Irrigation

The scarce and costly irrigation supplies need to be used with utmost care. While under irrigation to crops reduce their yields, over irrigation involves wasteful expense of water and can also cause serious problems of waterlogging, soil salinization and downward leaching of nutrients. All these ills are bound to reduce the productivity of soils. The problem of a rising water table in the south west districts of the state is a matter of serious concern. Unlike the central districts, the problem here would be difficult to manage since the use of underground water for irrigation purposes is restricted due to their saline nature.

To check the rising water table, crops should be irrigated at proper timings and with appropriate amounts of water that are absolutely essential for optimum plant growth and permit a minimum water loss below the root zone. The schedules of irrigation to crops were based on pan-evaporation, a reliable index of evaporative demand of the crops. Wheat requires a heavy pre-sowing irrigation and at the most four post-sowing irrigation for optimum grain yields. Application of four post-sowing irrigations to wheat according to the improved irrigation time-table gave as much yield as six to seven irrigations according to the farmers practice thus economizing 2-3 irrigations without sacrificing crop yield (Table 8.5).

Table 8.5: Effect of improved irrigation schedule on mean wheat yield

Irrigation Schedule	*Mean number of irrigation*	*Grain yield (kg/ha)*
Improved	4	3022
Irrigation table Farmer's practice	7	3039

The effect of irrigation on physiology of the crop and hence on its yield is considered more crucial in cotton than in other field crops. Timings of various irrigations need to be managed in relation to the climatic environment and available soil water in the root-zone so that excessive vegetative growth is avoided and the opening of the bolls is not hindered or delayed. The timing of last irrigation is particularly important for affecting boll opening under the suitable climatic condition. If this irrigation is given beyond September 30, it delays the opening of the bolls and affects the yield adversely (Table 8.6).

Table 8.6: Effect of timing of last irrigation on kapas yield of cotton

Time of last irrigation	*Irrigation applied*	*Mean kapas yield (kg/ha)*
September 30	4-5	1843
October 25 (Farmers' practice)	5-7	1717

In the 1980s, canal water supply in the distributory used to be withdrawn during May and June thereby delaying the sowing of cotton which led to lower yields of the crop. The irrigation authorities were apprised of the situation along with the data generated in the command area. The canal authorities were convinced and this subsequently assured the farmers supply of water for the timely sowing of cotton. The proposal was implemented from 1993 onwards along with adjustment of capacity factors keeping in view the cropping pattern in the region. Since then there has been no closure during May/June except during 2001 when the water storage in the Bhakra system went below the critical level.

Almost the entire area in the canal command as well as in the tubewell command in Punjab is flood irrigated. This is a highly inefficient method of water use. A need has long been felt in the region for more efficient methods of water application to be tested and propagated.

Irrigation Methods

Furrow Method

This method is generally followed in vegetables and certain fruit crops planted in rows or beds. Water from the main lateral channel is directed into small parallel furrows drawn along the slope in between the small ridges or broad beds. Water in the furrows moves both laterally and vertically to moisten the ridges and subsoil. The method is particularly useful for soils that crust on irrigation, exhibit a low infiltration rate, contain a high content of salts or need surface drainage. On lands with appreciable slopes, furrows are made along the contour. Furrow irrigation involves higher labour costs, but is generally more efficient than border and check methods. The dimensions of the furrows depend on several factors, *e.g.* soil type, land slope, irrigation depth, and nature of crop and implements available for cultural the operations. The lengths of the furrows generally increase with an increase in irrigation depth, fineness of soil and slope up to a certain limit (generally 3 per cent). Field demonstrations on furrow irrigation in cotton revealed that the mean yield was 70 kg/ha more under furrow irrigation than under the border method. It saved about 25-30 per cent of irrigation water (Table 8.7). Likewise in the case of sunflower, the grain yield was 100 kg/ha more under furrow irrigation with a similar saving in irrigation water. Thus it has been found to be useful even in light textured soils. Even if the crop yield is not improved, the water saved by the system more than compensates the yield decrement, if any.

Table 8.7: Performance of furrow irrigation in cotton and sunflower

Crop	*Yield, kg/ha*		*Irrigation water (min/ha)*	
	Furrow	*Border*	*Furrow*	*Border*
Cotton	1902	1832	280	381
Sunflower	1611	1511	386	513

Drip Method

In drip irrigation, the water is applied more frequently closer to the stems of plants through suitably spaced emitters attached to plastic or metallic pipes spread above or below the soil along crop rows. The pipes are hooked to the source of water supply through a storage tank or pressure device, which provides the necessary hydraulic head or pressure for movement of water to the drippers. In this case, a part of the soil in the vicinity of the plant roots is wetted and kept close to field capacity. Physically, the system is adaptable to a wider range of soils, topography and crops. The method not only ensures the highest economy in water use owing to reduced evaporation and seepage losses, but also provides an ideal moisture regime for high yields especially in vegetable crops. Increase in area under fruits and vegetables should be possible using efficient methods of irrigation as it will not put an additional strain on the water resources.

The evaluation of drip irrigation in kinnow revealed that for the net open pan evaporation from 1 January, 1993 to September 30, 1995 of 4464 mm the kinnow plants received 16,571, 26,069, and 35,55? litres of water for 35, 55 and 75 per cent level of irrigation, respectively. The plants under the basin method received 40 irrigations; but the pipe and drip system saved about 58.4 per cent of irrigation water required for the basin for 60 plants irrigated at mean 55 per cent of pan evaporation (Table 8.8). The yield of kinnow did not vary significantly with irrigation methods as well as irrigation levels.

Table 8.8: Amount of water applied under different irrigation levels and amount saved under drip and pipe systems during 1 January, 1993 and 30 September, 1995

Pan evaporation, mm	*Irrigation level*	*Water applied*	*Water saved (000'litres)*	
	% pan evapn	*litres*	*Pipe*	*Drip*
4464	35	16571	2763	2768
	55	26069	2193	2198
	75	35552	1624	1629

Agro-economic Feasibility

The cost of cultivation under drip, pipe and basin systems was found to be Rs. 7200, 6100 and 6900, respectively (Table 8.9). The drip and pipe systems used 9,114 litres as compared to 21,874 litres used under the basin method. The net seasonal income under the drip and pipe systems was Rs. 4355 and Rs. 875 more than under basin method. With this saved water 140 per cent additional area could be irrigated enabling additional income up to Rs. 60,000 and Rs. 49,008 under the drip and pipe systems. The benefit cost ratio under the drip system was about 25 per cent more than under the basin method. This leads to a net profit per litre of water to Rs. 3.21, 2.13 and 0.93 respectively under drip, pipe and basin methods. The water use efficiency was 175 and 125 per cent more under drip and pipe systems than under basin method.

Narang (1995) reported that drip and sprinkler irrigation systems were economically viable only under "real and acute water scarce situations" and where the underground water is saline/sodic. A study at Bathinda revealed that a synthetic sodic (6 me/L) water used through drip for two and a half years in kinnow did not have any adverse effect on its yield. The economy of the systems was caused by the increased coverage of irrigated area and planting of an additional crop and boosting of the net returns. The total installation cost of the drip system depends upon crop or tree type, the row to row and plant to plant spacing which determined the number of laterals and emitters per lateral row, area covered, source of water supply and pressure head required. The cost tends to decrease in proportion as the area is increased.

Table 8.9: Agro-economic feasibility of drip irrigation in Kinnow at RRS, Bathinda

Variable	*Drip*	*Pipe*	*Basin*
1. Total annual cost of the system	2790	700	—
2. Cost of cultivation	7200	6100	6900
3. Water used (L/tree/annum)	9114	9114	21874
4. Income (Rs./ha)	25000	20420	20345
5. Additional area cultivated, per cent	140	140	—
6. Additional net income, Rs.	24920	18508	—
7. Gross b/c ratio	2.47	2.02	1.94
8. Net profit/L water	3.21	2.13	0.93

A preliminary survey of drip irrigated orchards in south-west Punjab under National Agricultural Technology Project

PAA-007 revealed that benefits of the technology in terms of water and fertilizer saving are not being exploited fully because of a lack of technical know-how/advisory system and uncertainties associated with canal water/power supply in the region.

Irrigation Water Quality

Characterization

Tubewell waters collected from villages as well as those brought by the farmers themselves were tested for their quality.

Table 8.10: Mean values of water quality parameters in different blocks of south-west Punjab

Block	*Sample (Nos.)*	*Mean value of the parameter*		
		EC (dS/m)	*RSC (me/l)*	*SAR (me/l)*
Bathinda	64	3.3	-3.2	10.2
Sangat	56	3.6	-6.3	8.6
Talwandi sabo	28	3.6	-4.7	10.3
Kotbhai	19	3.8	-6.1	12.1
Nathana	19	2.9	-0.3	10.9
Muktsar	6	2.1	-4.8	7.3
Rampuraphool	8	1.8	4.0	7.8
Lambi	36	4.2	-13.3	7.6
Malout	9	3.4	-4.6	10.7
Kotkapura	8	2.5	-1.9	7.3
Abohar	13	2.9	-7.1	5.3
Khuian sarvar	10	3.3	2.1	11.1
Arniwala	14	3..8	-11.3	—
Mansa Khurd	11	2.9	3.2	—

These are mean values, the coefficient of variation in such data are very high and waters have to be used as per individual recommendations and no generalization may hold good appraisal. The water samples from Rampuraphool and Muktsar blocks had the electrical conductivity in the normal range (<2000 μmho/cm).The samples from Bathinda, Sangat, Talwandi Sabo, Kotbhai, Nathana, Malout, Kotkapura, Abohar, Khuian sarvar, Arniwala, and Mansa Khurd were all in the marginal range (2000-4000 μmho/cm). The samples of Lambi block were in the unfit category. The residual sodium carbonate values of water of Kotkapura and Khuian sarvar were <2.5 me/l; of Mansa Khurd, Bathinda, Talwandi sabo, Rampura, Muktsar, Malout were 2.5- 5.0 me/l; of Sangat, Kotbhai, Abohar were 5.0-7.5 me/l; and of Lambi were >10.0 me/l. The sodium adsorption ratio of waters in Muktsar Rampuraphool, Malout, Kotkapura, and Abohar were less than 8 me/l. All other blocks had a sodium adsorption ratio of 8-10, the maximum being 12.1 (me/l) in Kotbhai block. The range and mean values vary with the number of samples collected in each block, some values being highly divergent.

Conjunctive use of Saline and Canal Water

All marginal waters need to be used with caution and require frequent monitoring of both soil and water so that the soils remain productive and do not turn barren. Alternating poor and good quality irrigation water helps in reducing the salinity/sodicity hazards of the poor quality waters. It is preferable that poor quality waters only supplement the good quality water. Good quality water should be used in the early stages of crop growth and poor quality water later when the crop can tolerate high salinity/sodicity level.

The survey on conjunctive use prevalent in the region revealed that 50.7, 47.2, 37.5 per cent of irrigations were given with canal water in cotton, rice, and wheat, respectively. Irrigations with tubewell water alone accounted for 25.4, 24.2 and 38.9 per cent, respectively. Since a rise in the water table persists and greater use of tubewell water alone may salinize the soil, there is need to increase the conjunctive use of the canal and tubewell water to contain a rise in the water table. Wider adoption of these practices will help to contain the rising water table in this region.

A seven year study revealed that a gradually improving water could be safely used alternatively with canal water in the cotton-wheat system (Table 8.11). The best option will be not to apply brackish tubewell water in the early stages of the crop, but use it as a supplement for canal water in the later stages only when the crop can tolerate higher salinity/sodicity.

Since the groundwater can vary considerably even within small commands, experimentation over all soil, water and crop combination situations may not be feasible; and modelling attempts may be useful and explored.

Table 8.11: Mean yield of *kapas* and wheat grain under different treatments under different water quality

Irrigation treatment	*Mean yield, kg/ha*	
	Kapas	*Wheat*
All canal	1575	2757
All tubewell	1263	2651
Alternate	1388	2775

Ancillary Practices

Besides efficient management of water through irrigation technology, water expense efficiency of the crops can also be increased by the adoption of associated improved cultural practices. In fact, any operation which helps increase the crop yield by sustaining or even reducing the evaporation and deep percolation components of the field water budget is beneficial in improving the water use efficiency of the crop.

Cropping Sequence

Irrigation needs of crops may differ when studied individually. Irrigation supplies may be inadequate for a particular crop sequence and hence may be uneconomic. Little attention has been devoted to developing an appropriate cropping pattern, which can suit the irrigation supplies available at the farm. The role of crop planning in making more efficient use of water is important because:

1. Crops differ markedly in timings and amounts of the irrigation needs;
2. They exhibit a wide range of photosynthetic efficiency for similar water requirements; and
3. They exhibit differential sensitivity to water stress imposed at various growth stages.

Variation in the ET requirements of crops arise from differences in the duration of growth and severity of weather during the growing season and extent of foliage and morphological characteristics of their leaves. Plant species show genetic variation in photosynthetic efficiency and structure of canopy. The crops that develop their covers rapidly permit more efficient use of water through reduction of direct evaporation. In the summer, for crops growing during the hot and dry months, the irrigation needs are usually high while the yields are low resulting in low water efficiency of crops. The influence of the use of irrigation water at high rates for summer crops on availability of water for main season crops needs to be considered while deciding the cropping pattern under limited water availability situations.

The mean *kapas* yield was found to be 2019, 2161, 805 and 1911 kg/ha when cotton allowed wheat, raya, sunflower and barley, respectively. The mean yields of rabi crops were 3646, 1492, 1551 and 1653 kg/ha. These rotations required 11, 9, 14, and 8 irrigations (including) pre-sowing so that the net returns per irrigation were Rs. 2659 under cotton wheat, Rs. 1521 under cotton barley. The mean yield of sugar cane was 5267 kg/ha consuming 15 irrigations with a net return of Rs. 2150/ha per irrigation. The cost of irrigation is usually considered in crop production, but there is a need to quantify the returns per irrigation for different crops and crop sequences for efficient use of the irrigation water.

Table 8.12: Mean crop yields and net annual returns under different crop sequences

Crop sequence	*Mean yield, kg/ha*		*Returns (Rs.)/irrigation*
	Kharif	*Rabi*	
Cotton-wheat	2019 (4)	3646 (5)	2659
Cotton-raya	2161 (5)	1492 (2)	3180
Cotton-sunflower	805 (4)	1551 (8)	918
Cotton-barley	1911 (3)	1653 (3)	1521

Note: Figures in parenthesis indicate number of irrigations.

The yearly variations of *kapas* yields under different

rotations were significant, with maximum variation in sunflower. However, the experience of the farmers shows that except cotton-wheat, any other cotton-based rotation for more than two years reduces the productivity of the soil. Hence it has been difficult to persuade the farmers to follow other rotations for longer spans. Even otherwise, the stability of wheat yields is much higher than those of other rabi crops.

Balanced use of Fertilizers

The amount, time and method of fertilizer use are crucial aspects of fertilizer management in crops. In the selected command area, irrigation with respect to these aspects has been observed in the use of fertilizers—both nitrogen and phosphorus. Appropriate and balanced use of fertilizers causes a significant increase in crop yields with proportionally less or no additional use of water, thus, leading to a higher water use efficiency of crops. Different crops require different amounts and proportions of different nutrients for their optimum growth. An imbalance of essential nutrients causes growth patterns that are not conducive to optimum production and water use. The supply of a nutrient in excess of its requirements for a crop cannot obviate the adverse effect of the deficiency or excess of other nutrients, although it may help increase/decrease the availability of nutrients already available in the soil. The response of crops to fertilizers depends on soil type, source of fertilizers, method of application of fertilizers, use of inorganic manure and other cultural practices adopted by the farmers.

Table 8.13: Effect of judicious use of fertilizers on *kapas* yield of cotton

Fertiliser	*Kapas yield (kg/ha)*	
	Zn 0	*Zn 25*
N_{100}	1600	1585
N_{70}	1575	1708

In the command, the cotton crop requires only 70 kg N/ha fertilizer nitrogen instead of the recommended 100 kg/ha

(Table 8.13). No phosphorus and zinc fertilizer was needed for cotton. Later this was included in the package of practices thereby saving huge quantities of fertilizer N in the region. Raya needed 100 kg N along with 30 kg P_2O_5 (Table 8.14) and berseem needed 25 kg N and 75 kg P_2O_5 per ha (Table 8.15).

Table 8.14: Performance of raya under various fertilizer rates

Fertilizer rate	*Yield (kg/ha)*	*Water expense efficiency, kg/ha/cm*
FP N_{25}	1000	38.8
N_{100} P_{30}	1400	57.5

Table 8.15: Effect of balanced fertilizer use on the fodder yield of berseem

Fertilizer practice	*Mean green fodder yield (t/ha)*
Improved (N_{25} P_{75})	95
Farmer's (N_{100})	85
Irrigation applied	11

Application of Gypsum

The high atmospheric evaporativity of the region and occasional irrigation with tubewell waters that have higher amounts of sodium and other salts contribute to an excessive accumulation of soluble salts in the surface soils layers. With leaching down of these salts during the rainy season, the sodium ion gets absorbed on clay particles of soils. Because of its toxic effects, the adsorbed sodium deteriorates the physical condition of the soil, with an accompanying adverse effect on crop growth and yield. The addition of gypsum to the soil is helpful in replacing sodium with calcium and containing the detrimental effects. Moreover, gypsum also supplies sulphur, which may benefit a crop if it is limiting. Application of gypsum at the rate of 1.5 tonnes/ha increased the grain yield of wheat by 250 kg/ha and also increased the yield of cotton slightly (Table 8.16).

Irrigation is merely a component of the production

technology. The success of the crop, and the response to irrigation may be vitiated by any other component in this production technology. Therefore, package demonstrations were laid out in blocks of 10-15 acres of farmer's fields.

Mass adoption of improved and allied practices: The mean *kapas* and wheat yields under the improved package of practices were 202 and 373 kg/ha more than under the existing package of practices, respectively. The improved practices not only used less water, but also require less fertilizer. Only the timely application of the inputs was important. In the standard command area only 19 per cent of the farmers got *kapas* and wheat yield more than 1250 and 3000 kg/ha, respectively in 1983-4 whereas now about 80 per cent farmers are getting the yields.

Table 8.16: Effect of gypsum on mean yield and net income in cotton and wheat.

Rate of gypsum (t/ha)	*Mean cost of gypsum (Rs.)*	*Cotton*		*Wheat*		*Total net Rs./ha*
		Mean Kapas yield	*Addl. income Rs./ha*	*Mean yield Rs./ha*	*Addl net Rs./ha*	
0.0	000	1985	—	2925	—	—
1.5	167	2060	156	3147	200	357
3.0	334	2040	103	3228	166	63

Thus, the timely use of inputs is sufficient to effectively use water and other inputs without an extravagant input of money.

Table 8.17: Mean (5 years) crop yield and number of irrigations under mass adoption of improved technology

Crop	*Improved*		*Existing*	
	Cotton	*Wheat*	*Cotton*	*Wheat*
Border width, m	14	12.8	8.5	7.0
N-applied, 70 kg/ha	70	80	80	109
P_2O_5	-	40	-	70
Seed rate, kg/ha	25	90	16	120
Spray interval, days	10	-	10-15	-
Yield, kg/ha*	1561	3634	1398	3380
Irrigations, nos.	4.6	4.6	5.5	5.7

Recently, the Directorate of Agriculture, Punjab has initiated the emphasis on a scientific use of irrigation water in five more districts of the commands of the same canal system. For the project, the following suggestions deserve attention:

1. The study of the levelled sand dunes and possible effects of the changed strategy on the ecology of the region with respect to the productivity of the soil, its salinization, change in the water table regime and water quality in the region.
2. The increase in area under paddy despite the poor quality of the groundwater entails bringing more of the salts to the surface soils and the potential development of soil salinity particularly on the land between the Sirhind feeder and the Rajasthan canal. There have been instances of failure of paddy introduced in the cotton belt and the reversal to cotton production. A close monitoring of the soils, water table and water quality and the quality of the crops produced is essential to prevent a further deterioration of the situation.
3. The area under fruits and vegetables should increase using the efficient means of irrigation without constraining the water resources. The factors causing a decline in fruit production need to be studied and remedied.
4. The insect pest control measures advocated so far in the region have failed. There is a need for testing the alternate means of insect control for the successful production of cotton in the region. One such alternative is the trial of fly ash dusting on cotton which has been successfully used in south India (Narayanasamy, 2001) and plenty of ash is available from its thermal power plants.
5. With the ridge/bed planters becoming available (Chauhan *et al.*, 1995), furrow irrigation of different crops needs to be encouraged to control deep percolation and hence a rise in the water table, and through decreased pumpage in the falling water table zone.

6. The efficiency of the existing canal system should be increased through operation and management and through efficient use of water on farm as detailed in the paper. The farmers need to be advised about the amount of water and fertilizer to be applied through the drip to fruit plants at different stages.
7. Under the falling water table, an early transplanting of rice, replacement of rice with maize, mustard and groundnut, improved methods of irrigation, sowing of sunflower in December and transplanting in February can be helpful.

References

Badari Prasad, P.R., A. Naganagoud, and B.V. Patil. (2001). Evaluation of fly ash in the management of cotton insect pest. Proc. II National Seminar on use of Fly Ash in Agriculture, Annamalai University, Annamalainagar, March 5-6, 2001.

Chauhan, A.M., B.S. Bhatias and H.S. Dhingra. (1995). Multi crop planter of row crops. *Progressive Farming*, March issue.

Dhillon, G.S. (1992). "Canal system in Punjab". *PAU, Bulletin*. PAU, Ludhiana.

Dhingra, J.S., R.P.S., Chopra and S.K. Bansal. (1998). Groundwater resources scenario of Punjab — An overview. Ninth national symposium on hydrology. Amritsar. Nov. 26-27.

Hira, G.S., R. Singh. (1998). "Crop evaporatranspiration and rise or fall in water table" in Punjab. *Ibid*.

Narang, R.S. (1995). *Different systems of irrigation and their economics*. Inter-disciplinary scientific group on water resources management. Punjab Agricultural University, Ludhiana.

Narayanasamy, P. (2001). Fly ash pesticides in the new Millenium. In "Agricultural applications of fly ash" Proc. II National Seminar on Use of Fly Ash in Agriculture, Annamalai University, Annamalainagar, March 5-6.

Prihar, S.S., S.D. Khepar, R. Singh, S.S. Grewal, and S.K. Sondhi. (1993). *Water resources of Punjab: A critical concern for the future of its agriculture* (second edition), Punjab Agricultural University, Ludhiana.

Sandhu, B.S. and S.S. Prihar (1983). *Irrigation scheduling to crops in Punjab*. Department of Soils, PAU Ludhiana.

Sandhu, B.S., C.J. Singh, S.S. Prihar and K.K. Vashisht (1993). Field evaluation of improved water management and allied technology in a canal command of Punjab. Dept of Soils, Ludhiana and Regional Research Station, Bathinda.

Singh, C.J. (1996). Final report of the Punjab Irrigation and Drainage Project. Phase II. Credit 2076/IN/LOAN 3144 IN from the World Bank. Regional Research Station, Bathinda. Directorate of Research. Punjab Agricultural University, Ludhiana.

Singh, C.J., M.S. Aujla, K.S. Saini, G.S. Buttar, and J.S. Brar (2002). "Conjunctive use of fresh and salty water in cotton and wheat in South West Punjab". Presented 17th World Congress Soil Science, Symposium 34, Paper 1237-1-7, Bangkok, Thailand.

9

Groundwater Recharge

Need of the Day

ALPANA KATEJA

Introduction

Groundwater has been a reliable source of fresh water for human consumption, irrigation and industrial uses. It assumes the even more significant role of supplying water for almost all the development activities in water deficient arid areas in Rajasthan. The short spell of monsoon coupled with its erratic behaviour and scanty rainfall make Rajasthan the most water deficient state in the country. The severity of water scarcity in Rajasthan can be best understood by the following data as percentage of all a India indices:

Geographical land mass	10.4
Population	5.2
Availability of water resources	1.04

Ensuring access to safe drinking water to all has not been among the successes of planning in this state. Of a total of 93,946 habitations, 7864 still remain to be covered for drinking water supply in the state. In the face of an acute shortage of surface water, groundwater is the only source of this elixir. Almost 70 per cent of the population in the state is dependent upon groundwater for drinking and irrigation purposes.

The paper presents a profile of the water scenario in

Rajasthan. Considering 1984 as base year, it evaluates the trend in the states groundwater level. It also attempts to project the stage of development in the next quarter century. It underlines the need for a revival and strengthening of water harvesting systems in the state.

Drought

Drought, famine and scarcities are the most familiar words in the state. The scourge of nature visits the state in its unrelentless march almost year after year. After the 1998, 1999 and 2000 monsoon failed in 2002 as well, it completely eroded the coping capacity and economic potential of the people. The drought has a serious impact on the bare survival needs of the people including those of safe drinking water and adequate nutrition, leading to large scale out migration. It is therefore quite amazing to learn that in the several hundred years old history of the city of Jodhpur, there has never been such an occasion when it had had to be evacuated because of water shortages. People, quite intelligently and resourcefully made the best use of their environment. By developing wonderful water harvesting systems they had tried to reduce the fury of nature. Communities of this fragile eco-region had established a balance between the carrying capacity of nature and their needs. But gradually with the introduction of tap water, with an increasing population and several other development activities these traditional systems got neglected and the gap between water use and availability widened. Now it is being widely admitted that the recurring drought conditions are the result of bad water management in the past. Failure to harvest rainwater, excessive extraction of groundwater and failure to ensure the recharge of the aquifers has led to the water table falling sharply over the years (Table 9.1). Now in a deficient rainfall year there is no subsurface reserve to draw upon.

Trends in Water Level

Water levels vary with rainfall, physiography, drainage and

climatic environment. These parameters show wide variations in Rajasthan. Eastern Rajasthan has more rainfall and it decreases westward.

Table 9.1: Change in water level (1984-2000)

S.No.	*Districts*	*Depth of Water Level*	
		1984	*2000*
1.	Ajmer	-5.59	-2.67
2.	Alwar	-5.12	-19.03
3.	Banswara	+0.001	-19.45
4.	Baran	-2.01	-5.80
5.	Barmer	-1.33	-17.90
6.	Bharatpur	-1.11	-19.38
7.	Bhilwara	-4.71	-7.45
8.	Bikaner	+2.47	-12.60
9.	Bundi	-3.30	-12.85
10.	Chittorgarh	-4.09	-18.75
11.	Churu	+0.04	-19.70
12.	Dausa	-2.18	-13.56
13.	Dholpur	-1.35	-16.50
14.	Dungarpur	-2.02	-7.75
15.	Ganganagar	+5.62	-8.20
16.	Hanumangarh	+4.42	-1.25
17.	Jaipur	-5.74	-3.67
18.	Jaisalmer	+0.23	-21.25
19.	Jalore	-8.00	-7.98
20.	Jhalawar	-1.22	-21.45
21.	Jhunjhunu	-6.22	-6.60
22.	Jodhpur	-6.60	-14.81
23.	Karauli	-3.43	-54.26
24.	Kota	-1.51	-29.75
25.	Nagaur	-6.50	-13.17
26.	Pali	-6.69	-27.50
27.	Rajsamand	-4.69	-28.86
28.	Sawai Madhopur	-1.74	-16.80
29.	Sikar	-4.77	-9.00
30.	Sirohi	-6.49	-16.82
31.	Tonk	-2.37	-24.99

Source: Groundwater Department, Jodhpur.

For evaluating trends in the water level, 1984 has been considered as the base year. The change in water level between the pre-monsoon 1984 and pre-monsoon 2000 has been referred to as the water level trend. Out of 237 development blocks of the state, only 26 blocks show a rising trend in water level whereas the remaining 211 blocks covering 76 per cent area of the state registered a depleting trend of water level (GWD, Jodhpur). Twenty-six districts remain under depletion (Table 9.1). Nineteen districts, namely Jalore, Pali, Sirohi, Jodhpur, Nagaur, Sikar, Jhunjhunu, Alwar, Bharatpur, Dholpur, Jaipur, Dausa, Karauli, Sawai Madhopur, Tonk, Ajmer, Bhilwara, Rajasamand and Chittorgarh have become critical where more than 90 per cent area has registered a decline in the water level. Seven districts fall in the semi-critical category, namely Baran, Barmer, Bundi, Jhalawar, Kota, Dungarpur and Udaipur with more than a 50 per cent area under depletion. Slight deterioration has been observed in the peripheral area of Dungarpur, Banswara, Barmer, Kota, Bundi and Dholpur districts. The average depletion in the state is 0.18 metres per year (GWD). During 1984-1998 only 65 per cent area of the state was under depletion, while this has has swollen up to 76 per cent during 1984-2000, which is quite alarming.

The northern part of the state comprising the districts of Ganganagar, Hanumangarh, Bikaner and parts of Churu show a rising trend in the water level. A rising trend has also been observed in patches in south and south-eastern Rajasthan covering parts of Banswara, Kota, Bundi, Baran and Jhalawar. (Fig. 9.1) A rise in the water table in the Ganganagar, Banswara and Kota area can be attributed to the fact that these lie in the canal command area. The rising trend in Banswara, Churu and Jaisalmer are marginal, indicating almost stable conditions. Similarly, the depleting trend in Barmer, Bharatpur, Dholpur, Jhalawar and Kota is also marginal referring to more or less stable conditions in these districts.

A moderate decline with a depletion rate between 0.11-0.17 metre per year has been observed in Baran, Dausa, Dungarpur, Sawai Madhopur and Udaipur districts.

However, a maximum number of districts show a considerable depletion with a declining trend of more than 0.20 mt/year up to an extent of half metre per year. These include Ajmer, Alwar, Bhilwara, Bundi, Chittorgarh, Jaipur, Pali, Jalore, Jhunjhunu, Jodhpur, Karauli, Nagaur, Rajsamand, Sikar and Sirohi. Thus, 76 per cent area of the state is under depletion.

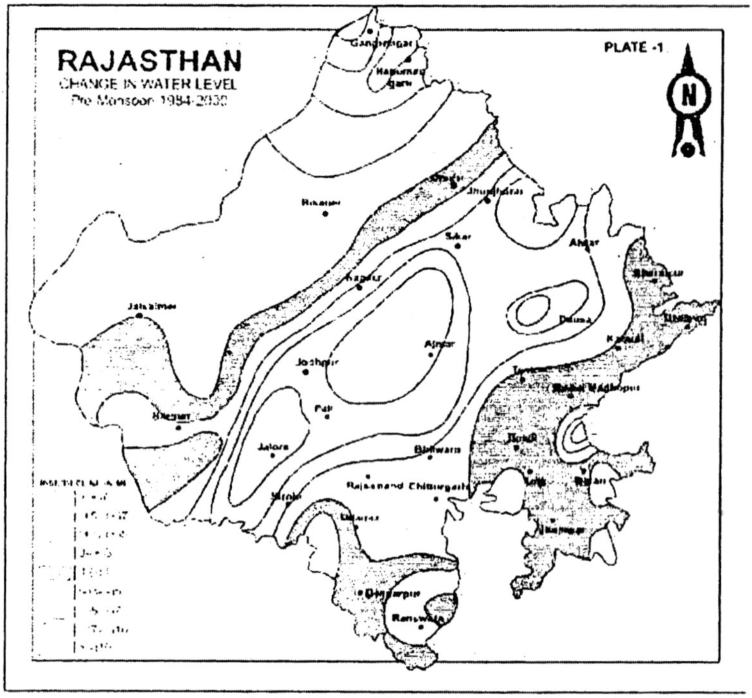

Fig. 9.1: Change in Water Level, Pre-Monsoon 1984-2000 in Rajasthan.

The present analysis of the period 1984-2000 reveals a very grim situation in the state. The monsoon pattern over Rajasthan has been dwindling for the last seventeen years causing regular droughts in many parts of the state. As a consequence, groundwater reserves in a major part of the state are not sufficiently replenished and are constantly under strain of heavy withdrawal through an ever-increasing number of energized wells and tubewells. Suspension in an unbalanced state of uncertain recharge through rainfall on the one hand and increased draft through wells on the other, has ultimately resulted in a depletion of groundwater reserves in Rajasthan.

Water Quality

Groundwater, which used to be a guaranteed pollution-free,

is not free from contamination, now-a-days. On the one hand, because of the over-exploitation of groundwater, inorganics like fluorides and arsenic present below the ground find their way into the extracted water, and on the other hand pollutants from surface sources leach into the aquifers below. They cause fluorosis and arsenic poisoning. The extent of the water quality problem in Rajasthan may be judged from the fact that the number of fluoride affected habitations is as high as 16,560. Similarly, salinity affected habitations are 14,415.

In contrast to humid climates, effluents arising in arid regions pose a danger not only to the surface water but also to the groundwater because here most waters ultimately feed into the subsoil. In this way pollutants contained in one body of water can disperse in all directions so contaminating valuable groundwater resources. This makes wastewater treatment in arid regions even more important. In Rajasthan, economic activities like mining, animal husbandry, farming, chemical and dyeing industries are posing a great danger to the quality of groundwater.

Groundwater Resources of Rajasthan

Its Past, Present and Future Status

As per the estimation made by the Groundwater Department on March 1984 based on GEC (Groundwater Estimation Committee) guidelines, the annual groundwater recharge, net draft and surplus were estimated as 16,223.60 mcm, 4,926.82 mcm and 11,296.78 mcm, respectively. Accordingly, the stage of development in the year 1984 was 30.37 per cent reflecting almost normal conditions.

Water Harvesting Systems

Natural replenishment of a vast supply of underground water occurs very slowly, and therefore excessive continued exploitation may require from decades to centuries for natural replenishment. Artificial replenishment at calculated rates is needed to replace groundwater used by industries, by agriculture and by municipalities. Indiscriminate over-development or poor practices can dewater aquifers and cause compaction of sediments, loss of storage space, land subsidence and degradation of quality. The aquifers once polluted are difficult to bring back to their natural state.

As a large area of Rajasthan is threatened to fall in the category of the dark zone of overexploitation, proper management of water resources has become absolutely necessary. This can be achieved by reviving the traditional water harvesting systems such as *talab, tanka, nadi,* roof water harvesting and *khadin*. The other water harvesting systems which may be reactivated and put to use are *bawari* and *jhalara*.[1]

Talab

Talab is the name given to big pond/tank generally found in cities or big towns. In the city of Jodhpur itself, there are eight *talabs* but most of them are non-functional and abandoned.

Nadi

Nadi (open tank) is constructed for storing water available from an adjoining natural catchment during the rainy season. In arid Rajasthan, the construction of a *nadi* is one of the oldest practices and is still the common source of water for conjunctive use. The capacity of a *nadi* generally ranges from 1200 to 15000 m^3 depending upon physiograpic conditions and the rainfall pattern. A traditional *nadi* has as a limitation, high evaporation losses from the free water surface, seepage losses through porous sides and the bottom and heavy sedimentaiton due to degradation in the catchment. A *Nadi* should be located in areas with the lowest elevation in order to benefit from the natural drainage and its need of minimum excavation of earth. Surface of the catchment area should preferably be impermeable. The surface to volume ratio should be kept at the minimum possible (less than 0.3) for reducing the evaporation losses. Slit trap should be provided at the inlet point to prevent a sediment load entry in the structure.

Tanka

Tanka is a small circular or square underground tank constructed with lime mortar or cement plaster, constructed on an individual family basis. It is constructed normally on fallow ground where the surface runoff can be diverted to the structure by creating a clean catchment all-around. Plantation of suitable species of trees around the periphery of the catchment area of a *tanka* also helps in environmental improvement.

Roof Water Harvesting

In the ancient times, houses in western Rajasthan constructed with stone and lime were so structured that they had a self-contained roof water harvesting system. Rainwater from rooftop was regulated through an in-built drainage network in underground tank constructed in the premises of house.

Bawari and Jhalara

Bawari and *Jhalara* (step-wells) are constructed to mitigate the drinking water supply, particularly in cities and towns. Bawaris are made pucca providing a pully to draw the water through a rope and bucket. However, in case of *Jhalara*, more investments are made.

Khadin

Khadin is a unique practice of water harvesting and moisture conservation in suitable deep soil plots surrounded by some sort of natural catchment zone. In this system, runoff from uplands and rocky surfaces is collected in this adjoining lower valley formations. The plots are vigorously built and managed to make the entire system a self-contained unit for cultivation.

Suggestions and Conclusion

To control "water chaos" in Rajasthan, various measures can

be taken. Thus traditional water harvesting systems must be revived and strengthened which must have evolved as specific responses to ecology and culture used to satisfy local needs in an environment friendly manner. However, with the introduction of the present system of centralized supply and management of water, these age-old techniques have been thrown into disuse and even eliminated. Traditional water sources such as wells, ponds, *baoris* and hand pumps, which were used regularly a few years ago are of no use as the government is supplying water to almost every area of the city.

Rehabilitating traditional water systems may require levels of investment much beyond the community's capacity. On the one hand, community participation must be encouraged, on the other hand, state finances should be invested in the research and development of all the aspects of traditional systems. Conflicts over community management as witnessed over *Johad of Lava ka Bass* in Alwar, Rajasthan are not good signs.

Water rights must be clearly defined. Individuals and households ought to have the right to harvest the rain in systems owned and managed by them. In Rajasthan people withdraw an unlimited quality of sweet water from *beras* (wells) and sell this at very high rates to earn big profits. In that way rights of other individuals are compromised, groundwater recharge is affected and water sources get polluted.

Depending upon the geological conditions, different techniques of water recharge must be used. The physiographic condition of Rajasthan is too varied which suggests that different techniques suitable to local conditions must be used in different areas.

With the supply of tap water, water use per capita has increased manifold. Water considered as a free good is used liberally. Use of price policy and subsidy instrument is needed to conserve water. Price of water should reflect its scarcity value and environmental cost. The water price must be linked to the amount of the water use. More efficient use of water through public education may help. In agriculture, sprinkles, drip irrigation and hydrophonics should be explored while in industries, reuse of water must be encouraged. The sustainable use of water is equally important.

In the arid and semi-arid climate of Rajasthan sufficient water resources are not ubiquitous. At the same time there has been a rapid and continuous rise in the consumption of water. The long-term effects are that the water tables are dropping. One of the most important challenges of the immediate future is to stabilized them. Intergeneration equity requires sustainable harvesting of groundwater.

Note

1. *Dying Wisdom, State of India's Environment 4: A citizen's report*. Various reports, Groundwater Department, Jodhpur.

10

Water Resource Utilization and Disposal of Waste Water in Class-I Cities and Class-II Towns in Gujarat

B. M. JANI*

Introduction

In Gujarat, urban planning which includes land and water use planning, water resources planning, fair utilization of water along with an adequate disposal of waste water by the local self-government in class-I cities and class-II towns has drawn the attention of town and country planners, economists, environmental scientists and administrators in the state. It is most necessary for local self-government to have an adequate water resource planning from the supply side, a productive efficiency of water quality and efficiency in terms of investment made in the water sector. An effort is made in this chapter to study the leading river basins of the Sabarmati, Mahi, Narmada and Tapti rivers from this perspective as well as cities and town-wise wastewater generation and collection in Gujarat. An economic environmental policy is necessary for water resource planning and urban development in a fast industrializing state like Gujarat.

The present chapter is divided into two parts : the first one

* The author is grateful to the World Bank Capacity Building Project funded by the Environment Economics Research Committee of the Indira Gandhi Institute of Development Research, Mumbai. The present paper forms part of a chapter in the first draft of the report.

highlights the strategy of the Ninth Five-Year Plan for water resource planning in Gujarat *vis-à-vis* India, the second one reveals the wastewater disposal facility in the state, whereas, the third one offers environment economics policy and strategy for sustainable growth in the state.

I. Strategy of the Ninth Five-Year Plan

The Ninth Five-Year Plan 1997-2002, of the Government of India aims to manage the environment through social mobilization and participation of the people at all levels. The plan has emphasized a multi-pronged environmental strategy for the sustainable development of the country, which is as follows:

1. Empowering people through information generation, dissemination and access.
2. Involving the industry—both public and private.
3. Integrating environment with decision-making through valuation of environmental impacts; evolving market-based economic instruments as an alternative to the command and control form of environmental regulation, appropriate pricing of natural resources based on their long term marginal cost, appropriate fiscal reforms and natural resource accounting.
4. Involving rights for common property resources.
5. Inter-sectoral co-ordination and co-operation.
6. Ensuring scientific and technological inputs by :

   (*a*) participation of people (particularly) in the management and sharing of usufruct through joint forest management;

   (*b*) involvement of NGOs for awareness building and as an interface between forest department and the people would be encouraged during the Ninth Five-Year Plan;

   (*c*) Integrated development of villages in and around forests.

(The Ninth Plan Document, p.952)

The Eighth-Five Year Plan (1992-7) with its lopsided priorities spent forty per cent of its total expenditure (Rs.154 crore) on Antarctic research for monitoring ocean pollution. The programmes organized were directly beneficial to the socio-economic or environmental activities. The draft of Ninth-Five Year Plan (1997-2002) made financial provisions for Rs. 900 crore. The share of research in this has been reduced to twenty per cent. The proposed financial outlay aimed at direct investment for the welfare of the society, integrated coastal zone management, marine area management, coastal community development and information technology services on the ocean. It is most necessary to preserve the Arabian sea coast of Bharuch District in Gujarat from wastewater pollution.

II. Wastewater Pollution in Class-I Cities of Gujarat

Levels of urbanization and the rural-urban migration of people in the big cities of the state have increased the need for water, which has ultimately contributed to problems of wastewater disposal. Such a process also increased the demand for investment in both water supply and drainage for disposal arrangements of wastewater. Cities and towns in Gujarat have problems of both river pollution and ocean pollution found in the cities of the coastal Saurashtra region as also in the highly polluting industrialized centres in south Gujarat where many of its polluting industries are located. It is educative to draw an all-India comparison with cities of Gujarat for location of population, collection and generation of wastewater, basinwise and city-wise facility for sewerage treatment by the local self-government in the state.

Comparison of state-wise water supply in class-I cities, shows that Gujarat has 21 class-I cities, whereas, UP has 41 class-I cities followed by 27 in Maharashtra. Gujarat had a population figure of 84,43,960 as per 1991 Census report and availability of 48.80 million litres daily (MLD) groundwater resources. Surface-wise water sources were 248.96 MLD and when combined these come to 1006.14 MLD. Total water supply was 1303.90 in MLD. Per capita availability of water was

172 litres per capita daily (lpcd) and 90 per cent of the total population was covered under organized water supply as per the latest report of Central Pollution Control Board (CPCB).

Sewage farming is different from sewage treatment as it refers to the use of wastewater for farming purposes, namely to grow vegetables on the available basin area on the bank of a river. Sewage treatment however refers to the direct treatment of wastewater through a treatment plant by the local self-government.

Basin-wise Analysis of Class-I Cities: India and Gujarat

It can be seen from Table 10.1 that the basin-wise water supply situation in class-I cities reveals that 21 cities are located on the river basin areas for example Sabarmati (7), Mahi (3), Narmada (4), and Tapi (7). According to the category-wise per capita water supply in Gujarat state, while one city has < 100, 18 cities have between 100-200, anothes city has 200-300 while another city has >3000 lpcd. In the distribution of class-I cities as per catchment areas per capita water supply, it can be observed (*see* Table 10.1) that: Most of the class-I cities (65%) in Gujarat have between 100 to 200 lpcd per capita availability in the major catchment areas of the river basins. Tapi Basin has the highest location of class-I cities (37%) while the Narmada basin has 30 per cent of the total of such cities with 2 cities that have less than 100 lpcd supply of water per capita whereas, 2 cities have between 100 to 200 per capita supply of water in lpcd. However, the Tapi river basin has more cities with the same availability of per capita water supply. As against this, Mahi basin has a lesser number of cities but a higher facility of per capita availability of water. It is very clear from the table that a majority of the cities about 67 per cent have a per capita availability that is between 100 to 200 lpcd. Gujarat state has a problem of water resources for its need of drinking water, irrigation water and industrial water both for river basin towns and cities and non-river basin towns and cities.

It should be noted here that all class-I cities are getting grants and aids from the World Bank project-I in Gujarat to monitor and treat water resources. The same is the case with the municipal corporations of Gujarat. However, the resources are found inadequate for financing drinking water in the context of consecutive years of drought in Gujarat. Consequently disposal of wastewater could not be made a priority.

Table 10.1: Distribution of class-I cities as per catchment areas per capita water supply in India *vis-à-vis* Gujarat

Sr. No.	*River Basin*	*No. of Class-1*	*No. of Cities in Different Category of Per Capita Water Supply in lpcd*			
			<100	Between 100-200	Between 200-300	>300
1.	Mahi	3	1	1	1	—
2.	Narmada	4	2	2	—	—
3.	Sabarmati	7	—	6	—	1
4.	Tapi	8	3	5	—	—
Total	Gujarat	22	6	14	1	1
Total	India	299	109	138	36	16

Source: Central Pollution Control Board Report on Status of Water Supply in Class-I Cities, February, 2000, p. 17.

Table 10.2 provides details of the sources of wastewater and its mode of disposal for example, agriculture, river, both river and agriculture, agriculture and others including sea, ponds and drainage facility. The Narmada basin area has a moderate number of such cities as compared to the Sabarmati and Tapi, where wastewater is disposed of in the main body of ponds, river and sea. These cities are affecting the eco-system of the sources of disposal, which is a matter of great concern for environmental activists.

It will be very interesting to compare the status of wastewater generation, collection and treatment in class-I cities in India and in Gujarat. There are 299 class-I cities in India in which 72 per cent of the total wastewater is collected. Of these, 16,270.8 MLD is from domestic, and 391.7 MLD from industrial sources. 11,936.2 MLD is the total generated and 11,938.2 MLD is collected for treatment. Out of 299 class-I cities, 160 have a sewage system for more than 75 per cent of the population and 90 cities have more than 50 per cent of population coverage. This mean that 70 per cent of the total population of class-I cities have a sewerage facility compared to 48 per cent in 1988 at the all-India level. As far as waste treatment is concerned, of the sewage generated the level of treatment available ranges from 2.5 per cent (Chitradurg) to 89 per cent in Sholapur.

Table 10.2: Basin-wise distribution of class-I cities in Gujarat according to mode of disposal

Sr. No.	*Major River Basins*	*Number of Class-I Cities with varied mode of disposal*					
		No. of Cities	*Agriculture*	*River*	*Both River and Agri.*	*Agri. and Others*	*Others (Sea, ponds, drainage, etc.)*
1.	Sabarmati	7	—	2	2	—	3
2.	Mahi	3	—	1	1	—	1
3.	Narmada	4	1	1	—	2	—
4.	Tapi	8	—	1	1	3	3
	Total Gujarat	22	1	5	4	5	7
	Total India	299	63	41	44	33	11

Source : CPCB Report, 2000.

As against these estimates, cities with primary level treatment plant facility, in Gujarat state include, Anand, Bhavnagar, Bhuj, Mahesana, Nadiad, and Surat and cities like Ahmedabad, Gandhinagar, Rajkot, and Vadodara have both primary and secondary level wastewater treatment plants. The number of cities and population served in these cities of Gujarat are higher than in other states. The Tapi river basin area has the highest number of class-I cities with the least arrangement for waste water disposal. The local self-government of the cities heavily relies upon external sources for such investments.

Sewage farming is another area that affords a comparison between the class-I cities of Gujarat and those at the all-India level. The municipal corporations of Gujarat have sewage farms organized by private means/farmers and controlled by local self-government. The cost of sewage at the all-India level ranges from Rs. 75/ha/year (Hyderabad) to Rs. 400/ha/year (Jaipur) and the average sale price of sewage in some class-I cities varies according to the nature of the sewage. Table 10.3 provides details of the big gap between the collected quantity of the wastewater and the wastewater supplied to sewage farms. Surat is a city in Gujarat which may be seen as highly polluted in that the least amount of waste reaches the sewerage. In Vadodara the local self-government arranges to reach 22 per cent of total waste collected. The municipal corporation of Ahmedabad city is assumed to be supplying less than 25 per cent of total collected waste to the sewerage in the city. The remaining class-I cities appear cleaner than the cities discussed. As such, sewage treatment is an economic activity as, Gandhinagar, Nadiad and Vadodara Municipal Corporations could generate income to cover the cost of treatment. In the case of these cities, the wastewater collected is covered under sewerage found to be 100 per cent, 100 per cent and 75 per cent, respectively.

Table 10.3: Details of sewage farming in class-I cities in Gujarat

Sr. No.	*Name of the Class-I City*	*Quantity of Waste-water Collected (mld)*	*Quantity of Wastewater Supplied to Sewage Farms (mld)*	*Land used for Sewage Farming (ha.)*	*Cost of Sewage Rs./ha/Year*	*Total Annual Income in Rs.*
1.	Ahmedabad	445.00	N.A.	N.A.	N.A.	N.A.
2.	Anand	10.40	2.60	20.00	N.A.	N.A
3.	Gandhinagar	43.50	41.86	500.00	100.00	50,000
4.	Nadiad	10.00	2.00	140.00	80.00	11,200
5.	Rajkot	48.70	40.00	1000.00	N.A.	N.A
6.	Surat	112.00	1.30	2.60	N. A.	N.A.
7.	Vadodara	105.00	20.00	1000.00	120.00	1,20,000
	Total Gujarat	774.60	107.76	2662.60	300.00	1,81,200

Source: CPCB Report
N.A.= Data not available

The above table indicates that most of the municipalities and corporations have not provided data regarding the cost of sewage and the total income from sewage. However, it is very clear from the table that sewage has an economic value if it is properly treated and sold professionally by contract system. It needs to be optimally reused. It can stop main body of the river if directly let out to the wastewater. In many urban centres, the treated or untreated sewage is used for fodder cultivation by the government or non-governmental agencies. The economic value of sewage can be assessed through its nutrient value and also water value. In a developing country like India, it is especially necessary for local self-governments to consider sewage as a source of income so that sewage treatment plants became economically justifiable. As per the CPCB report, the economic value of sewage generated by 17 class-I cities of Gujarat, was 1175.8 in MLD in the form of wastewater while the concentration of nutrient in tonnes per day was found to be 35.3 nitrogen (30mg/l), phosphate 8.8 (7.5 mg/l), and potassium 29.4 (25mg/l) as per the formula adopted in assessing the fertilizer value of sewage.

The cost estimates for local self-government can be stated as per NRCD, Government of India as under:

Table 10.4: Cost of analysis of sewage treatment plants (STPs) in class-I cities

Cost Factor	*UASB*	*Technologies*		
		ASP	*TF*	*OP*
Land Requirement Acre/mld	00.42	00.50	00.50	2.5
Unit cost of plant (Rs.Lakh/mld)	29.65	32.00	31.00	7.00
Unit O&M cost (Rs. Lakh/mld/year)	00.52	1.05	0.75	0.5

Legend: UASB = Up-Flow Anaerobic Sludge Blanket, ASP = Activated Sludge Plant, TF = Trackling Filters and OP = Oxidation Pond.

Source: CPCB Report 2000.

The above mentioned table provides cost estimates in the case of class-I cities at the all-India level. However, the cost may vary as per the price of the land, size of the land and technology used for sewerage treatment in each class-I city of

Gujarat as per engineering estimates. Techno-economic planning is necessary at each municipality or local self-government level in Gujarat.

Table 10.5 reveals data for 21 class-I cities, volume of waste generated which is domestic and industrial, as well as waste collected and percentage of sewerage covered.

Table 10.5: Status of wastewater generation, collection and percentage of sewage covered in class-I cities of Gujarat

Sr. No.	*Name of the city/town*	*Municipal Population*	*Volume of Waste Water Generated*			*Volume of Waste Water MLD*	*Per-centage Sewage MLD*
			Domestic MLD	*Industrial MLD*	*Total MLD*		
1.	Ahmedabad	28,76,710	520.0	36.0	556.0	445.0	80
2.	Anand	1,10,266	14.0	0.4	14.4	10.4	50
3.	Bharuch	1,33,102	12.9	N.A.	12.9	9.8	76
4.	Bhavnagar	4,02,338	38.0	N.A.	38.0	32.3	85
5.	Bhuj	1,02,176	8.4	N.A.	8.4	6.3	75
6.	Gandhinagar	1,23,359	43.5	N.A.	43.5	43.5	100
7.	Gandhidham	1,04,585	8.4	N.A.	8.4	6.3	75
8.	Godhara	96,813	5.6	N,A.	5.6	5.6	100
9.	Jamnagar	3,41,637	45.0	N.A.	45.0	33.8	75
10.	Junagadh	1,30,484	7.6	N.A.	7.6	5.7	75
11.	Mehsana	88,201	7.4	N.A.	7.4	6.0	82
12.	Morvi	90,357	9.9	N.A	9.9	7.4	75
13.	Nadiad	1,67,051	10.0	N.A.	10.0	10.0	100
14.	Navdsari	1,26,089	18.0	4.0	22.0	16.5	75
15.	Patan/Veraval	1,20,178	10.6	N.A.	10.6	N.A.	N.A
16.	Porbandar	1,16,671	11.3	N.A.	11.3	8.5	75
17.	Rajkot	5,59,407	60.0	5.0	65.0	48.7	75
18.	Surat	14,98,817	140.0	N.A.	140.0	112.0	80
19.	Vadodara	10,31,346	120.0	20.0	140.0	105.0	75
20.	Valsad	57,909	8.2	N.A.	8.2	6.0	75
21.	Wadhwan	1,66,466	11.5	N.A.	11.5	11.5	100
	Total Gujarat	84,43,962	1110.4	65.4	1175.8	936.7	74

Source: Class-I City Status Report, CPCB, Delhi, 2000, p. 44.

The figure varies according to the population size of the city, amount of waste generated and collected in MLD. However, the level of wastewater pollution would be different as per the quality and capacity of treatment of wastewater by the treatment plant. The last column provides important details that Gandhinagar, Godhara, Nadiad and Wadhwan have a hundred per cent sewage facility for the volume of wastewater in MLD. Moreover, the Gujarat average (74) is higher than all-India level (70) in terms of percentage of wastewater for sewage in MLD. It is also interesting to note from the table that the wastewater pollution load is higher in the domestic sector rather than in industrial sector in leading industrial cities Gujarat. However, most of the cities have not provided data to CPCB.

Table 10.5 clearly indicates that the Gujarat average is higher than all-India average of the percentage of sewerage coverage. Similarly, an average of class-I cities is higher than different location of cities in Gujarat state. We have the Class-I cities in the state like Bhuj, Gandhinagar, Nadiad, and Wadhwan having a 100 per cent sewerage coverage. An effort is made by the local self-government in Gujarat to manage and treat wastewater for possible reuse by managing sewage farming and treatment plants.

The Position of Class-II Towns in Gujarat

There are 345 class-II towns in 1991 compared with 241 in 1988 with municipalities that have not provided details of the wastewater generated by them. An attempt has also been made by CPCB to measure the water supply, collection, treatment and disposal status of the wastewater pollution load in class-II towns periodically (Table 10.6). It has been observed that total wastewater generation in the class-II towns is 1649 MLD in 1994 and is projected as 1897 MLD for 1999.

Table 10.6 provides a picture of Gujarat state class-II towns which on an average are seen as superior to the all-India level. This has happened largely due to an awareness of the local self-government and awareness of the people. Moreover, Sincere efforts are made by local self-government of central Gujarat region than that of class-II towns of Saurashtra region in the state for collection of wastewater. As against this, very scant attention is paid by local self-government of north Gujarat. They argued that they do not have sufficient resources for collection and treatment of sewerage water approached Gujarat Municipal Finance Board, bond market and Global level agencies to fund treatment plants and sewerage arrangements.

Table 10.6: Class-II towns-wise waste water generation, collection in Gujarat

Sl. No.	*State / Town*	*Population*	*Volume of WW Generated (MLD)*	*Projected WW Generated (MLD)*	*Volume of WW Collected*	*Per cent of WW Collected (MLD)*
1.	Amreli	67,827	3.54	4.1	2.66	75
2.	Anjar	51,209	9	10.4	4.5	50
3.	Ankleshwar	51,739	6	6.9	3	50
4.	Bilimora	51,039	3.68	4.2	2.39	65
5.	Botad	64,603	6.25	7.2	4.69	75
6.	Dabhoi	50,641	9.08	10.4	6.81	75
7.	Deesa	62,435	3.66	4.2	2.75	75
8.	Dhoraji	77,748	5.16	5.9	3.87	75
9.	Dholka	49,855	3.60	4.1	2.16	60
10.	Dhrangadhra	57,961	36	41.4	27	75
11.	Dohad	66,500	4	4.6	4	100
12.	Gondal	80,584	7	8.1	5.25	75
13.	Himatnagar	51,461	2.63	3.0	1.97	75
14.	Jetpur	73,560	15	17.3	11.25	75
15.	Kadi	50,733	3.04	3.5	1.98	65
16.	Kalol	82,137	10.8	12.4	8.10	75
17.	Keshod	50,172	2.56	2.9	1.28	50
18.	Khambhat	76,746	10.42	12.0	5.21	50
19.	Mahuva	59,912	6.56	7.5	4.92	75
20.	Palanpur	80,657	7.12	8.2	5.34	75
21.	Patan	96,112	6.21	7.1	4.66	75
22.	Savarkundla	64,815	6.25	7.2	4.69	75
23.	Sidhpur	50,770	3.50	4.0	2.63	75
24.	Unjha	51,003	3.40	3.9	2.55	75
25.	Upleta	51,801	5.68	6.5	4.26	75
26.	Viramgam	50,698	6.30	7.2	6.30	100
27.	Visnagar	57,839	4.80	5.5	3.60	75
Total (Gujarat)		1,680,557	191.23	219.9	137.79	72

Source: CPCB Report on Class-II Towns, March 2000, pp. 25-6.

Gujarat state has 27 class-II towns with a population of 2,48,344 as per the 1991 census. Twenty-six per cent of total population of class-II towns lives in the coastal region of Gujarat and 32.5 per cent of the urban population in the river basin areas. As against this, there are 345 towns at the all-India level having a population of 2,36,45,614. 2.8 per cent of total population lives in the coastal areas and 10.8 per cent of the total urban population in the river basin areas. The percentage distribution of population in the major basins is 76.9; coastal 4.6; and in the non-coastal basin it is 18.5. Thus, wastewater treatment and sewage treatment have become important for the state government and local self-government as both industrial wastewater and municipal wastewater create a problem for the pollution regulatory authority. Gujarat Infrastructure Development Corporation has prepared a Master Plan for the state, Gujarat : Vision For 2010 under which disposal wastewater and outlay for the purpose is earmarked.

Table 10.7 indicates the per capita water supply in class-II towns of the Narmada basin in Gujarat as 33.9 MLD and the per capita availability of water supply as 108 litre per day. These figures are very low considering the non-perennial nature of a river like Sabarmati in the central region of Gujarat State. As against this, the Tapi river basin also has a very low availability of waste supply in MLD and per capita waste supply compared with other basins in Gujarat. However, the percentage of population covered by the organized water supply remains the same for all-India level which is 88 per cent.

Regarding the basin-wise mode of disposal of wastewater in the class-II towns of Gujarat, out of five towns of the Narmada river basin, two disposes this in sea, lake and pond, two in agriculture or other mode of disposal whereas, one town disposes this in both — river and agriculture. Gujarat has cities and towns located on the coast line. These include Okha, Porbandar, Jamnagar, Veraval, Patan, Mahuva, Cambay, Valsad and Vapi. However, this researcher is of the view that class-II towns outperformed the class-I cities of Gujarat in terms of disposal rates due to sewage farming along with treatment plants.

III. Policy Options

It is most necessary for environmental economists to prescribe

Table 10.7: Basin-wise per capita water supply in class-II towns in India *vis-à-vis* Gujarat

Sr. No.	*Name of the Major River Basin*	*No. of Towns*	*Population*	*Ground source (MLD)*	*Surface source (MLD)*	*Combined (MLD)*	*Total Water Supply (MLD)*	*Assessed Water Supply (MLD)*	*Per capital Water Supply (litre/day)*	*Population Covered by Organized Water Supply*	*Per cent of Population Covered by Organized Water Supply*
1.	Sabarmati	6	342,993	35.0	4.5	—	39.5	—	120	329,701	96
2.	Mahi	4	238,770	4.0	—	22.0	26.0	—	111	234,021	98
3.	Narmada	5	330,307	18.9	10.7	4.0	33.6	—	108	310,503	94
4.	Tapi	5	371,292	23.1	—	—	23.1	—	67	345,751	93
	Gujarat Sub-Total	20	1,283,362	81	15.2	26.0	122.2	—	406	1,219,976	95
1.	Coastal	16	966,375	13.4	12.5	27.2	53.1	27.0	100	533,599	77
1.	No major Basin	67	4,185,618	219.1	53.5	91.4	364.0	10.3	101	3,601,463	88
	Grand Total	103	6,435,355	313.7	81.2	144.6	539.3	37.3	607	5,35,5038	88

Source: As per CPCB Report of Class–II Towns, March 2000, Delhi, p.8.

the policy options to be followed by local self-government as well as by pollution regulating authorities.

Table 10.8: Basin-wise distribution of class-II towns according to mode of disposal in Gujarat

Sl. No.	*Name of the Basin*	*No of towns*	*No. of Towns as per mode of disposal*				
			Agri-culture	*River*	*Both River and Agriculture*	*Agriculture and Other*	*Others (sea, lake pond)*
1.	Sabarmati	6	3	—	—	—	3
2.	Mahi	4	—	--	—	—	4
3.	Narmadà	5	—	—	1	2	2
4.	Tapi	5	—	—	—	—	5
	Gujarat Sub-total	20	3	—	1	2	14
	Coastal (India)	16	2	—	2	1	11
	No Major Basin (India)	67	—	—	3	6	58
	Grand Total (India)	345	12	12	18	42	261

Source: CPCB Report.

1. Gujarat Pollution Control Board should constantly watch wastewater disposal by the local self-government both in the case of river basin cities and towns as well as non-river based cities and towns. The Central Pollution Board River basin states that municipal wastewater including solid wastes dominates river pollution. A periodical sample analysis of wastewater is absolutely necessary for municipal as well as industrial wastewater.
2. Constant technical guidance is necessary from the State Pollution Control Board for treatment plants run and managed by local self-governments in Gujarat. It is also necessary to monitor the quality of wastewater disposed after due treatment by the regulating authority.
3. Local self-government should approach the financial

institutions to run and manage treatment plants for public health and occupation at health safety so as to provide hygienic drinking water and to save drinking water getting polluted by the wastewater.

4. Local self-governments in Gujarat should go in for a privatization process for disposal after due treatment as per the norms of the Pollution Control Board rather than fully bear the expenses for such a process can increase the supplementary income of the body. For this, a contract system is better than to maintain technical staff on permanent basis.
5. It is necessary for the local self-governments in Gujarat to go in for a zero effluent discharge in drinking water sources like river, wells, tanks, ponds and other ground-water sources which are getting highly polluted. A study of Sabarmati and Narmada rivers in Gujarat by the Central Pollution Control Board reveals that the municipal waste load is higher than the industrial waste load in these rivers.
6. For this, it is most necessary to organize an awareness progamme through NGOs for both educated and uneducated people by means of different media sources not to pollute water resources. As such, this kind of pollution adversely affects occupational health and public health in the state.

Conclusion

It appears from the data analysis that industrial wastewater does not significantly dominate the total volume of wastewater. Domestic wastewater constitutes the greater share even in bigger industrial cities and towns. Environmental activists have never approached the local self-government for a proper disposal in cases where the people are more responsible than industrialists. As, such wastewater or effluent disposal have become a non-return economic activity for the local self-government. However, privatization of disposal initiated by the local self-government of Gujarat state is necessary if due weightage is to be given to the cost economics of wastewater. The Ninth Five-Year Plan should earmark 30 per cent of total expenditure for this purpose. The state government should also emulate the other state level initiatives. Due to lack of such an action, the local body needs external sources of finance not only in Gujarat but at the all-India level also. This study also represents that class-II towns have performed better than municipal corporations largely due to sewage farming facilities which involve lesser cost in the disposal of wastewater. The Rajkot Municipal Corporation has set an example in the privatization of solid waste disposal to be emulated by leading Municipal Corporations at the All India level.

11

Valuation of Safe Water Supplies for Urban Households

PURNAMITA DASGUPTA

Introduction

This paper looks at the valuation of water as an economic resource in the context of a low income, infrastructurally disadvantaged urban household. In the context of a rapidly growing population, a typical urbanizing economy increases the demands on the natural system to absorb the fall-outs of enhanced economic activity while reducing the per capita raw material (natural resource) availability. The demands on the ecosystem increase both in terms of its ability to act as a sink and its ability to regenerate and provide a pool of natural resources. This gives rise to the potential for creating stress on natural resources and leads one to focus on the concept of scarcity of a natural resource. Economics therefore seeks to put a value on the effects of changes in a natural resource. It is in this context that the question of economic valuation of water arises.

Given the divergences between the richer and poorer sections in the economy, typically the poorer sections of society suffer more from conditions of water-stress. In urban areas the problem becomes particularly acute, with the poor being concentrated in settlements spread over a small land area and, stress being created in terms of both access to safe and adequate water and, disposal of solid and liquid wastes. Ensuring adequate supplies of safe water sources, therefore, becomes an issue of paramount importance for environmental and health considerations. It is in this context of urbanization that the present paper addresses the problem of "water-stress" in terms of both water pollution and water scarcity.

Two alternative damage assessment exercises are conducted to assess the impacts of both the quantity and quality of water being accessed and consumed by households. Issues related to both scarcity and water pollution are thereby captured. An objective assessment of the damages is made by looking at the health costs borne by households reporting diarrhoeal illness while an alternative subjective assessment is made by conducting a contingent valuation exercise for these households. Subsequently, the study investigates the discrepancies between value judgements and damage assessment as revealed by the health costs approach, *i.e.* between the subjective and objective valuation results.

Delhi: The Poor and Access to Water

Among the major Indian metropolises with a population of over one million in 1981, Delhi has experienced the highest demographic growth during the post-independence period. Such demographic expansion and enhanced economic activity has had obvious implications for the provision of infrastructure, especially for basic services such as water supply. Today, a large proportion of the population is still dependent on public hydrants for their water needs. Moreover, inadequacy or non-availability of sufficient quantities of piped water supply, forces individuals to resort to alternatives which can be harmful. Further, there are certain situations when threats from such alternatives increase. During summer, the demand for water goes up while the supply of water is reduced due to a fall in the raw water made available from the river (Yamuna) upstream. The poorer households are hit hardest by shortfalls in piped supply and the risks of contamination faced by them are compounded since they can least afford the better means of storage. During the rainy season, the soluble nature of faecal and other waste matters implies greater and faster dispersion of wastes. As a result the groundwater is contaminated more easily than during the dry months, affecting all those who are dependent on it. Once again, the ability of the poor in taking steps to purify this water is the least. Thus, there are a number of settlements which suffer from the common problems of poverty, high population densities, frequent population movements, poor supply of (piped) safe drinking water, and, lack of sanitation and sewage facilities. It is in this urban environment that waterborne diseases such as diarrhoeal illness flourish.

Methodology for Contingent Valuation of Water

Natural resources yield flows of services while changes in the flow of these services give rise to benefits and costs related to these resources. Because of certain economic characteristics such as the presence of externalities, common property and public good characteristics (non-excludability, non-depletability) of such services, pure market forces cannot be relied on to achieve efficiency in allocation and use of these resources. Nor can the market be used in revealing prices that reflect the true social value of these goods. It is this failure of the market system that creates the need for developing specific non-market measures of value for what is commonly termed as the "resource-environmental system." The market for water resources is also an imperfect one with problems of imperfect information being predominant in both demand and supply aspects. What must be recognized is that the assessment of what type of technique is suitable to value which change in resources, has to evolve in tandem with the particular field situation to be studied. There is no one best method for evaluating the consequences of a particular alternative on human well-being.

Neo-classical welfare economics has provided the basis for deriving measures of the economic value of changes in resource-environmental systems. Such value measures have been expressed either in terms of willingness-to-pay or in terms of willingness-to-accept compensation. A model for estimating willingness-to-pay as a measure of benefit evaluation can be derived from a model of consumer behaviour.

Starting with the utility function, $U = U(X, P)$, the household maximizes utility subject to the budget constraint leading to the derivation of demand functions for X [$x_i = x_i(p, I, P)$]; and subsequently, the indirect utility function. Here P represents the pollution/contamination level, X is a vector of consumption goods and I is income. The indirect utility function is then inverted to obtain the expenditure function (by duality theorem). This leads to the Hicks compensated inverse demand functions, which is interpreted as the marginal willingness to pay of the household for an improvement in its water supply.

Adopting this line of approach the present study estimates a measure of the willingness to pay where willingness to pay should vary with income, prices, a vector of individual characteristics, and the pollution levels. The willingness to pay in question is a dependent variable elicited from the respondents through a contingent valuation survey. The contingent valuation method collects preference information by asking households how much they would be willing to pay for some change in the provision of a public good. Such a willingness to pay measure compares with the compensating variation measure in welfare economics.

The sample covered constituted a mix of respondents who already have individual home water connections and, respondents who do not have individual connections. Thus the elicitation procedure was tailored to suit the requirements of these two distinct types of respondents, in order to make the hypothetical scenario as realistic as possible. For respondents without individual home connections, a hybrid method involving three stages was adopted for eliciting responses from this group. This method can be equated to a direct open-ended method, followed first by a referendum model and further with an open-ended follow-up. For respondents with individual home connections, the method adopted was a single, direct open-ended question on what would be the respondent's maximum willingness to pay for such improvements.

Upon estimation, the model yielded a mean willingness-to-pay estimate for a representative household in each of the two groups of respondents. These estimates were extrapolated to the entire population to derive the total willingness-to-pay.

Methodology for Evaluating Health Costs

An economic evaluation of the health effects of a change in an environmental or resource good, would focus attention on the changes in individual behaviour stemming from the perceived adverse effects (reduction) on the individual's utility. A model for valuing the damages from contaminated water supplies, based on the theory of utility maximizing consumer behaviour was developed for estimating the probability of illness for a household.

Thus, a health production function is used in modelling individual behaviour. The monetization of health losses borne by households reporting diarrhoeal illness provides a useful measure of the minimum willingness-to-accept of the households for avoiding damages from contaminated water.

A utility function is defined as $U = U(X, L, S)$, where X represents expenditures on all non-health related goods (household's aggregate consumption) and has a price of unity, L is leisure time per period and S is time spent ill (number of days sick). The household derives utility from the consumption of X and L, while S causes disutility. Time spent ill (S) is modelled as a function of the exposure to contaminants (P) and averting or defensive behaviour (D) to reduce the likelihood of illness. The household maximizes utility subject to the household's budget constraint. This solves for a set of optimality conditions. At the margin, the wage loss (w) corresponding to the unit increase in time spent on defensive activities, must equal the net gains (in terms of non-health consumption expenditure and reduced sick time) from the increase in time spent on defensive activities. The specification appropriate for the model is based on binary observed dependent variables and the model is estimated as a bivariate probit specification. The estimation procedure therefore requires the estimation of a reduced form relationship between illness and the ambient pollution levels while controlling for other variables that affect health status such as physical and socio-economic characteristics of age, sex, income and education.

The estimation of this model provided an estimate for the predicted probability of observing illness in a household. The probability measure is used along with the data on expenditure on illness to derive the treatment costs for the representative household. The wage-loss arising from absence from work due to ill health is also computed. This in turn leads to an estimate of the opportunity cost of illness for the entire population, comprising of the treatment costs and the wage-loss as a monetized value of damages from illness. The results obtained on the cost of illness lead in turn to policy implications for the water sector, especially in view of the findings from the willingness-to-pay exercise.

Sampling Frame of the Study

Selection of Households

At the outset the present study recognizes that the complexity of transmission routes of the diarrhoea-causing pathogen has to be taken into account in assessing the impact of any single intervention to control the disease (Briscoe, 1984). Nevertheless, the case for interventions focusing on improvements in drinking water has withstood the test of time and empirical analysis (Esrey *et al.*, 1985; 1991; Anderson and Cavendish, 1992). While a wide assortment of organisms cause acute diarrhoea, the occurrence of cholera signifies serious levels of contamination of water as the infective dose to cause clinical disease by *Vibrio cholerae* is highest among all other pathogens causing diarrhoeal diseases. In view of this, the sampling procedure for the field survey adopted in the present study, was based on data for cholera cases occurring in Delhi.

The sampling technique was based on a two-stage sampling procedure. At the first stage, keeping in mind the epidemiological considerations for such a study, all the 14 localities in Delhi that had been reporting more than 5 cases of cholera annually, consecutively over the three years, 1996 to 1998 were selected. The survey was conducted during the summer months of 1999.

At the second stage, 600 households were to be selected from these 14 localities. Based on the data on the number of cases for these three years, the three-year period prevalence was first calculated for these 14 localities. Each period prevalence was converted into a proportionate prevalence by dividing it by the sum of the prevalence ratios. Thus, a set of weights was generated which took into account the disease load for each locality. The total number of households in each of these 14 localities was subsequently weighted by the corresponding weights. A simple proportional rule was thereafter applied to these weighted household numbers in order to distribute the sample requirement of 600 households across the 14 selected localities. On further investigation, these 14 localities were found spread across 23 specific colonies. In all such cases where a locality was spread out over one or more colonies, the designated number of households to be sampled from each locality were divided among the constituent colonies in the same ratio as that of the total number of households in the component colonies. The actual selection of households within each colony was done by random sampling.

Selection and Laboratory Testing of Water Samples

An important part of the field study involved the testing of water samples from the surveyed colonies. Water samples were collected from each colony and laboratory testing of these water samples was conducted at a government-accredited laboratory. All the samples were tested for faecal coliform counts. Among the available bacteriological tests for water quality surveillance the presumptive coliform test was the most relevant one for purposes of the present study. This test is based on the most probable number (MPN) of coliform organisms in 100 ml of water. *Based on the guidelines for drinking water quality recommended by the WHO in 1996, the ISO (IS 10500) prescribes that the coliform count of a drinking water sample should be less than or equal to 10 per 100 ml of water.* The WHO recommendation for drinking water quality lays down the primary standards as prescribed limits that must never be exceeded in water meant for drinking purposes.

For purposes of sampling, a total of 90 water samples were collected, *i.e.*, 15 per cent of the 603 households that were surveyed. This would be sufficient to ensure that epidemiologically sensitive results were obtained for deriving conclusions about the exposure to contamination through the water that was being accessed by these 600 households. It is important to clarify that the term water source is defined in this study with reference to the means by which water is being accessed. The means of access play the most important role in influencing as well as indicating the quality of water being consumed by a particular household. Means of access that have similar characteristics and implications for water contamination, are clubbed together and labelled as a particular source. Thus, for example, shallow handpumps would be one source, as distinguished from deep bore tube-wells, which would be considered to be another distinct source. In all there were nine such sources defined in the present study. The distribution of the 90 water samples was done in a manner ensuring that from each colony, at least one of each type of water source with similar characteristics was tested. For colonies where community level water sources (*e.g.* public hydrants, community handpumps) were being accessed, all the specific points that were being accessed by the surveyed households were sampled. For those with individual water sources, be it handpumps, tube-wells or illegal connections, 15 per cent of the total number of households surveyed were randomly selected and their water samples tested in each colony. For colonies with individual piped connections, while the same 15 per cent rule applied, in addition water samples were also taken from the entry points of the water mains to the colony.

Survey Instrument

The primary data for the analysis were gathered through a survey conducted as a personal interview between the surveyor and the head of the household. The survey questionnaire had

13 modules covering various aspects of socio-economic and demographic features of the household, apart from detailed sections relating to water and illness aspects.

Preliminary Data Analysis

Results from Laboratory Analysis of Water Samples

Table 11.1 reveals that out of a total of 90 samples tested for coliform counts, 53 per cent had coliform counts that were unacceptable for safe drinking water. While the proportion of samples failing to meet the required coliform standards for drinking water varies across the different sources of water, an overwhelming 93 per cent of water samples taken from shallow handpumps failed the test. All the samples taken from deep bore tube-wells also failed the test. In the case of legal piped water supply, for both individual connections and public hydrants, the picture is reversed with failure rates ranging from 31.6 to 37.5 per cent. The evidence from the raw data, therefore, points overwhelmingly in favour of piped water supplies as being less contaminated than groundwater supplies. For most of the colonies, especially the slum clusters, public hydrants turn out to be a better option than handpumps or tube-wells, which is of course an expected outcome. What is of more concern, however, is that while groundwater has come to be accepted as a potentially risky source with regard to contamination, several public hydrants too contain coliform counts that are above the acceptable limits. The implications for diarrhoeal and other waterborne illnesses for areas where large numbers are dependent on public hydrants are very serious.

Sources of Drinking Water, Household Income and Diarrhoeal Illness

Table 11.2 gives the distribution of the primary source of drinking water by income quintiles. While most of the water sources mentioned in these tables are self-explanatory, code 9 deserves special mention. This code is used in all those cases where there is a private arrangement for buying water from a neighbour's deep bore tube well. The piped sources constitute the bulk of the water sources being accessed for drinking purposes. Predictably, the reliance on individual piped supplies increases while that on groundwater options decreases as one moves up the income brackets. It would be interesting to see how this data compares with the data on the number of cases being reported by households. Examination of the data reveals that out of 264 households that reported cases, 163 households used piped water supplies (either individual connections or public hydrants) from the municipality. It becomes clear that there is a need for analysing issues relating to both groundwater as well as piped water.

Table 11.1: Source-wise distribution of water samples

Source	*Total samples tested (per cent distribution)*	*Proportion of unacceptable samples (per cent)*
1	22	93
2	6	0
3	0	—
4	6	50
5	23	37.5
6	28	31.6
7	8	60
8	6	100
9	1	100
Total	100	53

Note: *Codes* for source of drinking water: (1) shallow handpump, (2) tanker (municipal supply), (3) community tank, (4) deepbore tube well, (5) individual piped connection (legal), (6) public hydrant, (7) individual piped connection (illegal), (8) Mark II (deep bore) handpumps, (9) piped water from private deepbore well.

The households sampled can be classified into two groups as per the data on illness. Those households that have reported at least one illness are referred to as the affected households, while the rest are the unaffected ones. There are a total of 266 households that are affected by this classification. In all there were 1422 individuals in these affected households, comprising 544 children, 824 adults and 36 elderly. Among these 1422 individuals, 150 children, 168 adults and 9 of the elderly reported at least one illness episode during the preceding 15 days, at the time of the survey. In all, there were 327 individuals who had been affected by at least one episode of illness.

Table 11.2: Source of drinking water by (household) income quintiles

	5 Quintiles of Income					
Source	*1*	*2*	*3*	*4*	*5*	*Total*
1	19	12	16	13	7	67
2	10	6	18	11	5	50
3	0	0	0	0	2	2
4	6	2	13	13	17	51
5	30	19	42	38	61	190
6	75	32	39	12	15	173
7	13	9	15	4	4	45
8	6	3	4	1	4	18
9	0	0	0	2	5	7
Total	159	83	147	94	120	603

Note: *Codes* for source of drinking water: (1) shallow handpump, (2) tanker (municipal supply), (3) community tank, (4) deepbore tube well, (5) individual piped connection (legal), (6) public hydrant, (7) individual piped connection (illegal), (8) Mark II (deep bore) handpumps, (9) piped water from private deep bore tubewell. Income ranges (in Rs per month) ~ 1: 1000-2500, 2: 2600 – 3250, 3: 3300 – 4500, 4: 4600 – 6000, 5: 6400 – 25000.

Incidence of Illness

The definition of diarrhoeal illness includes a reporting of at least three loose stools per day, along with any other symptoms of gastro-intestinal discomfort such as vomiting, nausea, etc. for adults. In the case of children, the mothers reporting of illness is accepted. The incidence rates for diarrhoeal illness among the population surveyed, both clusterwise and overall, for adults, male and female children were calculated. All these incidence rates are very high as shown in Table 11.3. However, it must be understood that the present study covered the most vulnerable colonies during the most vulnerable season from the point of view of diarrhoeal illnesses. In the present study,

19.95 per cent of the children below 5 years had suffered from one or more episodes of diarrhoea over the previous two weeks at the time of the survey. This is directly comparable to the 24.6 per cent figure obtained by the Bhatnagar and Dosajh study (1986) on diarrhoeal disease morbidity in children below 5 years in urban slums in Delhi. The average duration of illness was 3.9 days per episode for the latter study while, in the present study the mean duration was 3.2 days per episode.

Table 11.3: Illness incidence rates for the sample households

Definition	*Incidence rate*
1. For entire sample (number of cases/total population)	113.10
2. Male children (total reported cases of male children ill/total number of male children)	128.55
3. Female child (total reported cases of female children ill/total number of female children)	135.36

Interruptions in Piped Supplies

It is also of interest to study the data on frequency of supply interruptions for those with piped water supply sources. Out of the 603 households surveyed, the majority (355 households) is dependent on piped water. Interruptions in piped water supply have been identified as a crucial variable in explaining diarrhoeal illness in earlier studies (Alberini *et al.*, 1996). The data reveals that about 70 per cent of households using either individual piped connections or public hydrants are subject to interruptions in the water supply.

Defensive Behaviour

The data on defensive behaviour reveals that only 97 households adopted purification measures for their drinking water. While 26 households boil water, 35 use filtration techniques and 36 use chlorine tablets. It becomes evident that purification measures are being primarily adopted by those using piped water supplies.

Access to Sanitary Latrines and Disposal of Waste

Among the sanitation and solid waste management variables, while most of the respondent households had access to sanitary latrines, there were a sizeable number especially among children, who used open drains within the cluster for defecation. What is of more concern is the fact that over 70 per cent of the households sampled were using local dumps for disposing their solid waste which were primarily in the form of empty plots from where there was no system of regular removal by municipal or private authorities. For groundwater contamination, this poses a serious threat, since the waste is ultimately buried or burnt, posing a major health hazard.

Other Socio-economic Characteristics

While more than 50 per cent of the household heads have completed up to secondary school and almost 70 per cent have done their primary schooling, a little less than 20 per cent of the mothers have completed primary education. The largest numbers are mothers who are either illiterate or have received no formal schooling. It was also found that the average time spent on collecting and storing water is approximately two hours per household.

Expenses for Procuring Water

An attempt was made to estimate what proportion of household income was being spent on procuring adequate water for the family members. Table 11.4 gives a distribution of the proportion of water-related expenses by household income slabs. The expenditure variable comprises three distinct items: the water bill (if any) paid by the household, the expenditure on maintenance of equipment to procure water from sources other than piped supply and, the expenditure on storage facilities arising from the lack of continuous or reliable water supply. It becomes apparent that the poorest households spend a proportionately higher amount of their income on procuring and storing adequate water in the present sample.

While those in the in-between slabs in terms of household income spend less than the poorest, the highest income group spends proportionately the least on this head.

Table 11.4: Expenditure on water as a percentage of household income, by income groups

Household Income Groups (Rs /month)	*Mean Water Expenses (percentage of household income)*
Less than equal to 2500	2.9
Greater than 2500 and less than equal to 4000	1.9
Greater than 4000 and less than equal to 6000	2.3
Greater than 6000	1.6
Overall (for entire sample)	2.19

Results from Econometric Estimation

Health Production Function Model

For purposes of estimation, the different sources of water are clubbed into two broad categories: piped (legal) and non-piped sources. This classification leads to a crucial variable for the model. Within the piped category, both individual connections and public hydrants are included. It may be recalled that while 72 per cent of the laboratory tests conducted on water samples from the non-piped category failed, in the piped category 34 per cent of the tested water samples failed to meet the required coliform count.

The source of drinking water is highly significant and positively related to diarrhoeal illness indicating that households that are dependent on non-piped sources have a higher probability of reporting diarrhoeal illness. It can therefore be inferred from this result that those consuming drinking water from non-piped sources such as, shallow handpumps and even the much-acclaimed—Mark II handpumps, are most vulnerable to suffer from diarrhoeal illness. This is in keeping with the hypothesis underlining the need to provide treated and piped water supplies to these infrastructurally deficient areas.

Certain other risk factors theoretically important for explaining diarrhoeal disease are found to play a much less significant role. Thus not having access to sanitary latrines does not emerge as a significant explanatory variable for diarrhoeal illness. On the other hand, educational attainment of the head of the household also does not play a role among the determinants of diarrhoeal illness in the estimation. Among other factors, irregular removal of garbage from dumps within colonies and non-availability of municipal disposal facilities for household solid waste, play a significant role in explaining the incidence of illness. The results also confirm the hypothesis that interruptions in water supply are a significant determinant of diarrhoeal illness relative to households who do not experience such interruptions in piped supply.

Per capita income is significant as a determinant of illness. Expectedly and in keeping with the findings of earlier studies (Bhandari, 1992; Panda, 1996), it is found to be negatively related to the probability of illness. It possibly captures the effects of related socio-economic variables, which have not been quantified otherwise. Better-off households are in a better position with regard to buying information, and have a greater capacity to adopt alternative choices that affect diarrhoea, *e.g.* better hygiene, better living conditions, etc.

As far as the adoption of defensive practices is concerned; both per capita income and the educational level of the head of the household play a significant and positive role in determining defensive behaviour. Not having access to a latrine and the presence of a foul smell in the water act as significant proxy risk factors for possible contamination of the water and induce defensive behaviour. The timings variable is also significantly and positively related to defensive behaviour. This indicates that among households with piped supplies, those who experience interruption in the supply are more likely to undertake defensive measures to purify water.

Estimating Cost of Illness

The estimation of the health model leads to the derivation of estimates for the predicted probability of observing illness in each household which has been surveyed in the study. The univariate (marginal) predicted probability of success in the outcome, which in this case is defined as the probability of observing diarrhoeal illness in a household, is estimated for the sample. The average value obtained for this predicted probability is 0.44. This probability value is used along with other statistical measures to arrive at the monetized cost of illness for the entire population under discussion. These costs can alternatively be interpreted as an indicator of the (minimum) willingness to accept compensation for the surveyed population. The cost of illness comprises of two components, the costs of treatment and the wage-loss arising from absence from work due to ill health.

The Cost of Treatment

For arriving at a representative household's treatment costs, the data on the individuals sampled in this survey was first divided into three age categories. Children were defined as the age group below 15, adults as age 15 to 60, and those above 60 were classified as the elderly, and the probability of falling ill and the average treatment expenses incurred by each age group was calculated.

The Wage Loss from Illness

The opportunity cost of an illness episode would comprise of the costs of treatment as well as the implied wage loss, arising from mandays lost due to ill health. The wage loss is computed for the adult population. The probability of an adult being ill is multiplied by the average man days lost, the average wage rate over all working adults, the rate of employment for the sample, and finally the average size of the household (s), in order to arrive at the total wage loss (w_1) for a period of fifteen days, for the representative household.

While the model predicts a mean probability of illness measure of 0.44 for the representative household, the opportunity cost of illness (comprising of Rs. 35.17 as wage-loss and Rs. 36.26 as treatment expenses) is computed at

Rs. 71.43 as the average cost for a representative household over 15 days during the peak period for diarrhoeal illness. The annual cost of treatment was subsequently arrived at by weighting the peak period costs by a scaling factor based on the monthly distribution of cases over the last three years.

Willingness-to-Pay (WTP)

What emerges strongly from the willingness-to-pay exercise is the positive perception about water as having an economic valuation, even among the poorest households. The analysis on the willingness of households to pay for safe water supplies yielded a mean monthly willingness-to-pay of Rs. 15.50 for households without individual connections and Rs. 54.52 for households with such connections.

Several factors influence the willingness-to-pay expressed by households. The household income plays a very significant role in determining the household's WTP. This is an expected outcome, higher the household income higher is the WTP. On the other hand, it is interesting to see that the expenses on annual maintenance of equipment related to procurement of drinking water for the household plays an important role in determining the WTP of those who already have access to household connections. The higher is the expense on maintenance, the higher would be the household's WTP for an alternative that would not require such maintenance-related expenditures. The result is in keeping with the basic theory of consumer demand in as much as households are willing to pay more for an improved supply, the higher are the financial costs of obtaining water from the existing system.

Illiteracy among both the head of the household and the mother/woman of the household exerts a strong negative influence on the WTP for households without connections. The educational level of the mother is found to be a significant explanatory variable even among households with household connections. The socio-economic implications of these results are an obvious argument for improving the educational levels of both the head and the spouse/mother in terms of valuation of water supplies to a household. Thus households where the head and/or mother is educated, are willing to pay more for the improved supplies. Again, household's with working mothers exert a positive influence on the WTP as compared to households where mothers do not work. This captures the positive influences of work and earnings on the mother's awareness levels, and her role in decision-making within the household. This could be of crucial importance in the typically low-income situation which is being studied here.

Households which have to reschedule their activities as per the water supply timings, typically have a higher WTP for water, as compared to those who do not have to do so. This highlights the household's willingness-to-pay for "reliability" of a source. This is a significant variable influencing the willingness to pay across both groups of respondents.

The costs involved in obtaining water from the existing sources in terms of maintenance expenditure, the quality of the water available and the reliability of the sources, are three important aspects contributing to how households perceive the characteristics of the existing scenario *vis-à-vis* the (proposed) improved scenario.

Conclusion

The results from the health model point towards the importance of overall infrastructural variables such as garbage removal facilities, water sources, interruptions in water supply, which may not always be within the control of the individual household. Community and government involvement is called for, the latter more so in the context of the low-income households which have been studied here. The results obtained indicate the importance of safe water supplies from the point of view of both the household and the civic authorities that are interested in interventions to control diarrhoeal illness. Comparison of the costs of illness with the costs that would be incurred in supplying an adequate quantum of water to these very same households could for example, be crucial inputs into policy making for the water sector. Again, the costs of illness could also be used for judging the willingness-to-accept compensation for the affected households.

On the other hand, the willingness-to-pay indicates the awareness and ability to pay of these households. Their positive perception of water as an economic good, and as a scarce resource, augurs well for policy planners. It also implies that there is scope for the water supply undertaking to take initiatives to generate revenues by levying appropriate tariffs on these households, while ensuring adequate supplies of safe water.

Willingness-to-pay measures are conditioned on the income distribution within a community. This makes the efficiency criterion a poor and inadequate one for judging the benefits of a water supply project when one is considering the desirability of a project in terms of its social value as much as its pure economic value. In the context of the present study of primarily low-income households that experience inadequate access to a basic infrastructural service, this point is a crucial one.

The findings from the present study, indicate that the total willingness-to-pay of the surveyed population can meet the long-run cost of production and distribution of water from treatment plants for a minimum assured norm of 70 lpcd. What could actually be achieved in terms of cost recovery would ultimately depend on what the political economy would permit. If the costs of illness are considered to be a better indicator of the economic benefits accruing from improvements in the water supply to these vulnerable households than the willingness-to-pay, then such benefits would justify provision even up to the 50 gpcd level.

A critical minimum, a life-line rate of water supply has to be ensured. As per the current targets of the water supply undertaking (Delhi Jal Board), such a rate should be defined at the 25 gpcd level. Under such circumstances, it becomes obvious that the state would need to take the initiative for upgrading and providing better services, while recovering the costs as far as politically and administratively feasible. While the production costs would be potentially recoverable as indicated by the WTP exercise, it would have to frame policies for financing the distribution costs. Given the present scenario, the case for cross-subsidization by levying higher charges on upper-income groups seems to be limited. Treating the distribution cost as a public good and considering that true preference revelation has not taken place, the state could opt for either of two financing alternatives. The distribution costs could be charged to the households as actual costs incurred for avoiding the damage costs. Alternatively, if such charges are considered to be too high from a social perspective, then a lumpsum transfer from the exchequer seems to be called for.

References

Alberini, A., Eskeland, G.S., Krupnick, A. and McGranahan, G. (1996). "Determinants of Diarrhoeal Disease in Jakarta", *Water Resources Research* 32(7): 2259-69.

Anderson, D. and Cavendish, W. (1992). "Efficiency and Substitution in Pollution Abatement: Three Case Studies", World Bank Discussion Papers no. 186, World Bank, Washington D.C.

Bhandari, N. (1992). "The household management of diarrhoea in its social context : Study of a Delhi slum", Unpublished Ph.D. Thesis, CSMCH, Jawaharlal Nehru University, New Delhi.

Bhatnagar, S. and Dosajh, U. (1986). "Diarrhoeal disease morbidity in children below 5 years in urban slums in Delhi", National Institute of Health and Family Welfare, New Delhi.

Briscoe, J. (1984). "Intervention studies and the definition of dominant transmission routes", *American Journal of Epidemiology* 120: 449-55.

Esrey, S., Feecham, R.G. and Hughes, J.M. (1985). "Interventions for the control of diarrhoeal diseases among young children: Improving water supplies and excreta disposal facilities", *World Health Organisation, Bulletin no. 63(4)*: 757-72.

Esrey, S.A., Potash, J.B., Roberts, L. and Shiff, C. (1991). "Effects of improved water supply and sanitation on ascariasis, diarrhoea, dracunculiasis, hookworm infection, scchistosomiasus and trachoma", *World Health Organisation Bulletin no.*69(5): 609-21.

Panda, P. K. (1996). "The effects of safe drinking water and sanitation on diarrhoeal diseases among children in rural Orissa", in K. Pushpangadan, (ed), *Proceedings National Seminar on Rural Water Supply and Sanitation*, Centre for Development Studies: Trivandrum.

12

Industrial Pollution and its Impacts on Soil, Water, Crop, Livelihood Status and Ecology

A Micro-Level Investigation in Vellore District of Tamil Nadu

C. SEKAR

Introduction

Leather industries in Tamil Nadu contribute about six per cent to the total of the world's leather production. These industries provide total employment to more than one million people (Ramasamy, 1997). It is understood that the leather industries let out huge quantities of effluents into the nearby Common Property Resources (CPRs), which affect the land and water sources seriously. George (1984) found a decrease in the fertility status of soils due to the irrigation of land by tannery effluents. Singaram in 1992 observed a reduction in the micro-nutrient status with an increasing concentration of effluents in the soil. According to Mahimairaja *et al.* (1997), the concentration of total chromium in bore well water from selected sites in Vellore district of Tamil Nadu ranged from 178 to 990 mg/kg. Teekaraman and Ahmed (1982) in their study observed poor germination of paddy, bajra, sorghum and sugarcane, when irrigated with tannery effluents. Rao and Nandakumar (1981) revealed that crop yield had declined when irrigated with effluents from tanneries. All these studies have reiterated the negative externalities imposed by the leather factories on the basic natural resource systems, *viz.*, land and water bodies. Due to the release of effluents directly into the Palar river (CPR), the well water (Private Property Resource—PPR) in most parts of the district has been inexorably polluted. Besides about 2500 hectares of land are reported to be severely affected and another 12000 hectares are being moderately affected in Vaniyambadi, Ambur, Vellore, Wallajah and Ranipet taluks of Vellore district.

The present investigation was undertaken, (*i*) to study the extent of dependence on common village lands and water bodies by the villagers, (*ii*) to analyse the externality impacts of industrial effluents on CPRs, cropland, water, crop production environment, livelihood status of the village inhabitants and the micro-economic behaviour of the village inhabitants depending on CPRs, and (*iii*) to estimate the respondents' Willingness-To-Pay (WTP) for replenishing the original status of the river and tank CPRs and the likely policy implications and strategies to be adopted for micro-level planning.

Data and Methodology

Karugambathur village in Vellore district of Tamil Nadu was purposively selected for the study. In this village the deterioration of the Palar river CPR and its consequent impact on crop land, irrigation and drinking water, animal production, labour employment, farm income, etc. were significantly pronounced. The study was conducted during 1999-2000. Both primary and secondary sources of information were used for the study. Three different categories of respondents, *viz.*, (*i*) farmers, (*ii*) non-farmers, and (*iii*) agricultural labourers were selected purposively. Owing to time and financial constraints, for each category, only 50 respondents were chosen. The households were categorized into three group sizes since the impact of industrial pollution on village inhabitants was not uniform across categories.

Damage Functions

There were two types of damage functions observed in

agriculture, *viz.*, agricultural damage function and aggregate value damage function (for details, refer Cropper and Oates, 1992). For the present study, both agricultural and aggregate value damage functions were developed and employed. While the agricultural damage function was limited to agricultural damages alone, the aggregate damage function has considered loss in productivity of cropland and labour, crop output, livestock production, change in quality of water, damage on human and animal health etc. Agricultural and aggregate value damages were computed as the differences in the value of each affected farm over the value of the each item before the occurrence of such damages. The agricultural and aggregate damage functions were estimated by employing log-log regression models separately for all the categories. The damage functions used for the study were of the following form:

CROP OUTPUT = f {AVERTLIE, LANDQUALI, WATQUALI, PCLCPR}

AGDMGE = f {FAVERTLE, AVERTDW, AVERTEXHH, AVERTEXAH, WATQUALI, PCLCPR}

where,

CROP OUTPUT	—	Yield damage (Rs./ha)
AGDMGE	—	Aggregate damage (Rs./household)
PCLCPR	—	Proximity of cropland to local CPRs (river/tank/forest)
AVERTLIE	—	Averting input expenditure for land (Rs./ha)
LANDQUALI	—	Land Quality Index*
WATQUALI	—	Water Quality Index*
FAVERTLE	—	Averting input expenditure for land (Rs./farm)
AVERTDW	—	Averting input expenditure for drinking water (Rs./family)
AVERTEXHH	—	Averting expenditure for human health (Rs./family)
AVERTEXAH	—	Averting expenditure for animal health (Rs./family)

* *Poor-1; Average-2; Good-3*

Hedonic Model

A hedonic model was chosen to study the influence of CPR degradation and its effect on the characteristics of cropland in deciding the final value of croplands. Miranowski (1984), Bartic (1987) and Palmquist (1989) in their studies have demonstrated the application of hedonic techniques to value croplands, cropland sales and land improvements, respectively. The model used for the present study was of the following form.

VCLN = f {LQI, WQI, PCLCPR, CCPRACL, SCPRITIE, SCATFA}

where,

VCLN	—	Value of cropland (rupees/ha)
LQI	—	Land Quality Index (scale*)
WQI	—	Water Quality Index (scale*)
PCLCPR	—	Proximity of croplands to local CPRs (distance in kilometre)
CCPRACL	—	Condition of CPRs adjoining to cropland (scale*)
SCPRITIE	—	Share of CPR input contribution to total crop input expenditure (percentage)
SCATFA	—	Share of cropped area to total cropland area (percentage)

* *Poor-1; Average-2; Good-3.*

Contingent Valuation Technique (CVT)

An attempt was made to examine the economic valuation of the CPRs, *viz.*, river and tank using CVT. WTP was used to assess the value of changes in the quantity and quality of CPRs in the study area. The survey schedule was designed in such a way that it did gauge the knowledge and attribute of the respondents towards CPR degradation, their perception and interest in the protection of the village CPRs. Information was also elicited to study the dependence of households on local CPRs for employment, income, irrigation and drinking water. The manure, fuel and fodder, problems faced by them in the collection of water, fuel, tank silt and fodder, as also measures, if any taken by the users when shortage occurred and their WTP for the protection of local CPRs. In order to get responses for the WTP, the following hypothetical situation was created. The households were asked to assume that some local agency manages the village CPRs. It was explained to the respondents that the new system would be adequate enough to meet the needs of the inhabitants. The capital cost of the project was taken to be Rs. 5000 for farmers and Rs. 700 for non-farm and agricultural labourers, while the cost of maintenance and running expenditure to be involved were assumed to be Rs. 1200 for farmers and Rs. 360 for non-farm and agricultural labourers. A part of the expenditures of the new CPR management system was said to be met by the government. For meeting the remaining expenses of the new systems of managements, each household's WTP towards its capital and maintenance cost in monetary terms as well as labour time was sought for. The WTP of the households towards capital and maintenance cost of the new system in money as well as imputed cost of labour time estimated in rupee terms were added after annualizing the capital investment.

The capital cost was annualized by taking a lifetime of 15 years for the capital investment at a 12 per cent rate of interest. Similarly, using the per day wage rate, as prevalent in the study area during the time of survey, the labour time or days the villagers were willing to contribute towards the project, was monetized. Hence WTP for good quality water, land based inputs like green leaf manure, fodder and ecological functions was the sum total of WTP for annualized capital and maintenance and WTP in labour days estimated in monetary terms. The explanatory variables used for analysing maximum WTP were age of the respondent, household composition, educational level, per capita agricultural land, per capita standard cattle unit, interest in protection of CPRs, perception of CPR degradation, and distance of CPRs locations from the location of the farm, etc. The structural model of the function was of the following form:

WTP	=	{AGE, HHSIZE, HHEDN, HHINCOME, PCAL, PCSCU, IPCPR, PCPRD, PCLCPR},
where		
WTP	—	Willingness-to-pay (Rs./household)
AGE	—	Chronological age of the respondent (in years)
HHSIZE	—	Size of the household (number)
HHEDN	—	Household head's education (scale)*
HHINCOME	—	Household income (Rs./annum)
PCAL	—	Per capita agricultural land (in hectare)
PCSCU	—	Per capita standard cattle units (number)
IPCPR	—	Interest in protection of village CPRs (scale) **
PCPRD	—	Perception of CPR degradation (scale) ***
PCLCPR	—	Proximity of cropland to local CPRs (in km)

* — *Illiterate-0; Primary -1; Secondary -2; Collegiate -3.*
** — *Not interested -1; Somewhat -2; Fairly -3; Very much -4.*
*** — *Very much -1; Somewhat -2; Not at all -3.*

Non-availability of an adequate period of farm employment, poor crop stand and livestock income, reduction in net farm and non-farm income, land sales, diversion of farm lands for non-agricultural activities, labour mobility towards non-farm sectors and poor economic status were the most important consequences of pollution externalities observed in the study village.

Major Findings of the Study

Land Holding Details

Land area details of the sample are presented in Table 12.1. The average area of cultivation by farmers is found to be only 0.15 ha. This might be due to a fragmentation of land holdings, industrialization (presence of tanning factory in the study area), etc.

Table 12.1: Land holding details

Sl. No	Particulars	1999-2000 Average Land Holding (ha)	1999-2000 Net Area cultivated (ha)
1.	Wet land	12.10 (0.24)	2.17 (0.04)
2.	Dry land	13.55 (0.27)	2.35 (0.04)
3.	Garden land	5.30 (0.10)	3.03 (0.06)
	Total	30.95 (0.62)	7.55 (0.15)

Note: Figures in parentheses indicate the per household possession of land holding.

Sources of Irrigation

The river Palar, open well and a non-system tank are the primary sources of irrigation in the village. There are three canals in the village. The average length of canal is 20 km. It is implicit from Table 12.2 that the total number of abandoned wells in the sample farms has been high. This is due to the non-availability of good quality water in shallow depth and an improper maintenance of the water table by poor conservation and abatement techniques. It is understood that the quality of water has been affected due to the presence of tanneries in and around the study village.

Livestock Status

Due to contaminated water in the river Palar, non-availability of community grazing lands, etc. the livestock possessed by farmers community is found to be low. The effluents let out into the river have also affected the domestic and irrigation wells (PPRs), which has altered the basic characteristics and the fertility status of the native soils. Previously, farmers used to rear bullocks for draught purposes but now their dependence on animals has been reduced because of the non-availability of fodder, restricted agricultural activities and also partly due to farm mechanization. The high overhead cost of maintaining bullocks and engaging them only for a four-to-five month period while feeding them throughout the year without a farm level availability of fodder and CPR support were found to be the major factors that restricted the large ruminant population.

Table 12.2: Details of well irrigation

Sl. No.	*Particulars*	*1999-2000*
1.	Total number of wells in the sample farmers	40
2.	Depth of open wells (in metres)	9.78
3.	Percentage of wells in use	14.30
4.	Percentage of wells abandoned	85.70
5.	Water quality (per cent)*	
	Good	—
	Medium	13.0
	Poor	87.0

* As perceived by the respondents.

With land and water deterioration, farmers had to spend greater amounts on purchasing fodder from outside. Farmers have also engaged themselves in non-farm activities, *viz.*, beedi-making (*rolling of tobacco leaves*) and pay only secondary importance to agricultural occupations. The increase in goat population was primarily due to the greater availability of area under rain fed cultivation and agricultural fallow lands. Hence farmers now allow their goats to graze on croplands, which were once the promising fields for bumper crop output.

Sources of Animal Grazing

The respondents have to depend on many sources for animal grazing. But the extent of dependence on various sources is determined by the category to which they belong to. Farmers who possess croplands, have a lessen extent of dependence on

CPRs. In fact the use of community land has been less than 50 per cent, which indicates a shrinkage of the CPR status (*see* Table 12.4). Since non-farmers and agricultural labourers did not possess land of their own, they had to depend entirely on the CPRs (Poramboke lands, tank bunds and river banks) to meet part of their animal feed requirements. In addition, they had also been dependent on other sources, *i.e.* getting their animals grazed in other farmer's land/field bunds, etc. The decline in usage during recent times might be attributed to the deterioration and degradation of river and tank CPRs. The village CPRs (tank and river) are also not in good condition as the available water has been heavily polluted by the nearby leather factories.

Table 12.3: Availability of livestock (*in numbers*)

		1999–2000		
Sl. No.	*Category*	*Farm*	*Non-farm*	*Agricultural labourer*
1.	Bullock	20 (0.40)	—	—
2.	Buffalo	11 (0.22)	3	2
3.	Local cow	39 (0.78)	10 (0.20)	—
4.	Cross bred cow	46 (0.92)	9 (0.18)	—
5.	Goat	40 (0.80)	—	11 (0.22)
6.	Poultry	94 (1.88)	20 (0.40)	6

Note: Figures in the parentheses indicate the possession of livestock per household.

Area and Productivity

The area and productivity of different crops grown in the study area are reported in Tables 12.5 and 12.6. It is obvious to note that there has been a gradual decline in the area and production of major crops, namely paddy, sorghum, sugarcane and tomato. The reason for decline in area is due to the effluents discharged from leather factories. The polluted water has also spoiled the quality of well water (PPR), which in turn reduced the crop area and yield. The gross cropped area was only 8.57 ha in 1999-2000.

Table 12.4: Sources of animal grazing

Sl. No.	*Particulars*	*Sources*	*1999-2000*	
			No.	*Per cent*
1.	Farmer	i. Own land	36	72.00
		ii. Own land + Tank and river CPRs	14	28.00
2.	Non-farmer	i. Tank and River CPRs	29	58.00
		ii. Other sources	21	42.00
3.	Agricultural labourer	i. Tank and River CPRs	29	58.00
		ii. Other sources	21	42.00

Table 12.5: Gross cropped area (*in hectares*)

S.No.	*Name of the crops*	*1999-2000*
1.	Paddy	2.51
2.	Ragi	0.48
3.	Sugar cane	1.09
4.	Groundnut	0.24
5.	Tomato	0.23
6.	Coconut	1.03
7.	Agathi	2.99
	Gross cropped area	8.57
	Gross irrigated area	3.60
	Cropping intensity (per cent)	113.50

It is reported that the average yield of paddy was 3575 kg/ha during early eighties and it declined to 2976 kg per ha during 1999-2000. The average yield of ragi has also decreased drastically. One important feature to be mentioned here is that ragi in spite of its saline resistant character, has not been cultivated due to the poor quality of water and also partly due to a change in food consumption behaviour. The low cropping intensity is obviously due to the conversion of cropland for non-agricultural purposes and also due to the occupation of more area under fallow agricultural lands. So the farmers are unable to make a considerable profit from agriculture. The area under irrigation to total cropped area has also declined by more than 30 per cent due to a deterioration of water, in wells in the Palar river as well as of the community tank. It is also implied that the *agricultural land value* has declined by more than 25 per cent due to the deterioration of soil and water qualities by leather factory effluents.

Table 12.6: Average productivity of crops

(kg per hectare)

S.No.	*Name of the crops*	*1999-2000*	*1980s*
1.	Paddy	2976	3575
2.	Ragi	1798	2670
3.	Sugar cane (tonnes)	38.15	46.58
4.	Groundnut	1746	2015
5.	Tomato	4591	7185
6.	Coconut (nos.)	8050	12513
7.	Agathi	2120	2850

The pollution of river Palar has also created changes in employment opportunities. The results presented in Table 12.7 indicate the poor cropland income. The declivity in cropland area since eighties was observed to be 79.50 per cent. This has brought a huge decline in crop income of Rs. 6820 per hectare. The foremost reasons attributed for a decline in crop income were change in cropping pattern and low crop productivity on account of pollution externalities.

Cropland and Drinking Water-Based Averting and Defensive Expenditures

Leather factory effluents have forced the farmers to undertake averting/defensive activities and expenditure on cropland and irrigation water. Table 12.8, the pragmatic steps the farmers took by spending a major proportion for soil amendments, *viz.*, application of organic manure, which amounted to Rs. 642.30 (54.74 per cent of total expenditures). Farmers also applied gypsum and tank silt for internalizing the externalities. Expenditures involved in getting drinking water from non-polluted areas amounted to Rs. 980 per annum (only implicit value for labour involved in fetching water from unaffected area was considered).

Table 12.7: Status of cropland and loss of crop income

Sl. No.	*Particulars*	*Observed changes*
1.	Total land area of sample farmers	30.95 ha.
2.	Cropland area of sample farmers	8.57 ha.
3.	Percentage of cropland	24.39 %
4.	Loss in crop income (Rs./ha/yr) due to pollution externalities	6820

Table 12.8: Cropland and drinking water-based averting and defensive expenditures

Sl. No.	*Particulars*	*Farm household*	
		Expenditure (Rs./ha)	*Percentage*
I.	Averting expenditure for land		
	(a) Additional cost of seed	65.00	5.54
	(b) Organic manure	642.30	54.74
	(c) Application of gypsum	126.00	10.74
	(d) Tank silt application	150.00	12.79
	(g) Inorganic fertilizers	190.00	16.19
	Total	1173.30	100.00
II.	Expenditure involved in getting drinking water from non-polluted areas (Rupees per household)	980.00	100.00

Externalities on Human and Animal Health (Expenditure in Rs./annum/household)

The effluents discharged from the leather factories not only impaired the quality of water and soil but also created anarchy in human and animal health. From the results exhibited in the Table 12.9, it is apparent that the farm category respondents had to spend a much higher amount than the other categories. The amount incurred was Rs. 684.55, Rs. 599.00 and Rs. 345.05, respectively, by the farm, non-farm and agricultural labourer categories. As farmers had been the direct users of polluted water for irrigating crop fields and feeding cattle hence they were the worst affected group. With respect to the deterioration of animal health, the farmers and non-farmers had incurred an expense of Rs. 728.45 and Rs. 598.90, respectively the farmers' incurring a larger expenditure.

Table 12.9: River CPR externalities on human and animal health

Sl. No.	*Particulars*	*Farmers (Rs.)*	*Non-farmers (Rs.)*	*Agrl. labourers (Rs.)*
1.	Incidence of human health disorder			
	Cost of medicine	262.38	249.75	226.30
	Physician cost	187.77	199.25	-
	Loss in time of employment (Rs.)	234.40	150.00	118.75
	Total	684.55	599.00	345.05
2.	Deterioration of animal health			
	Cost of medicine	487.20	379.15	79.85
	Physician cost	241.25	219.75	81.15
	Total	728.45	598.90	161.00
	Loss in value of livestock product (Rs./annum)	2150.00	755.00	281.00

Willingness-to-Pay and Changes in Employment Pattern

The willingness of the households to pay towards capital and preservation cost of the CPRs and for purchasing good drinking water from outside was estimated in terms of money. Some households expressed that income was a constraint disturbing their ability to pay. For such households, the respondents were asked whether they would be able to contribute in terms of voluntary labour or monetary terms. The average amount was calculated and the results are reported in Table 12.10. The farmers were willing to pay an average amount of Rs. 904 per annum. The non-farmers were willing to pay an average amount of Rs. 353 annum. In the case of agricultural labourers, comparatively a lesser amount of willingness-to-Pay (Rs. 182) was observed. It is also observed that some of the respondents had either changed their occupation or migrated to nearby areas due to the smaller returns in agriculture. The man-days of farm employment per year earlier and presently were calculated and reported (*see* Table 12.10). There is almost a three-fold decline in farm employment in the case of farmers and two-fold drop of employment for the non-farmers and three-fold reduction in employment for the agricultural labourers.

Table 12.10: Average WTP and changes in employment pattern

	Particulars	*Farmers*	*Non-farmers*	*Agricultural labourers*
1.	Willingness-to-Pay (WTP) (Rs./annum)	904	353	182
2.	Change in occupational pattern of the household due to the externalities			
	a. Earlier (man-days of farm employment/yr.) 1980s	168	17	144
	b. At present (man-days of farm employment/yr) (1999-2000)	57	9	49
	c. Percentage reduction in farm employment	66.07	47.06	65.97

WTP for Internalizing Externalities

The respondents were categorized based on their WTP. As data in Table 12.11 indicates the farmers were willing-to-pay greater amounts as they had been the worst affected category. Out of 50 farmers, a majority of 19 farmers had stated that they were willing to pay in the range of Rs. 500-1000 per annum. A majority of 40 per cent of non-farmers and 62 per cent of agricultural labourers were willing to pay only less than Rs. 250 per annum.

Attitude of Respondents towards Externalities

The externalities of leather factory effluents have created changes in the socio-economic orders of the inhabitants in the study village. It is observed that some farmers have migrated to other areas in search of employment and a small section of the population have disposed of at least a portion of their croplands during the last two decades as a result of poor agricultural operations. The results reported in Table 12.12 have implied that 20 per cent of the farmers migrated to other areas and 24 per cent sold at least part of their land and 48 per cent of agricultural labourers migrated out of agriculture in search of employment. The poor agricultural activities and problems due to industrial effluents were observed to be the key reasons for land selling and migration. Due to problems like high cost of fodder, non-availability of grass rich common grazing yards and poor water quality, farmers have reduced the livestock population significantly.

Table 12.11: WTP for internalizing externalities

Sl. No.	*WTP (Rs./annum)*	*Farmers*		*Non-farmers*		*Agricultural laboures*	
		No.	*%*	*No.*	*%*	*No.*	*%*
1.	<250	3	6.00	20	40.00	31	62.00
2.	250-500	5	10.00	12	24.00	7	14.00
3.	500-1000	19	38.00	2	4.00	—	—
4.	>1000	15	30.00	—	—	—	—
5.	Not willing to pay	8	16.00	16	32.00	12	24.00

Note: No. — Number, % — Percentage.

Estimates of Agricultural Damage Functions

The agricultural value damage function explains the value of agriculture related damages to averting expenditures of land, human and animal health and land and water quality indices. The estimated parameters of agricultural value damage functions are presented in Table 12.13. It is obvious from the table that the averting input expenditure for land and water quality indices and proximity of croplands to the local CPRs (PCLCPR) has influenced the farm-level agricultural damages significantly. It is implicit that a one per cent increase in land related averting expenditure *ceteris paribus* has decreased the agricultural damage by 0.64 per cent from the mean level. While the averting expenditure for land and proximity of river CPR were significant at the five per cent level, the land and water

Table 12.12: Attitude of respondents towards externalities on socio-economic condition

Details	*Farm*		*Non-farm*		*Agricultural labourers*	
	Number	*Percentage*	*Number*	*Percentage*	*Number*	*Percentage*
(a) Labour migration	10	20.0	—	—	24	48.0
(b) Land selling	17	34.0	—	—	—	—
Reason for land selling						
(i) Poor agricultural activities and off-farm income	6	33.3	—	—	—	—
(ii) Problem due to effluents discharged from industry	11	67.7	—	—	—	—
Total	17	100.0				
Reduction in livestock population	39	78.0	12	24.0	5	10.0
(i) Non-availability of CPR grazing land	17	44.0	7	58.0	3	60.0
(ii) Poor water quality	13	33.0	3	25.0	2	40.0
(iii) High cost of fodder	9	23.0	2	17.0	—	—

quality indices were significant at the one per cent level. The above facts and figures have clearly shown the intensity of the leather industry-related soil and water pollution in the study village.

Table 12.13: Estimates of agricultural damage function

Name of the variables	*Regression co-efficient*
Constant	8.852
	(1.517)
AVERTLIE	-0.712*
	(-1.947)
LANDQUALI	- 0.935**
	(-5.053)
WATQUALI	-1.012**
	(-3.765)
PCLCPR	-0.085*
	(-1.849)
R^2	0.642

Note: Figures in parentheses indicate t value; * significant at 5 per cent level; ** significant at 1 per cent level.

Results of Aggregate Damage Function (ADF)

The ADF was employed to study the influence of the explanatory variables, *viz*., FAVERTLE, AVERTDW, AVERTEXHH, AVERTEXAH, WATQUALI and PCLCPR on AGDMGE (value of farm level aggregate damage). The logic behind the selection of per farm ADF function was due to the inclusion of variables, *viz*., AVERTDW, AVERTEXHH and AVERTEXAH. A Log-log regression model was found to be the best choice for the analysis of the ADF variables. The parameter estimates of the model are presented in Table 12.14.

It is explicit from the results reported in Table 12.14 that the land-based averting expenditure was found to be negative. The estimates of the model have shown that one per cent increase in land-based averting expenditure would reduce the aggregate damage above mean level by 0.56 per cent, *ceteris paribus*. The AVERTDW, AVERTEXHH and WATQUALI were significant at five per cent level, while PCLCPR was significant at one per cent level. It could also be interpreted that about 74 per cent of the aggregate damage was explained by the variables included in the model. The preceding discussions have clearly demonstrated the necessity for undertaking some abatement measures apart from resource management activities at the farm level to curtail the aggregate farm level damage in the study village, where the extent of river CPR degradation and its resultant impact on soil, water and crop growth had been greatly pronounced.

Table 12.14: Results of the aggregate damage function

Name of the variables	*Regression co-efficient*
Constant	8.901**
	(4.239)
FAVERTLE	-0.561*
	(-2.049)
AVERTDW	-0.146*
	(-1.874)
AVERTEXHH	-0.034*
	(-1.953)
AVERTEXAH	0.109*
	(1.941)
WATQUALI	-0.309*
	(-2.031)
PCLCPR	-0.019**
	(-5.304)
R^2	0.738

Note: Figures in parentheses indicates t value; * Significant at 5 per cent level; ** Significant at 1 per cent level.

Hedonic Regression Model

The hedonic regression model elucidates the influence of cropland and other associated characteristics on land value. In the present study, the qualitative and quantitative characteristics, *viz.*, Land Quality Index (LQI), Water Quality Index (WQI) Proximity of Cropland to Local CPRs (PCLCPR), condition of CPRs adjoining to cropland (CCPRACL), cropping intensity (CI), share of CPR input to total crop input expenditure (SCPRITIE) and share of cropped area to total cropland area were expected to influence the land value and hence these variables were considered. The semi-logarithmic model was used for the estimation of the parameters in the model. Only farm category respondents were considered for running the hedonic regression. The results of the hedonic regression analysis are presented in Table 12.15.

Table 12.15: Estimates of the hedonic model

Name of the variables	*Regression co-efficient*
Constant	12.013** (6.267)
LQI	0.659* (2.056)
WQI	0.941** (3.685)
PCLCPR	0.071* (1.930)
CCPRACL	0.045* (1.902)
SCPRITIA	0.009 (0.807)
SCATFA	0.054* (1.969)
R^2	0.886

Note: Figures in parentheses indicate t value; * Significant at 5 per cent level; ** Significant at 1 per cent level.

In the study village, the land and water quality indices were found to have a significant role in deciding the land value. The LQI was significant at five per cent level, while WQI was significant at one per cent level implying the intensity of river CPR's degradation. The PCLCPR was positive, showing that the croplands adjacent to the river CPR had been fetching low values, *i.e.* the distance of cropland from the river (where the leather factory is located) if increased by one kilometre, *ceteris paribus*, increased the land value by seven per cent. The CCPRACL was also found to influence land value positively, which exhibited that the condition of the river CPR also played a major role in deciding the cropland value. The share of cropped area to total cropland area was also found to influence cropland price significantly at 5 per cent level. The analysis has also showed that about 89 per cent of the variations in land value had been taken care of by the explanatory variables included in the model.

The above findings clearly demonstrate the influence of CPR related qualitative attributes in deciding the cropland value. The result of the study is also in consonance with the outcome of the research conducted by Palmquist and Danielson (1989) and Miranowski (1984). In sum, the findings have ascertained the fact that qualitative parameters like the river and tank water quality, the nature of CPRs present in the vicinity of the cropland, etc., should be protected and given importance for getting a higher land value.

Factors Influencing WTP

The factors influencing the WTP of the sample respondents were analysed using the constant elasticity model for all the three categories across the study zones. The explanatory variables used in the analysis included age (AGE), family composition (HHSIZE), education (HHEDN), income of the household (HHINCOME), per capita agricultural land (PCAL), per capita standard cattle unit (PCSCU), interest in protection of village CPR (IPCPR), perception of CPR degradation (PCPRD) and proximity of crop lands from household to local CPR. The regression results for the aggregate sample categories across categories are presented in Table 12.16.

Estimates obtained through regression for the farm, non-farm and agricultural labour categories have shown the expected signs. The HHSIZE co-efficient was significant and positive for all the categories, implying that as the number of family members increased, the dependency for water, fodder, fuel also increased which results in an enhanced WTP. One per cent increase in HHSIZE *Ceteris paribus* increased the WTP from the mean level by 0.11 per cent. HHEDN was yet another important variable considered. It was expected that a person with higher educational qualification would be more concerned with the protection of river CPR. The HHEDN co-efficient was significant and positive for farm and non-farmers. As expected, the variable HHEDN was not significant for the agricultural-labourer category. The HHINCOME was positive and significant for non-farm and agricultural labourer categories. The PCAL which showed a positive and significant co-efficient indicates that as the per capita farm size has increased, the WTP would also increase because of pollution externality and enhanced dependence of river CPR by this category of respondents. The PCSCU was found positive and significant only for the agricultural labourer, which indicated the greater dependence of CPRs by this category for rearing cattle and also to meet part of the fodder requirements of cattle. The IPCPR was positively significant, implying the interest shown by the farmers in conserving the river CPR, which had been one of the vital sources of input for agricultural activities. The PCPRD variable was significant only for non-farm respondents. The PCLCPR was negative but significant for farm and agricultural labourers, indicating that as the distance increased, the WTP had also declined since the people living very close to the river CPR were realizing more benefits (or externalities ?) rather than the respondents living away from the river CPR.

CPR Management, Public (Institutional) Failures and Erosion of Local CPRs

It is unfortunate to note that other agencies than the users were the source of externality problems associated with the management of village CPRs. Upon discussion with the respondents, it is understood that the traditional systems of management of village CPRs had distributed the benefits equally and ensured a sustainable exploitation of the existing resources in an environmentally benign manner. The respondents stated that the modern system of government maintenance of village CPRs broke down the community system of resource management and increased the dependence of the government. The absence of proper monitoring and control mechanisms at the grassroot level had indirectly encouraged the leather industries to let out the effluents in the river Palar, which had been the lifeline of the villagers, supplying water not only for irrigation but also for household applications.

Table 12.16: Factors influencing WTP-results of contingent valuation models

Name of the variable	*Partial Regression co-efficient*		
	Karugambathur		
	Farmer	*Non-farmer*	*Agricultural labourers*
Constant	6.795**	5.905*	5.194**
	(2.658)	(1.865)	(2.687)
AGE	0.051	0.048*	0.584
	(1.023)	(1.935)	(0.256)
HHSIZE	0.105*	0.063*	0.198**
	(1.759)	(1.810)	(3.125)
HHEDN	0.042*	0.116*	0.018
	(2.078)	(1.871)	(0.853)
HHINCOME	0.069	0.035*	0.087**
	(0.450)	(2.005)	(3.958)
PCAL	-0.041**	—	—
	(2.912)	—	—
PCSCU	0.031	0.031	0.095**
	(0.645)	(0.167)	(2.538)
IPCPR	0.043*	0.142	0.035*
	(1.862)	(0.056)	(1.865)
PCPRD	0.005	0.023*	0.081
	(0.665)	(2.145)	(0.712)
PCLCPR	-0.081*	-0.106	-0.128*
	(-1.904)	(-0.043)	(-2.012)
R^2	0.716	0.735	0.681

Note: Figures in parentheses indicate t value; * Significant at 5 per cent level; ** Significant at 1 per cent level.

CPR-PPR Inter-relationships

CPR input requirements and output flows are qualitatively, functionally and temporally different from those of the PPRs. This forms the basis of complementarities of the production systems of CPRs and PPRs. The critical examination of these aspects has highlighted the complementarity effects between CPRs and private farm. CPR-land-water management has remained a complex challenge in the study village. This challenge has involved among other things, the distribution of benefits of CPRs in terms of water, silt, sand, fodder, fuel, etc., and ensuring the appropriate use. The degraded river CPRs have changed the water quality of the private wells, besides affecting the soil health and village ecosystem by creating human and animal health disorders. Due to CPR degradation, the private cultivated land to total land area is found to be less than 25 per cent, which was mainly due to the discharge of tannery effluents. CPR degradation due to industrial population has ultimately led to reduction in animal population, declivity of crop area, yield and also reduction in farm labour employment.

Conclusion and Policy Implications

The outcome of the study underlines the fact that the area and the yield of important crops have dwindled over the years due to the discharge of effluents from leather tanning factories, which spoiled the quality of river and tank CPRs as well as the well water. It is also evident from the foregoing analysis that the pollution due to industrial effluents has altered the cropping pattern, employment potential, income distribution and has therefore created social instability. The huge averting expenditure incurred by the respondents towards human and animal health disorders for internalising externalities has unambiguously stated the seriousness of the existing external effects. It is also understood that the intensity of soil and water pollution is significantly high being reflected by the significance of land and water quality indices. The results reveal the following key policy issues for future planning and adoption:

(*a*) Installation of effluent treatment plants should be made mandatory in the leather factories to reduce the externalities and the non-compliers should be dealt with severely to reduce the degradation of river CPRs, which are considered to be the life-line for agricultural production and rural prosperity.

(*b*) The polluter paid principle should be advocated so as

to internalize the external effects based on the actual damage caused to the human, animal and other natural resource endowments including CPRs and local eco-system. To achieve this, a well-netted enforcement institution should be developed.

(c) It is strongly advocated that the CPRs should be protected so as to enhance the production potential of private resources as it is evident that the degradation of one resource (CPRs), would drastically influence the other (PPRs) in many ways. The impact of influence is either physical or biological or socio-economical or environmental or in combination. Hence the future policy should be oriented towards simultaneous strengthening and development of both the CPRs and PPRs productivity to ensure complementarity, maintain social harmony and to attain environmental sustainability.

(*d*) As far as possible, at least a portion of the resources for the initial investment to be made for internalizing the externalities should be collected from the villagers or actual beneficiaries to get the people totally involved in the system and make the project sustainable. The exact details and modalities of financing and cost recovery should be left to the community institutions themselves. The local communities must also be made to get themselves involved at all stages of planning, execution and implementation.

(*e*) The principal role of managing the village CPRs should be vested with the communities and the role of the government should be kept at a minimum. The government institutions' involvement should be restricted to the extent of providing guidance for community participation not on community governance, *i.e.* the government's role should be through promotion and extension, like the counselling unit.

References

Bartik Timothy, J. (1988). "Evaluating the Benefits of Non-Marginal Reductions in Pollution Using Information of Defensive Expenditure", *Journal of Environmental Economics and Management*, 15: 111-127.

Cropper, Maureen L. and Wallace, E. Oates, (1992). "Environmental Economics: A Survey", *Journal of Agricultural Literature*, 30 (2): 675-740.

Mahimairaja, S., S. Sakthivel, K. Ramasamy, P. Thangavel and R. Naidu (1997). "Chromium Contamination in Soil and Groundwater due to Disposal of Tannery Wastes", Paper Presented During *National Symposium on Environment V1*, held at Tamil Nadu Agri. Univ., Coimbatore, 7-9 January.

Miranowksi, J.A. and Brian D. Hammes, (1984). "Implicit Prices of Soil Characteristics for Cropland in Lowa", *American Journal of Agricultural Economics*, 66: 745-49.

Palmquist, B. Raymond and Leon E. Danielson, (1989). "A Hedonic Study of the Effects of the Erosion Control and Drainage on Cropland Values", *American Journal of Agricultural Economics*, 7 (1): 55-62.

Ramasamy, K. (1997). "Treated Tannery Effluent Related Pollution on Land and Water Ecosystem in India". Proceedings of the Extended Abstracts from the *4th International Conference on the Biochemistry of Trace Elements* held at University of California, Berkeley, USA: 1. 771-12.

Rao, Guruprasad M. and N.V. Nandakumar (1981). "Analysis of Irrigation Reservoir Contaminated by Tannery Effluents", *Indian J. Envir. Health*, 23: 239-41.

Singaram, P. (1992). "Study on Tannery Effluent Collected from Various Sources and its Effect on Soil Physico-chemical Properties". *Report of Research Project Submitted to Tamil Nadu Agriculture University*, Coimbatore.

Teekaraman, G. and N. Farooque Ahmed (1982). "A Study on the Influence of Tannery Effluents on Soils, Water and Crops in North Arcot District". *Special Report No. 7*. Soil Survey and Land Use Organization, Vellore.

13

Water Use in Indian Industry

Estimates of Value of Water and Price Elasticity of Demand

BISHWANATH GOLDAR

1. Introduction

It is widely recognized that an appropriate pricing of water for Indian industries, coupled with effluent charges, can lead to substantial savings in industrial water consumption. This is further possible through an application of water recycling and water conservation technologies,[1] which are important for the sustainable use of water resources in the country.[2] However, notwithstanding the significance of this matter, there has been very little econometric research on water use in Indian industries.[3,4]

It is unnecessary to emphasize that an econometric analysis of industrial water demand can provide useful inputs for the formulation of water pricing policy for industries. Clearly, for taking a decision on the water price to be charged to industries, one needs to take into account the price elasticity of demand. One has also to consider the contribution that water makes to industrial production, *i.e.* the value of water in industrial use. Estimates of these and other such parameters, useful for pricing policy formulation, would be provided by econometric research on water use in industries.

Attention may be drawn here to another important aspect

of water pricing policy for industries, namely the pricing of water for cost recovery by the water supply agencies. Inasmuch as the poorer households are often unwilling or unable to fully pay the supply cost of their water requirements, one may have to turn to industry (and commercial establishments) for revenue generation (cross-subsidization) so as to meet the cost of water supply systems. This raises the question: how much is the industry willing to pay for water? While there have been a number of studies in the Indian context (and in the context of other developing countries) on the willingness of households to pay for water, often using *contingent valuation techniques*, there has hardly been any study on the industries' willingness-to-pay for water. The contribution that water makes to industrial production, that is, the value of water in industrial establishments, will determine industrial establishments' willingness-to-pay for water. Evidently, an econometric analysis of industrial water use can help in estimating the value of water to industries and this in turn can provide an estimate of the industries' willingness-to-pay for water.

In view of the above, this paper addresses itself to the following two issues: (1) what is the value of water for the industrial sector in India, and (2) what is the price elasticity of demand for water in Indian industries. These obviously have important implications for the water pricing policy for Indian industry.

For the valuation of water for Indian industries, a marginal productivity approach is taken in this study, based on an estimated production function. Earlier this has been done by Wang and Lall (2001) in a recent World Bank study on the valuation of water for Chinese industries. To get an estimate of the price elasticity of demand for water in Indian industries, a demand function for water has been estimated, as several earlier studies on industrial demand for water have done.

The next section describes the models used for the econometric analysis. The data source used for the study and the measurement of variables are discussed in Section 3. The empirical results are presented in Section 4. The main findings of the study are given in Section 5.

2. The Models

A four-input production function is used for the analysis, in which capital, labour, energy and water are taken as the four inputs. Accordingly, output is defined as the value of production minus intermediate inputs, except value of energy and water used.[5] This is basically the value added production function extended to include energy and water along with the primary inputs.[6]

Two specifications of the production function have been used: the Cobb-Douglas function and the translog function. Let Y denote output, L labour, K capital, E energy and W water. Then, using the Cobb-Douglas functional form, the production function may be written as:

$$\ln Y = \alpha + \beta_L \ln L + \beta_K \ln K + \beta_E \ln E + \beta_W \ln W \qquad (1)$$

And, using the translog functional form, the production function may be specified as:

$$\ln Y = \alpha + \beta_L \ln L + \beta_K \ln K + \beta_E \ln E + \beta_W \ln W$$

$$+\beta_{LL} [½ (\ln L)^2] + \beta_{KK} [½ (\ln K)^2]$$
$$+ \beta_{EE} [½ (\ln E)^2] + \beta_{WW} [½ (\ln W)^2] +$$
$$\beta_{LK} (\ln L)(\ln K) + \beta_{LE} (\ln L)(\ln E)$$
$$+ \beta_{LW} (\ln L)(\ln W) + \beta_{KE} (\ln K)(\ln E) +$$
$$\beta_{KW} (\ln K)(\ln W) + \beta_{EW} (\ln E)(\ln W) \qquad (2)$$

For estimating econometrically the two equations (1) and (2) above, a random error term is added to them. Both have been estimated by the Ordinary Least-Squares (OLS) technique.[7]

From the estimate of the production function, the marginal product of water is derived in the following way. In the first step, the elasticity of output with respect to water, denoted by σ, is obtained. For the Cobb-Douglas production function, the elasticity is given by the coefficient of water in equation (1), *i.e.*

$$\sigma = \beta_W \quad (3)$$

For the translog production function, the elasticity of output with respect to water is obtained as:

$$\sigma = \beta_W + \beta_{WW} (\ln W) + \beta_{LW} (\ln L) + \beta_{KW} (\ln K) + \beta_{EW} (\ln E) \quad (4)$$

In this case, the elasticity is not constant, but varies with the level of different inputs.

Given the elasticity of output with respect to water, σ, the marginal product of water (MP_W) is obtained as:

$$MP_W = \frac{\partial \ln Y}{\partial \ln W} \times \frac{Y}{W} = \sigma \frac{Y}{W} = \frac{\partial Y}{\partial W} \quad (5)$$

Thus, the ratio of Y (output) to W (quantity of water used) is computed from the production and input use data of industries, and this is multiplied by s to get the marginal productivity of water.

Price Elasticity

For studying the price response of water demand, a water demand function for Indian industries has been estimated. Water use in industries has been taken as a function of output and input prices.

For estimating the water demand function, a double-log functional form has been adopted. The explanatory variables are output, the price of water and prices of other inputs. The demand function used for the analysis may be written as:

$$\ln W = \phi + v \ln Y + \mu_W \ln P_W + \mu_L \ln P_L + \mu_K \ln P_K + \mu_E \ln P_E + \mu \quad (6)$$

In.this equation, P_i is the price of the i'th input and μ is the random error term. The equation is estimated by regressing ln (W) on ln (Y) and the logarithm of prices of inputs.[8]

It should be pointed out here that the above equation does not correspond to the translog production function given in

(2) above. Nor does it arise from a translog cost function. This form is, however, convenient to estimate and interpret. Also, the double-log functional form has been used for estimating water demand function in several earlier studies.

3. Data and Measurement of Variables

The basic data for the study have been drawn from the *Annual Survey of Industries* (ASI), 1995-6. ASI data are available at present also for 1996-7 and 1997-8. But, as detailed data on water and energy consumption are available only for 1995-6. ASI data for this period was chosen for the study. Thus, the estimation of the production function and water demand function has been done from industry-wise cross-section data for one year.

It should be pointed out here that the ASI data relates to the organized industrial sector. It covers factories employing 10 or more workers with power or 20 or more workers without power. Consequently, very small units, particularly the household units, are not covered under ASI. Yet, an analysis of industrial water use based on ASI is useful because factories covered under ASI account for a dominant part of the aggregate industrial production and also of the aggregate water consumption by the industrial sector.

Estimation of the production function and water demand function has been done from cross-section data for 165 three-digit manufacturing industries. ASI results for 1995-6 are available for 177 three-digit manufacturing industries. Of these, in 12 cases, water consumption (purchase) is reported to be zero. These industries have therefore been excluded from the analysis. For each industry, the reported figures on output and inputs (persons working, capital invested, etc.) have been divided by the number of factories. This gives, for each industry, the output and inputs of a typical factory. These enter as observations to be used for the estimation of production function/water demand function.

The total number of persons working (including working proprietors) is taken as the measure of labour input, and fixed capital stock as the measure of capital input. To get a measure of energy input, the quantities of coal, firewood, fuel and electricity used in the industries have been expressed in coal replacement units and then combined. As regards the fourth input, namely water, some difficulty occurs in using ASI data. ASI does not report water consumption — it reports water purchased by the factories. In the absence of water consumption data, the data on water purchased by factories had to be used for the estimation of the production function. Needless to say that this is a limitation of the study. Since "water purchased" is used as an explanatory variable in the production function, instead of "water consumed", this may affect the estimate of elasticity of output with respect to water, and may lead to an overestimation of the marginal productivity of water. Therefore, while working out the estimates of marginal productivity of water, some adjustment has been made to make up for the difference between consumption and purchase of water by industries.

Turning to the measurement of variables for an estimation of the water demand function, water used (purchased) per factory is taken as the dependent variable while output per factory is taken as one of the explanatory variables.[9] These variables are the same as those used for the estimation of the production function. The other explanatory variables are the prices of inputs. ASI does not provide data on the prices of inputs directly, but proxies for the prices have to be computed or derived from the data reported on the quantity and value of inputs used. Thus, the reported figure on emoluments has been divided by the reported figure on the number of persons working to obtain emoluments per person, and this ratio has been taken as the price of labour input. The value of water purchased has been divided by the quantity of water purchased to compute the unit value of water purchased, and this has been taken as the price of water.[10] It has been mentioned above that the quantities of various types of energy inputs have been expressed in coal replacement and then combined to get a measure of energy input. The value of energy inputs used by an industry has been divided by the quantity of energy used (expressed in coal replacement) to compute the price of energy input. From the available data, it is not possible to work out a good measure of the price of capital input. Therefore, the water demand function has been estimated without the capital price variable in the equation. The exclusion of the capital price variable may have affected to some extent the estimates of the water demand function. But, there are no reasons to expect a strong correlation (across industries) between the price of capital input and the price of water. Therefore, the exclusion of the price of capital input from the water demand function may not have caused a serious bias in the estimate of the coefficient of the water price variable, and therefore in the estimate of price elasticity for the demand for water.

4. Empirical Results

Preliminary analysis of the data reveals wide inter-industry variation in water consumption per factory as well as the ratio of water consumption to output, invested capital and employment. In about 60 per cent of the industries, water consumption (purchase) per factory is less than five million litres per year. Taking all the 165 industries together, the average water consumption (purchase) per factory is about 26 million litres per year. Of the 165 industries, 29 may be regarded as relatively high water consumption industries. In these industries, water consumption (purchase) per factory is more than 20 million litres per year. These industries account for 88 per cent of total industrial consumption of water. Among these 29 industries, there are 11 industries in which the level of water consumption (purchase) is over 100 million litres per factory per year. These include (*i*) fertilizers and pesticides, (*ii*) plastic in primary forms and synthetic rubber, (*iii*) man-made fibres, (*iv*) petroleum refinery products, (*v*) iron and steel in primary and semi-finished forms, (*vi*) paper, and (*vii*) sugar.

Water consumption (purchase) per employee ranges from less than 10 kilolitres per person per year in about one-sixth of the industries to over 500 kilolitres per person per year in about one-tenth of the industries. The average is about 320 kilolitres per person per year.

The value of output per kilolitre of water consumed (purchased) varies widely across industries. The average for all the 165 industries taken together is about Rs. 760 per kilolitre of water purchased. But, there are industries in which the relevant ratio is below Rs. 100 per kilolitre of water. On the other hand, there are industries in which the ratio is above Rs. 100,000 per kilolitre of water purchased. For about 80 per cent of the industries, the relevant ratio falls in this range of Rs. 500 to Rs. 27,000 of output per kilolitre of water purchased.

Estimates of Value of Water

Table 13.1 presents the estimates of the Cobb-Douglas production function. The first column presents the estimates based on data for all the 165 industries. In this regression, the coefficient of water is found to be statistically insignificant. Also, the coefficient of energy is statistically significant only at the 10 per cent level. Since the results of this regression are not very satisfactory, especially in regard to the estimate of the coefficients of energy and water, the residuals were examined for possible outliers. From an examination of the residuals, four outliers (industries) were identified, and when these were excluded the results improved appreciably. This set of results based on 161 industries is reported in the second column of the table (Regression 2). In this regression, the coefficients of all the four inputs are statistically significant. Also, it is seen that the estimated parameters are of the correct sign and plausible magnitude.

The third column shows the results obtained when a dummy variable was introduced for the industries in which water consumption is over 100 million litres per factory per annum. The purpose in introducing this dummy variable is to capture in the estimation of production function the technological differences between the eleven highly-water intensive industries and the rest of the industries. The coefficient of the dummy variable is found to be negative. But, it is not statistically significant. Nonetheless, Regression (3) is arguably a better specification than Equation (2), since in Regression (3) an attempt is made, albeit in a limited way, to incorporate in the estimation of the production function, technological diversity among industries in regard to water requirements. The estimates obtained in Regression (3) have therefore been used for the estimation of marginal product of water.

Table 13.1: Estimates of Cobb-Douglas production function for Indian industry, 1995-96

(Dependent variable: ln V)

Explanatory variables	*Regressions*		
	(1)	(2)	(3)
ln L	0.388***	0.522***	0.525***
	(6.246)	(10.113)	(10.141)
ln K	0.558***	0.495***	0.498***
	(9.687)	(10.667)	(10.693)
ln E	0.084*	0.085**	0.085**
	(1.746)	(2.261)	(2.255)
ln W	0.021	0.042**	0.048**
	(0.817)	(2.054)	(2.225)
D100			-0.129
			(-0.887)
Constant	-0.170	-0.470	-0.489
Adjusted R-squared	0.877	0.925	0.925
No. of observations	165	161	161

V= output, L = labour, K = capital, E = energy, W = water
D100 = dummy variable taking value one for those industries in which water consumption per factory is more than 100 million litres per year, zero otherwise.
The t-ratios of coefficients are shown in parentheses.
*, **, *** statistically significant at ten, five and one per cent level, respectively.

Estimates of the translog production function are presented in Table 13.2.[11] The estimates shown in the first column correspond exactly to the specification in Equation (2) above. The estimates shown in the second column, *i.e.* Regression (2), have an additional explanatory variable. This additional variable is the dummy variable for industries consuming more than 100 million litres of water per factory per year. The coefficient of the dummy variable is found to be statistically significant at the ten per cent level. The sign of the coefficient is negative as in the case of Cobb-Douglas function estimates.

It is seen from Table 13.2 that most of the estimated coefficients in Regressions (1) and (2) are statistically insignificant. Therefore, in Regression (3) some of the second order terms have been dropped. With the dropping of some second order terms, the coefficients of other terms become

Table 13.2: Estimates of translog production function for Indian industry, 1995-96

(Dependent variable: ln V)

Explanatory variables	*Regressions*		
	(1)	(2)	(3)
ln L	1.393 ***	1.358***	1.511***
	(3.252)	(3.193)	(4.230)
ln K	0.018	-0.0014	-0.209
	(0.048)	(-0.004)	(-0.824)
ln E	0.197	0.235	0.346**
	(0.659)	(0.791)	(2.158)
ln W	0.108	0.094	0.132
	(0.988)	(0.861)	(1.593)
½ (ln L) ^{2}	-0.293**	-0.304**	-0.307**
	(-2.091)	(-2.182)	(-2.429)
½ (ln K) ^{2}	0.030	0.0305	
	(0.362)	(0.370)	
½ (ln E) ^{2}	0.035	0.029	
	(0.505)	(0.423)	
½ (ln W) ^{2}	-.0197	-.0075	
	(-0.980)	(-0.355)	
(ln L) (ln K)	0.149*	0.161**	0.173**
	(1.848)	(2.002)	(2.950)
(ln L) (ln E)	-0.035	-0.034	-0.066**
	(-0.578)	(-0.559)	(-1.907)
(ln L) (ln W)	-0.058	-0.057	-0.047**
	(-1.601)	(-1.597)	(-2.489)
(ln K) (ln E)	-0.045	-0.048	
	(-0.738)	(-0.787)	
(ln K) (ln W)	0.023	0.029	0.023**
	(0.561)	(0.735)	(2.066)
(ln E) (ln W)	0.012	0.0098	
	(0.477)	(0.385)	
D100		-0.373*	-0.387**
		(-1.794)	(-2.121)
Constant	-1.668	-1.677	-1.902
Adjusted R-squared	0.929	0.930	0.932
No. of observations	161	161	161

V = output, L = labour, K = capital, E = energy, W = water

D = dummy variable taking value one for those industries in which water consumption per factory is more than 100 million litres per year, zero otherwise.

The t-ratios of coefficients are shown in parentheses.

*, **, *** statistically significant at ten, five and one per cent level, respectively.

statistically significant. This indicates that the estimates of the translog production function presented in the first and second columns of Table 13.2 may have been affected by the problem of multi-collinearity. While the estimates of the translog production function presented in Col. 2 are not very satisfactory, as most coefficients are statistically insignificant, this estimated equation has been preferred over the estimated equation in Col. 3 for deriving the marginal product of water.

Estimates of the marginal product of water (value of water) are presented in Table 13.3.[12] Separate estimates have been made for high water consumption industries (consuming more than 20 million litres per factory per year), and for other industries, *i.e.*, low and medium water consumption industries (consuming less than 20 million litres per factory per year).

For all the 165 industries taken together, the estimate of marginal product of water is Rs. 19.29 per KL (in 1995) based on the Cobb-Douglas function and Rs. 17.40 based on the translog function. These are much higher that the (average) price actually paid for water, Rs. 1.94 per KL. In the high water consumption industries, the estimated marginal product of water is about Rs. 11 per KL, while the price paid for water is Rs. 1.52 per KL. In other industries (low and medium water consumption industries), the estimated marginal product of water is about Rs. 62 per KL (based on the translog production function) or higher, while the price paid is about Rs. 5 per KL. Since the marginal product of water or value of water for industries is quite high as compared with the price charged, it may be concluded that there is ample scope for raising the water price.[13] The high value of water in industries also indicates that water shortage in industries has a significant cost in terms of lost industrial output.[14]

It would be interesting to make a comparison here with Chinese industry. According to the estimates of Wang and Lall (2001), the industry-wide average elasticity of output with respect to water is 0.17 in Chinese industry. By comparison, the estimates present here indicate that the elasticity for Indian industry is much lower at about 0.43 to 0.48, depending on the production function used. The marginal value of water for Chinese industry

Table 13.3: Estimates of marginal product of water in Indian industries, 1995-96

Water consumption per factory (million litres per year)	*Number of industries*	*Elasticity-CD*	*Marginal product (Rs./KL) based on CD function*	*Elasticity-Translog*	*Marginal product (Rs./KL) based on translog function*	*Price paid (Rs. per KL)*	*Water consumed*	*Per cent*
Up to 20	136	0.0481	80.84	0.0366	61.51	5.05	604082	12
Above 20	29	0.0481	11.10	0.0466	10.76	1.53	4544376	88
All	165	0.0481	19.29	0.0434	17.40	1.94	5148459	100
Water consumption per employee (KL/year)	*Number of industries*	*Elasticity CD*	*Marginal product (Rs./KL) based on CD function*	*Elasticit Translog*	*Marginal product (Rs./KL) based on translog function*	*Price paid (Rs./KL)*	*Water consumed*	*Per cent*
Below 50	91	0.0481	148.11	0.0238	73.28	5.62	111832	2
50 to 200	44	0.0481	53.23	0.0435	48.14	3.91	692117	13
Above 200	30	0.0481	10.56	0.0673	14.78	1.53	4344510	84
All	165	0.0481	19.29	0.0434	17.40	1.94	5148459	100

estimated by Wang and Lall is about 2.45 Yuan per cubic metre (for 1993). This comes to about US$ 0.4 per KL. The estimate for Indian industry presented in Table 13.3 is about Rs. 17 to 19 per KL of water (for 1995), which comes to about US$ 0.5 per KL. It appears therefore that even though the elasticity of output with respect to water is lower in India, the value of water in industrial use in India is slightly higher than that in China.

The bottom part of Table 13.3 presents estimates of the marginal product of water for industry groups classified according to the ratio of water consumption to employment. It is seen that for the most water intensive industry group (water consumption per person over 200 KL per year), the estimated marginal product of water is about Rs. 11 to Rs. 15 per KL of water (depending on the production function used), which is well above the price of water of Rs. 1.94 per KL. This again shows the high marginal product of water in Indian industries in comparison with the price paid for purchase of water.

Price Elasticity of Industrial Demand for Water

Estimates of the water demand function are presented in Table 13.4. Regression (1) is based on data for all the 165 industries. The coefficient of water price is found to be negative (as expected) and statistically significant. The estimated price elasticity of demand for water is about 0.9. In Regression (2), a dummy variable is introduced for industries in which water consumption per factory is 20 million litres or more per year. Both intercept and slope dummy variables are used. The estimate of price elasticity of demand based on this regression is about 0.4 for high water consumption industries, and 0.64 for other industries. To examine this aspect further, the water demand function has been estimated separately for the high water consumption industries and other industries. The results are reported in the last two columns of the table. The results suggest that the price elasticity of demand for water is about 0.37 for high water consumption industries and 0.64 for other industries.

From the results reported in Table 13.4 it appears that the price elasticity of demand for water in high water consumption industries is about 0.4. Since these industries account for about 88 per cent of total water consumption in the industrial sector, the price elasticity at the aggregate industry may be expected to be only a little above 0.4.

Table 13.4: Estimates of water demand function for Indian industry, 1995-96

(Dependent variable: ln W)

Explanatory variables	*Regressions*			
	(1)	(2)	(3)#	(4) ##
ln V	0.646***	0.404***	0.409***	0.391**
	(6.264)	(4.024)	(3.589)	(2.005)
ln P_W	-0.857***	-0.640***	-0.638***	-0.371***
	(-7.903)	(-4.626)	(-4.344)	(-2.857)
ln P_L	1.180***	1.095***	1.111***	0.944*
	(5.125)	(5.259)	(4.839)	(1.863)
ln P_E	-0.267	-0.249	-0.311	-0.121
	(-1.394)	(-1.430)	(-1.367)	(-0.626)
D20		1.798***		
		(4.819)		
(ln P_W.D20)		.243		
		(1.062)		
Constant	0.922	1.249	1.288	2.918
Adjusted R-squared	0.648	0.714	0.435	0.430
No. of observations	165	165	136	29

V= output, P_L = price of labour, P_E = price of energy, P_W = price of water

D20 = dummy variable taking value one for those industries in which water consumption per factory is more than 20 million liters per year, zero otherwise.

The t-ratios of coefficients are shown in parentheses.

*, **, ***statistically significant at ten, five and one per cent level respectively.

Estimated for industries in which water consumption per factory is less than 20 million litres per year.

Estimated for industries in which water consumption per factory is more than 20 million litres per year.

Attention may be drawn here to the fact that the dependent variable in the regression equation presented in Table 13.4 is the quantity of water purchased. The elasticity estimated thus shows the effect of the price hike on water purchase, and not on water consumption. Increases in price of water may lead to an increase in self-extraction of water by factories, or it may have no effect on the component of water consumption that is not purchased. In consequence, the effect of the water price hike on water consumption is expected to be lower than what the estimated elasticity indicates. Considering this fact along with the estimates of the water demand function, it may be inferred that in water-intensive industries, which account for the bulk of water consumption in the industrial sector, the demand for water will not be particularly responsive to hikes in the water price. Rather, the price elasticity may be low, say somewhere in the range of 0.2 to 0.3. Since the results of the analysis indicate the price elasticity of industrial demand for water to be low, it may be concluded that an increase in the water price for industries will not reduce much water consumption in industries, but will lead to an increase in revenue for the water supply agencies.

Conclusion

The paper presented estimates of the value of water in industrial use and the price elasticity of demand for water in Indian industries. Cross-section data for 165 industries for 1995-6 were used for the estimation. The results indicated that the value of water is quite high in Indian industries compared to the price paid for water. At the aggregate industry level, the value of water (or marginal product of water) was found to be about Rs. 17 to 19 per kilolitre of water (in 1995-6), much higher than the average price of water of about Rs. two per kilolitre of water. Accordingly it may be concluded that Indian industries have a high willingness-to-pay for water and there is ample scope for raising the price of water for industries. Another conclusion that may be drawn from the results is that water shortage in industries has a significant cost to the economy in terms of lost industrial output.

The estimates of water demand function indicated a statistically significant effect of water price on water consumption in industries. However, the estimates of price elasticity of demand were found to be low. Based on the results it may be inferred that a hike in the water price will reduce to some extent the consumption of water in Indian industries. But, water consumption in industries is unlikely to respond much to increases in water price. It appears therefore that the scope of reducing industrial consumption of water though price hikes is limited.

Notes

1. See, among others, Bhatia *et al.* (1994).
2. A recent study that has examined the issue of sustainable use of water resources in India is by Chopra and Goldar (2000). They have noted the importance of water conservation and recycling in industries for sustainable use of water resources.
3. A large number of studies have been undertaken on industrial demand for labour and energy in India. Some of them have used a cost function for studying industrial demand for labour, capital, energy and intermediate inputs. In such studies, intermediate inputs were taken as one composite input, and a separate analysis for water demand was not carried out. For a survey of this literature, *see* Goldar (1997). One study on industrial water demand undertaken in the Indian context is by Bhatia *et al.* (1994).
4. Wang and Lall (2001) note that there are very studies on the structure of industrial demand for water. This situation stands markedly in contrast to the exhaustive analysis that has been applied to industrial demand for capital, labour and energy.
5. One would notice a similarity of this approach to the approach taken is some earlier studies on substitution possibilities between energy and non-energy primary inputs. In such studies, a cost function was estimated taking labour, capital and energy as three inputs, and separating materials input from them (*e.g.* Pindyck, 1979).
6. Wang and Lall have used a five-input production function, which they have estimated from plant-level cross-section data. They take labour, capital, water, energy and materials as inputs. However, they did not have good data on energy and materials, which compelled them to use COD load in water as a proxy.
7. It may be mentioned here that in some production function studies, the translog production function has been estimated jointly with factor income share equations to gain greater efficiency in the estimation of parameters (see, Goldar, 1997). This, however, requires

the assumption that all factors of production are paid according to their marginal product. In this study, a single-equation approach is taken to the estimation of the Translog production function, as done by Wang and Lall (2001), which has the advantage that one does not have to make the assumption of competitive markets and factors being paid according to their marginal product.

8. A sophisticated procedure that could have been adopted was to estimate a translog cost function along with cost share equations for water, labour, energy, etc. One reason why such a framework could not be used is that this analysis is based on cross-section data, and from such data it is difficult to obtain a good measure of the price of capital input. Besides, such a framework would require the assumption of cost minimization.
9. In the case of water demand function, it is meaningful to relate water purchased to price of water, rather than to relate total water consumption to price of water. An increase in water price may have quite a different effect on self-extraction of water by factories as compared with its effect on the quantity of water purchased. Therefore, the fact that ASI reports only water purchased, and not the quantity of water consumed, does not appear to be a serious limitation of the econometric analysis.
10. The inadequacies of using unit value in place of price variable for the estimation of demand function are well known. Note, in particular, that errors in the measurement of water quantity (dependent variable) cause an error in the measurement of water price (one of the explanatory variables), and this would affect the estimate of the coefficient of the water price variable. The inadequacies of using unit value or average water price for estimating the water demand function has been noted by Wang and Lall (2001).
11. These estimates have been made using data for 161 industries, as in the cases of Regressions (2) and (3) of Table 13.1 (Cobb-Douglas function).
12. In making estimates of marginal product of water, adjustments have been made for the difference between water consumption and water purchase in factories. Chopra and Goldar (2000) have compared the water requirement in some water-intensive industries and the water purchase figure reported in ASI for them. The comparison revealed a large gap between water consumption and water purchased. The ratio of water consumption to water purchase was found to be 1.9, which was used by Chopra and Goldar (2000) to make an estimate of total water requirement of the industrial sector. In this study, this ratio (1.9) is applied to work out the average product of water from the data on output and water purchased, and then the marginal product of water is computed by multiplying the average product of water by the elasticity of output with respect to water.

13. It should be noted that the cost of using water in factories should include the cost of procuring water as well as the cost of treating wastewater. The gap between marginal product of water and price of water is partly explained by the cost of wastewater treatment. It seems, however, that even after allowing this factor the marginal product of water in Indian industries is significantly higher that the cost of using water.
14. The serious adverse affect that water shortage has on industrial production has been noted by Bhatia *et al.* (1994) in the context of India and some other developing countries.

References

Bhatia, R., P. Rogers, J. Briscoe, B. Sinha, and R. Cestti (1994). "Water Conservation and Pollution Control in Indian Industries: How to use water tariff, pollution charges and fiscal incentives", *Currents*, UNDP-World Bank Water and Sanitation Programme.

Chopra, K. and B. Goldar (2000). Sustainable Development Framework for India: The Case of Water Resources, Report prepared for the United Nations University, Tokyo, as a part of its project on Sustainable Development Framework for India.

Goldar, B. (1997). "Econometrics of Indian Industry", in K.L. Krishna (ed), *Econometric Applications in India*, Oxford University Press: New Delhi.

Pindyck, R.S. (1979). "Inter-fuel Substitution and the Industrial Demand for Energy: An International Comparison", *Review of Economics and Statistics*, 61(2), pp. 169-79.

Wang, H. and S. Lall (2001). "Valuing Water for Chinese Industries: A Marginal Productivity Approach", Paper, World Bank, New Ideas in Pollution Regulation.
available at www.worldbank.org/nipr.

14

The Water Supply in Imphal City and its Suburbs

Its Economics and Environmental Implications

KSH. JHALJIT SINGH

Introduction

Manipur is a small state of India situated in the north-eastern part of the country with a total land area of 22,327 sq km and a population of 23,88,634 (*Provisional figure from 2001 Census*). Physiographically, the outlying areas of the state are covered with rugged hills and narrow valleys. The inner area represents the features of the flat plain topography. The valley which forms the inner part, constitutes just 10.02 per cent of the total land area of the state.

Located in the vicinity of the highest rainfall spot in the world, Cherapunjee in Meghalaya, Manipur receives fairly high rainfall. The average annual rainfall in the hills, in the central valley and Jiribam, a small valley in the south-west region of the state are respectively 1734 mm, 1238 mm, and 2637 mm (*based on records of 4 decades ending 1997*) with some year to year fluctuations. When we study the meteorological data of Imphal (ICAR), Manipur for the period 1990-2000, with regard to monthly figures of rainfall, relative humidity, maximum and minimum temperature, we find that each of these variables possess quite a low co-efficient of variation. The significance of the above finding is that the monthly mean figure may be looked upon as a good proxy value of the variable concerned for any normal year. Having said that, we may conclude that it rains for

8 months in a year in Imphal (a rainy month here is a month with rainfall of 100 mm or more). (Refer to Tables 14.1 and 14.2.)

Table 14.1: Meteorological data of Imphal (Tulihal), Manipur

[Monthly average of 10 years (1980-89)]

Month	*Rainfall in mm*	*Total No. of rainy days*	*Temperature in °C*		*Relative humidity*
			Max.	Min.	*(per cent)*
January	15.04	3.5	21.76	3.97	68.07
February	36.53	5.0	23.96	7.09	60.28
March	93.37	11.7	27.96	11.15	61.22
April	181.01	16.5	28.16	15.79	65.75
May	150.3	17.7	29.98	18.16	72.29
June	234.73	21.6	29.78	20.69	79.10
July	249.00	24.3	30.41	21.09	80.71
August	163.02	20.3	29.97	19.02	79.55
September	145.96	17.4	27.99	16.97	81.71
October	122.83	9.2	26.99	10.07	80.86
November	37.25	3.8	22.47	8.47	77.11
December	10.60	1.8	20.89	5.26	74.33
Annual	1439.64	152.8			
Mean	119.97	12.73	26.69	13.15	73.41

Source: Imphal (Tulihal).

Table 14.2: Meteorological data of Imphal (ICAR), Manipur

[Monthly average of 15 years (1985-99)]

Month	*Rainfall in mm*	*Temperature in °C*		*Relative humidity*
		Max.	Min.	*(per cent)*
January	10.66	21.3	4.4	70.4
February	36.03	23.3	7.4	64.2
March	110.43	26.3	12.2	61.4
April	123.9	27.9	15.7	67.1
May	177.2	28.4	18.6	75.1
June	237.0	29.1	21.2	80.5
July	276.4	28.4	21.0	83.3
August	190.6	28.9	21.1	81.7
September	164.4	28.7	20.3	83.0
October	119.7	27.7	16.7	79.8
November	44.5	25.5	11.3	78.1
December	17.7	22.5	5.6	75.8
Annual	1501.34			
Mean	125.11	26.5	14.68	75.03

Source: Imphal (ICARI).

Practically, more than 90 per cent of the precipitation takes place during the period of March–October. Seasonwise the distribution pattern of the total annual rainfall is as follows : 2 to 5 per cent in the winter (January and February), 20 to 30 per cent in the pre-monsoon (March–May), 55 to 65 per cent in the monsoon (June–September) and 10 to 15 per cent in the post-monsoon (October–December).

Till quite recent times, *i.e.* a few decades immediately after independence, the hills which form the catchment areas of numerous rivers and streams flowing through the valley, were covered with dense forests which regulated the flow of water into these rivers and streams (by first absorption of rainwater during rain and then slowly releasing it). This assured a reasonably good flow of water the year round and also prevented soil erosion apart from a host of other eco-protective functions. As population pressure on land increased, large scale deforestation took place and intensified till the Supreme Court ban on tree-felling in 1998 when the devastation had already touched an alarming level with far reaching socio-economic and ecological consequences. A direct fallout of this in recent times is that the rivers and streams that flow from these hills become very lean and almost dry up during the dry season but are dangerously turbulent during the monsoon. Even 2-3 days of continuous normal rainfall in the hills makes them overflow and breach their embankments thereby creating flood havoc and dry and drought conditions between the rains. The occurrences of flash flood and drought conditions have become more frequent in recent times. This phenomenon may be attributed more so to an increase in the run-off rate of surface water and the raising of the river bed due to siltation rather than to a significant change in the climatic conditions of the region. Thus, the people of Manipur live amidst the unenviable paradox of *'rain, rain and more rain, but not a drop of water for use'*. The situation could not have come at a worse time than when the ever-increasing population and developmental process, has sharpened the urgency to develop a scientific water harvest, storage and distribution system.

The demand for water for domestic consumption forms a

very small fraction of the total demand for it for various uses in any region. According to a UNEP study estimate, by the year 2000, the exploitation of water will be 69 per cent for agriculture, 23 per cent for industry and 8 per cent for domestic uses. For India, as per estimate of Ministry of Water Resources (1999) the present water use in different sectors are respectively, 84 per cent for agriculture, 12 per cent for industry, thermal power and other sectors and 4 per cent for domestic. Clearly the principal use of water is in agriculture and only a small proportion of it is being used in people's homes. But whenever the demand for water exceeds its availability, the requirements of domestic use and livestock must have a first claim on available resources. In this paper, we shall deal with issues relating to water for domestic consumption in Imphal City and its suburbs (more compactly, known as Imphal).

Demand-Supply Balance of Water in Imphal

Imphal, the capital city, is situated at the centre of the Valley. Imphal comprises: (*i*) Imphal municipal area (29.57 sq km), and (*ii*) non-municipal area (105.72 sq km) consisting of 77 surrounding villages. Thus, the total area of Imphal is 135.29 sq km, a mere 0.6 per cent of the total land area of the state whereas the projected population as per 2001 census is 5.95 lakhs which is almost one quarter of the state population. The gross density of population works out to be 43.98 pph for Imphal whereas the corresponding figures for municipal and non-municipal areas are 108.63 and 25.89, respectively (100 hectares = 1sq km).

The Public Health and Engineering Department (PHED), Government of Manipur[1] has estimated the daily water requirement of Imphal to be 98.84 mld. The present installed capacity is 86.62 mld out of which a capacity of 83.03 mld comes under the Urban Water Supply Schemes (UWSS) and the remaining capacity of 3.59 mld comes under the Rural Water Supply Schemes (RWSS). Thus the present installed capacity falls short of the total demand by 12.22 mld.[2]

Even though the installed capacity of the water treatment

plants for Imphal city under UWSS is 83.03 mld, the actual production is about 65 mld (see *Appendix-I*). Reasons cited for this under-performance are non-availability of regular power supply, low voltage problem and aging of the treatment plants and machineries, etc. Normally, the operation and maintenance of the micro plants under RWSS runs into more problems than the plants under UWSS basically because of the less proportionate allotment of funds and the lower priority given to it. Assuming a production level at 60 per cent of its installed capacity, the production under RWSS comes to a meagre 2.15 mld. Under the RWSS, water is supplied freely but only at public hydrants. There is no provision for individual/private domestic connection under this scheme. Many of the micro-plants under this scheme perform much below the installed capacity and at times, some of these come to a grinding halt for want of maintenance and repair of machinery, shortage of inputs and running capital cost. This is despite the fact that most of them were started with installed capacity to provide the rural population under its coverage a basic minimum of 40 lpcd.[3] No wonder, the people of these areas have to fall back partially or completely on the river or ponds for their water requirements.

If the present water requirement of 98.84 mld (as estimated by PHED) is fully met, the water availability of Imphal works out to the respectable figure of 167 lpcd. The estimated figure of the actual production is 67.15 mld (65 mld under UWSS and 2.15 mld under RWSS). Assuming a 40 per cent system loss[4] the actual amount reaching the residents is just 40.29 mld which translates to 67.71 lpcd, indeed a very low figure for a class-I city.

Even as per the reduced supply norm as recommended by Ministry of Urban Development (CPHEED), New Delhi[5] the total water demand for Imphal works out to be 72.82 mld. This means that the existing water supply falls short by 32.53 mld even to meet the reduced norm. The deficit in supply works out to be 80.7 per cent of the present supply. Thus, we find that the water supply situation in Imphal, to say the least, is indeed quite dismal and gloomy.

What is the PHED Doing?

Presently Imphal city is gravity fed from high level service reservoirs located over hills which are fed by the water treatment plants either by gravity or through pumping. Since these service reservoirs are located in a few places, they fail to create pressure to reach the distant consumer points. Therefore, the zonal supply system has been introduced in the water supply scheme of Imphal. Supply to zonal water reservoirs is fed from the untapped conveying mains and supply from these reservoirs has been started to create an equitable distribution of drinking water and to extend the water supply to uncovered areas. Old pipelines are replaced or new lines are laid as and when required. Efforts are also made to improve the supply by controlling leakages and wastage of water by the repairing of leaking joints. Works for the installation of more zonal reservoirs are in progress for a further improvement in the water supply of the scarcity area in Imphal city (Table 14.3).

At present, almost the entire raw water being used in treatment plants is from the surface water sources. The ever-increasing demand of the growing population has put these sources under great stress in recent times. The need to explore and tap new sources of raw water is being greatly felt. Limited success has been met in its attempt to install public hydrants in the form of tube-wells being dug up at various points in and around Imphal. The water available from these tube-wells was found to be of high quality and free from all kinds of health hazards.

Also a project undertaken with the assistance of French experts for the exploration of groundwater, the North and West of Imphal including *Potsangbam* area has already been completed. The project was to determine the potentiality of groundwater and the available recharge capacity particularly in *Awang Potsangbam* area. The preliminary findings have been very positive on both fronts. Also, there being no industrial pollution of any kind in the area concerned, the quality of groundwater may be expected to be very high.

Table 14.3: Surface water quality of river Imphal (Mahabali Point)

Characteristics	*1993*	*1994*	*1997*	*1998*	*1999*
*p*H	7.2	7.2	7.43	7.4	7.54
Temperature	19.4	23.1	23.9	23.65	22.63
Turbidity	64	65	121.09	106.3	83.17
DO	8.2	8	7.58	7.18	5
BOD	2.1	2.7	2.13	1.63	2.7
Total Coliform	6	6.36	106.56	33.25	471.29
Hardness	29	39	37.63	38.2	32.86
TSS	NA	NA	156.31	198.73	171.9
TDS	0.8	NA	76.12	57.24	76.89
COD	6.4	5.8	6.04	4.75	8.96

Surface Water Quality of River Iril (Poromapat Point)

Characteristics	*1993*	*1994*	*1997*	*1998*	*1999*
*p*H	7.1	7.1	7.79	7.79	7.81
Temperature	19.5	22.2	23.68	26.63	24.24
Turbidity	86	71	101.28	106.3	91.88
DO	8.1	8.1	7.73	7.5	5.33
BOD	2.8	3.7	1.4	1.49	1.97
Total Coliform	5	6.23	70.83	148	327.33
Hardness	54	51	71.46	69.2	64.86
TSS	NA	NA	176.4	273.34	304.63
TDS	0.6	NA	94.37	91.93	120.61
COD	4.3	6.3	3.86	5.4	4.88

Source: Manipur Pollution Control Board.

At the policy stage, there is a proposal for the construction of a network of community ponds to augment the raw water availability specially during the lean seasons. These ponds will help in harvesting and storing the rain water and are also expected to aid in running the treatment plants closer to full installed capacity all year round.

Given that a majority of the areas and the population of Imphal is presently devoid of a reliable water supply system, the afore mentioned steps and works taken up by the PHED for improvement in the water supply seem too inadequate in terms of scale and pace. Though the general public impression of PHED may be that of a non-performing or non-functioning department, the fact that it has to work under many constraints should not be overlooked. Some of its major problems are: (*i*) inadequate infrastructure at its disposal to meet the growing demand, (*ii*) paucity of funds for investment in new infrastructure and improvement of the existing one, (*iii*) occurrence of rampant theft of water, (*iv*) disturbance and distortion in its distribution system arising out of people's apathy towards public property, (*v*) tariffs not paid in time and also no effective mechanism in hand to enforce it, and so on.

The Public's Response

Only a few areas of Imphal receive adequate water supply to meet the domestic requirements. For a majority of the population the water supply situation is a hopeless case. Some of common problems faced by them include: (*i*) low pressure and erratic supply, (*ii*) inadequate supply, (*iii*) pipe lines not present for many areas for houses within the city and even amongst the places with pipe connections, distribution is not equitable, (*iv*) cases of contaminated water supply, (*v*) tariff bills not issued regularly, etc. The people of Imphal have completely given up the hope of any possible improvement in the situation in the near future. They have instead devised their own ways to meet their daily water requirements.

In areas of low pressure supply, people are rampantly using pumps directly linked to the supply line to draw as much water as possible during that limited and erratic supply period and store it as reserve for future use. In the process some households have been completely deprived of their share of water.

People have been driven to this course of action because of the uncertainty and unreliability of the service. Water has become such a prized item for the individual in such a situation that he/she as a matter of fact is able to ignore the ethical aspect of the present act which consists in consciously taking away the rights of the others over water. J. Fletcher in *Situation Ethics* has rightly said, "The morality of an act is a function of the state of the system at the time it is performed".

There are others who in order to extract even the last drop of water in the supply line have bored holes in the lowest part of the pipe and made illegal connections from those points. There are another set of people who are making illegal connections from the untapped conveying main which are there to carry water from the main service reservoirs to the zonal reservoirs. No doubt the illegal and selfish aspect of the above activities is disturbing but even more disturbing is the fact that the said people have no qualms in admitting to their acts. On the face of it, these people seem to be less concerned about the moral side of their living but are happy just to carry on living in the face of the struggle for such a basic need as water. The insensitivity of the people is evident from their failure to take cognizance of their act as theft of someone else's property. It seems that the minimum thing that needs to be done in any effort to sensitize the feelings of citizens towards their fellow-beings is to ensure the provision of their basic minimum needs.

Another way in which people meet their water requirement is by purchase of water. It is seen that many among the newly constructed houses have provision a for large water storage tanks and even among many old houses, new water storage structures have been constructed if they were not there earlier. The above observation is true for all parts of Imphal, irrespective of whether the house is located in the few areas where the water supply is presently adequate. It seems that a majority of the population does not have much faith in the water supply system.

There are now a good number of private players besides the PHED, that provide tanker supply service on request. The rate is Rs. 200 for a tanker with a 4500 litre capacity, Rs. 350 for 9000 litre and above, a trailer tanker of 1000 litre capacity without jeep is Rs. 20 per trip and supply for private tanker from treatment plants is Rs. 2.50 per 1000 litres. Those with large enough storage spaces and located in the water scarce area of Imphal acquire and store water from a 9000-10000 litre capacity tanker on their own for their domestic uses. Of those with much smaller storage spaces, a few of them come together and acquire a 9000-10000 litre tanker supply and share this amongst themselves according to the available storage space and the requirement of each member. There are still others mostly with very very small or with almost no storage facilities who purchase water in small amounts daily from persons who are in the business of sale of water on a retail basis. The entrepreneurs who are in this line, purchase bulk water from tanker supply, store it in large storage tanks and sell it on a retail basis. People may purchase water from the retail outlet set up near the storage tanks or obtain home delivery service with a good service charge on it. The current rate is Re. 1 for 10 litres at the selling point and home delivery at the rate of Rs. 20 for 100 litres or Rs. 3 per 10 litres. Clearly to these people who depend on small purchases of water for their daily requirement, the popular notion of water being a free resource will sound far-fetched.

For an entrepreneur in the water business with his own water tanker of capacity say 10,000 litres, his gross revenue will be Rs. 300 for tanker supply. If he indulges in retail sale this figure will lie between Rs. 1,000 to Rs. 3,000 depending on how much of it was sold at the retail outlet and how much of it is home delivery. Obviously a person without his own tanker can enter only in the retail business. The margins in all the above cases definitely look attractive for anyone thinking of venturing into this business.

Now let us get the story from the consumer side. The total demand of water for a family of say 5 members per month (working with the reduced supply norm of 150 lpcd) is 2,25,00 litres. Whereas the average total water supply for this family per month works out at 10,156 litres (as per the average water availability in Imphal being 67.71 lpcd). Thus the deficit in supply is 12,344 litres.

Now assuming that this gap is filled up by water delivered through tanker supply (this being the next cheapest way to get water after the organized water supply), then the family will have to spend approx. Rs. 370 for this. This is over the Rs. 75 tariff for domestic pipeline connection.

This simple workout shows that if the department is able to improve its water supply to meet the minimum daily water requirement and even charge a stiff monthly tariff in the range of Rs. 170 to Rs. 440, the new situation will be *pareto-superior* to the existing one. In the new situation, the family will be spending less than Rs. 440 per month on water whereas the family currently spent Rs. 445. On the other hand, the PHED will be getting an additional amount of Rs. 95 in the tax collection whereas the additional cost to the department in availing the residents 150 lpcd at the rate of Rs. 75 for 67.71 lpcd comes to Rs. 91.15 only.

Through this excercise we have shown the vast scope for bringing about a *pareto-improvement* to the existing situation. This *pareto-improvement* could be brought about by increasing the production level or improving the level of operation or improving the distribution system or plugging the system loss, etc. All these obviously will require an improvement in the existing infrastructure of the department and this requires an additional investment. The fact that the state government's coffers are presently almost empty and with no improvement expected on this front in the near future, suggests that it is time to look for private interventions in the area of water production and distribution. We should also encourage and welcome the development of water markets which serve the people to meet their rising water demand. This is not to say that the state completely withdraws. Water being a social good and a basic human and animal need (and a sustainer of environment), very vital issues of equity, social justice and sustainability are linked with it. One should not expect market forces to satisfactorily address the above issues. Hence the necessity for institutions which address the social and basic need aspect of water should not be forgotten.

There are many houses in Imphal with structures on their roofs to harvest the rain. Since Imphal receives a fairly good amount of rainfall spread over almost 8 months, these structures if properly designed and planned can provide a sufficient amount of directly harvested rain water which will cater to all kinds of domestic needs and can productively meet the water shortage problem of the residents. Also, this measure can significantly bring down the pressure on the existing water supply system.

Another very important source of fresh water which meets a large proportion of the daily requirement of Imphalites is ponds.

The Economics of the Community Ponds

Traditional water harvesting systems such as ponds, locally known as *pukhri* of various sizes are still actively in use even today in this part of the country and are very strongly linked to the life of the majority of the people in the valley region of Manipur. These ponds have proved to be very efficient in capturing and storing rain water. However, ponds as a mode of collecting water for meeting the day-to-day requirement may not be compatible with the present urban lifestyle. No doubt it would be much more convenient to have a running water tap for meeting domestic requirements. Failing this, in context of Imphal[6], a pond within a short distance, which is well kept and maintained in a hygienic way and which can meet the domestic requirement of the residents is indeed a very good option.

Imphal city is dotted with a great many ponds of this kind. They may be broadly classified into two categories on the basis of ownership: (*a*) community ponds, and (*b*) private ponds. In general, private ponds are smaller in size than the community ponds. They are also not so well-maintained or kept hygienically. But it is seen that wherever private ponds exist the people largely depend on these because of the proximity convenience to the user even though the quality of its water is likely to be inferior to the community pond[7]. Nonetheless, these people do fetch water from the public hydrants or the community ponds at least for cooking and drinking purposes.

The ever-worsening land population ratio especially in municipal areas and on some of its fringes has taken its toll on the private ponds of these areas. Almost all the private ponds have been filled up to maximize the building area available. Imphal is a low rise and low-density city though the municipal area along with some of its fringes have quite a high density, in fact to the tune of about 10,863 per square km. This high density along with the feature of low rise residential houses have shot up the premium of land for building purposes in these areas and this has led to the disappearance of private ponds from there. This development along with the inadequacy in the water supply system has increased the dependency of these people on the community ponds of their area. Though private ponds have disappeared from certain areas of Imphal because of the pressure on land, fortunately none of the existing community ponds have been disturbed. On the contrary, the institution of community ponds have been strengthened in recent times.

The local club of the area where the community pond is located looks after its well being and maintenance. The local club in charge of the pond normally collects voluntary donations in money or in kind from the residents of the locality for the upkeep of the pond. The works in this regard taken up by the club or under its supervision may be listed as follows:

(*i*) Keeping the pond free from waterweeds, water hyacinths, algae, etc. by removing them from time to time.
(*ii*) Prohibiting all polluting activities like washing clothes, or livestock or taking bath, throwing garbage, etc. within the pond.
(*iii*) Cleaning and clearing the surroundings of the pond.
(*iv*) Getting the water quality of the pond tested regularly and if need be steps to be taken to improve the water quality (with the help/advise from PHED) say by using lime, bleaching powder, alum, etc.
(*v*) In lean season period, if the water level goes very low the club may get the embankment strengthened, may take up desilting and deepening work, make arrangement of getting water pumped from a river if possible or from a bigger water body or bulk supply from PHED through pipe connection and fill the pond.[8]

Another factor which is greatly in favour of ponds as a system of capturing rainwater and storing it, is the climatic condition prevailing in Manipur. It is found that for a normal year the water level in a pond will be quite comfortable round the year considering the average annual rainfall, spread of the rainfall, average day temperature and observed rate of evaporation (*see Appendix*-2). Also adequate precaution can be taken up for a year of adverse climatic condition so that the pond can survive even during that year.

Table 14.4: Physico-chemical characteristics of some selected community ponds (1999)

Characteristics	*Sampling Sites No.*			
	P1	*P2*	*P3*	*P4*
pH	6.5	7.5	8.5	7.02
DO	5.6	8.9	8.2	8.83
BOD	9.4	8.8	8.2	8.24
COD	31.8	12.6	625.3	314.14
Total Hardness	400	146	821	126
Nitrate	0.4	33.4	50.7	2.91
Calcium	25.6	512.0	216.8	321.65
Chloride	44.0	244	251.1	266.75
Magnesium	36.8	25.3	312.6	614.13
Phosphate	0.38	0.25	0.5	0.49
Sulphate	0.75	0.2	0.11	0.08

P1 =Langthabal Community Pond, Imphal (Drinking and Domestic Use).
P2 =Khurai Heikru Makhong Community Pond (Drinking and Domestic Use).
P3 =Ningthem Pukhri, Imphal East (Drinking and Domestic use).
P4 =Thangmeiband D.M. College Gate Pond, Imphal (Drinking and Domestic use).

Source: Ecology and Environment Wing, Department of Environment, Govt. of Manipur.

In fact it has been observed that it does not require much of an effort in terms of labour or money to maintain these community ponds in a environmentally healthy and clean state. All one needs is to avoid the "tragedy of commons" situation taking place. Every resident of the area has a stake in the upkeep of the pond irrespective of the quantum of pond water used by him daily or his degree of dependence on it for the daily water requirement. The value of a well-kept community pond in the vicinity seems to be so well ingrained in the minds of the people of Imphal (whether it is in the form of used value or existence value or aesthetic value or some other form) that they are ready to cooperate in any manner to safeguard it. People have developed a kind of social responsibility towards the institution of the community pond and have no inclination to be free riders as they value the future service of the pond.[9] The need-based factor has also contributed to a great extent to the existence of the community ponds.[10] Going by the primary water quality criteria, most of the community ponds are found to belong to class 'C' category, *i.e.* drinking water source with conventional treatment. (*See* Table 14.4 and *Appendix-3*).

It has been shown in the literature of pollution that a subsidy per unit of emission reduction establishes the same incentive for abatement activities as a tax of the same magnitude per unit of waste discharge. The club/association that looks after the upkeep of the community ponds seems to have combined both the options in order to curb the pollution of the pond. It prohibits pollution activities within the pond such as washing of clothes, taking a bath or giving a bath to livestock, throwing in garbage etc. Anyone caught performing these acts is fined. On the other hand, in some cases as a part of the subsidy to the potential polluters the club has provided facilities for a public bath, and washing of clothes just outside the pond complex. With this form of subzidisation more people may be attracted to avail of the services of the pond and in the process, the utility of the pond gets enhanced. On the other hand as more people start using the pond the club may have to be more vigilant about any possible polluting activities which may harm the health of the pond. Also it has been observed that the people are quite knowledgeable about the natural assimilative capacity of the pond. Pictures of some community ponds are attached in Appendix-4. Public facilities (like place for public bath, washing cloths, water collection points, etc.) built around the ponds can be seen in these pictures. The banks of the ponds are also found to be raised higher than its surroundings. This stops direct discharge of rainwater drains and/or domestic sewages into the ponds.

Lastly, the community ponds perform a lot of ecological functions. Besides the aesthetic value of a well-kept clean water body, it moderates the weather condition of the area in its proximity, recharges the groundwater and indirectly helps in the growth of vegetation in the area. Even if an improved and dependable water supply is made available to the area (which is unlikely to take place in near future) it is worthwhile for the sake of ecological functions and contrasting health implications for the residents — of a clean water body *vis-à-vis* a dirty water body to continue to maintain the community ponds in pollution free and environmentally hygienic condition.

Conclusion

It is quite clear that the population of Imphal is facing a problem in meeting with its domestic water requirements. The need to set up more zonal reservoirs in the water scarce and uncovered areas of Imphal to solve the low-pressure problem and increase its area coverage is urgent. The PHED should take adequate care and be vigilant to prevent the seepage of sewage and contaminated water into the supply line through leakage joints. For mobilizing more revenue, the PHED should take care to regularly bill the water tax due and to reach these bills to the customers in time. Another measure, which may be explored, is pegging the water tax to the amount of water used rather than the existing flat rate; this measure is likely to bring reduction in demand or an increase in revenue generation. First, one will directly reduce the existing gap between demand and supply and second, indirectly the same purpose is served by investing the additional revenue to water supply related infrastructure. The distribution system could also be streamlined to result in more equitable distribution and improvement in its efficiency by minimizing the system loss. At the present juncture when the state government is reeling under an acute financial crunch while the private sector in general has proved more efficient than public sector in providing utility services, it will not be out of place to explore private sector entry on partnership between private and public institutions in water supply. Last but not the least, one should try to develop some formal organizational structure or incentive schemes to strengthen the institution of community ponds. This should be undertaken not only because it is a source of water but also because it has a social, environmental and health impact on the residents around it.

It is often said that the water supply problem faced by

Imphal is that of quantity and not of quality. Turbidity has been considered to be the main problem in the quality of the available raw water used in treatment plants. Even today Imphal city, as such, does not face a problem of raw water availability considering the amount and the distribution of rainfall over a year. The problem is that of the management of available raw water. Because of an increase in the run-off rate, unless there is some scientific way of harvesting and harnessing the available raw water, there may be the problem of its availability at certain months of the year. We are still to explore and tap the vast potential of groundwater availability in and around Imphal. The preliminary findings of the survey work on the exploration of groundwater availability and its recharge rate has indicated a good prospect on both fronts. Even the quality of the raw surface water available scores quite high in terms of its physical and chemical characteristics (*see* Table 14.3). The rivers and ponds under study exhibited physical and chemical characteristics well within the Government of India permissible limits. The availability of good quality raw water within Imphal city as well as at points which are close to the city are favourable factors for the future augmentation of capacity and from the operation and maintenance cost point of view. It will be quite in place to point out that the likelihood of these sources of surface water getting polluted in future is very high. The pressure of increasing population, the changing lifestyle of the people especially the aspect of people's fatal attraction towards consumerism and non-existence of a proper drainage and sewage system in Imphal are the prominent factors which may be responsible for the above. Hence it is very important for the institutions (government as well as NGOs) involved with water to consciously and consistently work towards their maintenance through physical measures as well as educating the public in this regard.

APPENDIX-1

THE EXISTING TREATMENT PLANTS (UNDER IMPHAL UWSS) ALONG WITH INSTALLED CAPACITY (IN MLD) AS ON 1-4-2001

	Kangchup (Old)	14.53
	Kangchup (New) constructed under French assistance	9.08
	Chinga	1.14
	Minuthong	1.14
	Porompat (Old + New)	9.08
	Canchipur	6.81
	Koirengei	2.27
	Singda	18.16
9.	Ningthempukhri	4.54
10.	Tube Well (180)	1.36
11.	Porompat Phase-II	6.81
12.	Potsangbam Phase-I	6.81
13.	Kiosk (Ultra Filtration)	1.30

THE EXISTING TREATMENT PLANTS (UNDER IMPHAL RWSS) ALONG WITH INSTALLED CAPACITY (IN GALLONS PER DAY) AS ON 1-4-2001

Imphal East

1.	Ningomthong Kitnapanung	60,000
2.	Akampat	50,000
3.	Gangapat	50,000
4.	Khomidok	30,000
5.	Heingang	60,000
6.	Kontha Ahallup	40,000
7.	Kanglatombi Top	40,000
8.	Irilbung	60,000
9.	Leimaching	1,00,000
10.	Kiyamgei Bashikhong	60,000

There are 10 other plants (under RWSS) operating in Imphal West located at Patsoi, Lairenjam, Changangei, Chajing Karam, Hiyangthang, Wakching Khullen (Govindagram), Malom, Ghari, Moidangpok, Lamdeng, each with an average installed capacity 40,000 gallons per day.

Source: PHED, Govt. of Manipur.

APPENDIX-2

PAN EVAPORATION LOSSES

Month	*Losses in cm (Singda Project)*	*Losses in cm (Thoubal Project)*
January	—	4.47
February	5.08	6.53
March	8.89	10.41
April	10.16	11.53
May	—	14.68
June	—	12.17
July	—	11.68
August	—	11.1
September	—	11.02
October	—	9.88
November	4.06	4.93
December	3.56	4.24
Total		112.64

Source: PHED, Govt. of Manipur.

APPENDIX-3

PRIMARY WATER QUALITY CRITERIA

Criterion	Class A	Class B	Class C	Class D	Class E
Dissolved Oxygen (mg/l) minimum	6	5	4	4	
BOD (mg/l) maximum	2	3	3		
Total coliform count (MPN/100 ml) maximum	50	500	5000		
pH	6.5-8.5	6.5-8.5	6-9	6.5-8.5	6.5-8.5
Free ammonia (mg/l) maximum				1.2	
Conductivity (micro mho/cm) maximum					2250
Sodium absorption ration maximum					26
Boron (mg/l) maximum					2

BOD : Biochemical Oxygen Demand; pH : measure of the acidity or alkalinity of water: MPN: Most probable number; Mho: unit of conductivity.

Class A: Drinking water source without conventional treatment.

Class B: Water for outdoor bathing.

Class C: Drinking water with conventional treatment.

Class D: Water for wildlife and fisheries.

Class E : Water for recreation and aesthetics, irrigation and industrial cooling.

APPENDIX-4

Chanam Pukhri, Singjamei

Kakwa Pukhri, Kakwa

Tera Pukhri, Tera

Sayang Pukhri, Tera

Ningthem Pukhri, Wangkhei

Khuyathong Pukhri, Khuyathong

Notes

1. This department deals with planning, investigation, execution and operation and maintenance for both urban and rural water supply schemes in the state.
2. For an idea of the comparative scale of things we have the following figures for Delhi. The estimated population of Delhi is 135 lakhs with the corresponding daily water requirement being 3028 million litres. On an average, 2422.4 million litres are produced everyday. Hence the resulting shortfall is 605.6 mld.
3. The Seventh Five-Year Plan had envisaged providing safe drinking water facilities to the entire rural population and also to enhance the per capita norm for water as from 40 lpcd to 70 lpcd during the plan period. The progress has been tardy on both fronts in a majority of states due to the lack of adequate funds available with the respective state governments (who are primarily responsible for these schemes).
4. The reported system loss for Delhi is 40 per cent though the Delhi Jal Board claims it to be just 20 per cent.
5. In consideration of the existing situation of restriction in raw water availability in Imphal area, the Ministry of Urban Development recommended a reduced supply norm of 150 lpcd for the municipal area and 90 lpcd for the non-municipal area.
6. A water well or a handpump at home could have been a better option than a pond but this option is not available in Imphal. Till date almost all attempts at constructing a water well or in stalling a hand pump in the valley region have failed. A few hand pumps which have successfully been installed are in the foothills of hillocks, situated within and around Imphal.
7. It is not that people are not aware of the risk to health involved in using unhygienic and unclean water but that their lack of conviction has resulted from years of tolerance of these very waters. In fact, it may not be so much a lack of conviction as the people are quite well aware of water quality and its health implications but it may have more to do with the non-availability of a readily accessible source of clean water to meet their domestic requirement.
8. Manipur had experienced a severe drought like condition in the year 1996-7 when most of the community ponds had gone dry and even the rivers had dried up for a few weeks. During that time, the Manipur Housing and Urban Development Department, Government of Manipur (MAHUD) with Planning and Development Authority (PDA), Manipur as its implementing agent, had taken up the embankment and improvement work of quite a few community ponds.

9. Incidentally factors like compactness of the community ponds, the non-exclusivity nature of its services, users residing around it, vitality of its services to the users' daily life, the smallness of the number of users, noticeability of rule-breaking free riders and finally the reluctance of the state to enter the scene seem to be helping not only in its survival but also in strengthening the institution of community ponds in Imphal.
10. There is a very interesting case where the need-based factor and the practical approach to the water problem resulted in the creation of a community pond ; Wangkhei Thambalkhong is a village in suburban area of Imphal East. Organized water supply has not reached this village though it has reached its doorsteps. The village did not have any community pond worth till quite recently and faced the problem of getting water for their domestic requirement. There was a large private pond in the area, conveniently located from the point of view of accessibility by the villagers. Some elders of the area got together and approached the owner with a request and suggestion of converting it into a community pond for the welfare of the village. The owner readily agreed to the proposal and surrendered the title of the land to a local body which was formed to look after the pond. It was agreed that the head of the family of the original owner of the pond will be the president of this body and other members will be elected/appointed from the families of the area who are directly dependent on the pond. A monthly collection of the nominal amount is done from each family in the area covered by the pond. And in this way a fund has been created to be used for the upkeep of the pond. This community pond goes by the name of *Thambalnu Pukhri*.

References

Coase, Ronald, (1969). "The Problem of Social Cost", *The Journal of Law and Economics*, Oct. Reprinted in Economics of the Environment, Ed. W.W. Norton & Company, pp. 23-49.

Gilbert M. Masters, (1998). *Introduction to Envirc nmental Engineering and Science*, Prentice Hall of India: New Delhi.

Government of Manipur, Secretariat, MAHUD Department, 1994, *Manipur Gazette, No.198(A)*.

Government of Manipur, PHED, *Annual Administrative Report, 1999-2000*.

Hardin, Garett, (1968). *"The Tragedy of the Commons"*, Science *163:* 1243-48

M.N. Murty, A.J. James, Smita Misra, (1999). *Economics of Water Pollution*, Oxford University Press.

Wade, Robert, (1987). *The Management of CPR* : Collective Action as an Alternative to privatization or State Regulation, *Cambridge Journal of Economics*, 11, 95-106.

SECTION-III

INSTITUTIONS FOR WATER MANAGEMENT

15

Groundwater Markets and Institutional Mechanism in Fragile Environments

DALBIR SINGH

Introduction

Rights in groundwater belong to the landowner, since it forms part of the dominant heritage and land ownership that is governed by the tenancy laws of the state. The Transfer of Property Act necessitates that rights to groundwater can be given to anyone else only if the land is transferred. The consequence of such a legal framework is that only the landowners can own groundwater in India. It leaves out all the landless and tribes who may have community rights over land but not private ownership. It also implies that rich landlords can be water lords and indulge in openly selling as much as they wish (Singh, 1991). As a result access to this resource is privy to well do farmers and is beyond the reach of resource poor farmers, who cannot install their own wells. The only option left with these farmers is to purchase water for irrigation from well owners. In varying conditions, these practices operate differently. In water scarce regions, the

* The author is thankful to Prof. M.S. Rathore for valuable comments on an earlier draft of the study and to the editors of this volume for their valuable comments that helped to improve the present version.

farmers who have excess water sell it under monopoly power by charging a high price from their buyers. In areas where substantial groundwater deposits exists, there is the emergence of efficient and competitive markets, buyers would be able to buy groundwater at a lower price. Thus, they may make intensive use of their land for growing water intensive crops (Barah, 1993; Meinzen-Dick 1996; Shah, 1985; 1993). The market structure of groundwater is not only determined by its supply but also by the institutional mechanisms both formal and informal that have a strong hold on the operation of water markets. The main difference between formal and informal set-ups is the way in which the water trade is conducted. Under the formal framework, government establishes legal tradable water rights to retain and extend the advantages of water markets. Informal arrangement users contract for water on their own and no legal recognition is granted to the water trade. In the absence of formal rules, a dense social network leads to the development of customs, laws, thrust and normative provisions that constitute an informal framework that has a pervasive influence on the economy (Aron, 2002). Similarly, the market structure of groundwater is ecologically and socially embedded (Dubash, 2000).

The government's supportive policies made subsidies and credit facilities available to the farmers. This has resulted in intensive groundwater utilization. There are strong and direct linkages between power pricing, technological intervention, equity and efficiency in groundwater development. In the recent past, a sharp increase in groundwater use is recorded. This expansion of groundwater use has resulted in a speedy decline in the groundwater table in several parts of the country (Dhawan, 1995; Moench 1992; Bhatia, 1992). The evidence indicates that the lowering of the water table is quite rapid in water scarce regions. In Rajasthan this decline is recorded at the rate of one to five metres per year in varying conditions. This has resulted in over-exploitation of groundwater. If this trend continues then there will be irreparable loss to the society as evidence shows that socio-economic and environmental challenges have emerged or are about to do so. It is high time for immediate attention to be given to this serious problem.

Rajasthan is a water deficient state in the Indian subcontinent. Only one per cent of the country's water is available have for 5 per cent of its population living in 10 per cent of the total geographical area. Within the state as well, there are significant variations in the regional and temporal distribution of rainfall as is evident from the fact that a larger proportion of the area is classified as arid and semi-arid, mainly the western and northeast regions, respectively. The uncertain and uneven distribution of rainfall creates imbalances in the water availability for different uses. The agriculture sector is the major consumer of water where 90 per cent of the total availability is used for irrigation purposes. The domestic demand is fulfilled by about 9 per cent of the total supply while only one per cent is allocated to other uses. Groundwater is the major source of irrigation in the state. More than two-thirds of the total irrigation demand is met by groundwater. The scarcity of groundwater along with the presence of a large section of poor farmers make, the role of water markets crucial to the process of agricultural development in the region.

The literature shows that water markets play an important role in correcting the inequalities between poor and rich farmers in the access to groundwater. Markets provide water for irrigation to resource poor farmers, who cannot install their own wells, because this involves heavy investment. In such circumstances, the only option available with these farmers is to purchase water for irrigation from well owners. Markets also help to make sustainable use of water albeit at different prices. Experience shows that there are various institutions both formal and informal that influence water selling. While developing the groundwater markets it is important to pay attention to these institutions. To understand these institutions and their in role in water trading, this study was launched in Rajasthan with the specific objectives to: (*i*) understand the operations of groundwater markets in fragile conditions, and (*ii*) examine the institutional mechanism for the development of groundwater markets.

Selection of Villages and Approach

This study was carried out in the arid and semi-arid regions of Rajasthan. Access to groundwater and its marketing practices were the major criteria for the selection of two typical villages. Accordingly, Kukanwali village in Nagaur district and Srichandpura village in Alwar district of Rajasthan falling in arid and semi-arid regions, respectively were selected (*see* Map. 15.2).

The selected villages differ in respect of climatic conditions, *i.e.*, annual rainfall, land and water quality and aquifer condition. In the arid region, the average rainfall is considerably lower than that of the semi arid area that has a strong implication on groundwater availability (*see* Table 15.1). Kukanwali village has an undulating topography with sandy soil while Srichandpura village is surrounded by mountain with loamy soil. In the arid village, the farmers use sprinkler irrigation, while flow irrigation is used in the semi-arid village. In arid regions, the aquifer is depleting at a faster rate than that of the semi-arid regions (Singh, 2000). The terms and conditions of water trading also vary across the villages.

Table 15.1: Average rainfall in the selected areas (five years average in cm)

Year	*Kukanwali (Arid Region)*	*Srichandpura (Semi-arid Region)*
1951-55	39	54
1956-60	33	60
1961-65	31	68
1966-70	31	64
1971-75	62	72
1976-80	39	66
1981-85	31	64
1986-90	31	46
1991-95	36	60
1996-99	55	80

Note: District level figures are used to observe climatic variation.
Source: Statistical Abstract, Directorate of Economics and Statistics, Government of Rajasthan (Various Issues).

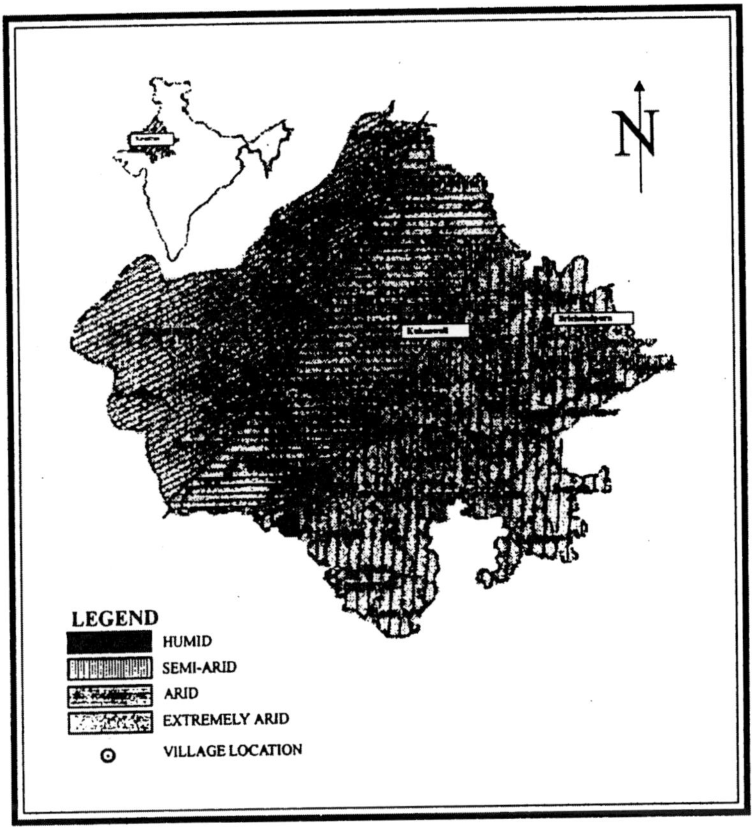

Map 15.1: Location of Selected Villages.

From the selected villages, a list was prepared and categorized into three groups, namely self-users, sellers, and buyers. Among the water users a representative sample size of 40 and 25 in Kukanwali and Srichandpura villages, respectively was drawn (*see* Table 15.2).

Village-level information was obtained through group discussions, participatory methods and the collection of primary data by using a structural schedule. The socio-economic and demographic information was obtained from

Table 15.2: Category-wise distribution of water users in the selected villages

Category	*Kukanwali*			*Srichandpura*			*Overall*		
	Total Water Users	*Sampled*	*Per centage to Total Water Users*	*Total Water Users*	*Sampled*	*Percentage to Total Water Users*	*Total Water Users*	*Sampled*	*Percentage to Total Water Users*
Self Users	155	13	8.39	25	6	24.00	180	19	10.56
Sellers	150	19	13.67	38	6	15.79	188	25	13.30
Buyers	80	8	10.00	75	13	17.33	155	21	13.55
All	385	40	10.39	138	25	18.12	523	65	12.45

Source: Field Survey.

published and unpublished village records. An intensive survey of households using groundwater for irrigation was carried out to obtain detailed information with regard to all aspects of groundwater markets for the year 1998-99.

Profile of Sample Villages

The present section deals with a profile of the sample villages, *viz.*, the demographic and social set-up, and utilization of land and groundwater resources. Mainly, the focus of attention was the sample respondents and their resource endowment. It is in fact, appropriate to begin with the land and water resource inventory and their utilization pattern across the water users and villages.

Socio-Demographic and Infrastructural Set-up

Kukanwali village has a population of 3962 persons (638 households) while Srichandpura has a population of 1408 person (230 households) as per the 1991 census. During the decade 1981-91, the population increased at the rate of 2.97 and 3.06 per cent in Kukanwali and Srichandpura villages, respectively (*see* Table 15.3). It may be stated that a faster growth of population not only resulted in many problems but also influenced the per capita benefits of irrigation facilities.

The caste system plays its own role in natural resource use particularly groundwater for irrigation. In Kukanwali village, the *Jat* community which constitutes 30 per cent the total households belong to the farming community and exercise leverage in managing the natural resources both private and commonly owned. About 14 per cent of the *Rajput* households and 10 per cent that are *Brahmin* and *Bania* are mainly engaged in the service sector and other non-farm activities. The remaining 46 per cent households belong to the scheduled and other castes. Some of the households, with a limited holding size and not being in a position to install their own well, met their demand for water by purchasing this from well owners. In Srichandpura village only 11 per cent of the total households belonged to the *Jat* and *Gujjar* community. *Rajputs* and *Brahmins* accounted for 23 and 27 per cent of the total households respectively. In this village, the scheduled castes and scheduled tribes are dominant, *i.e.* 39 per cent of the total households. They also have a very limited size of holding. They meet their water requirement through water marketing practices that prevail in the village.

Table 15.3: Demographic and social featurt. of the sample villages

	Particulars	Kukanwali (Arid Region)	Srichandpura (Semi-arid Region)
1.	*Demographic Characteristics*		
	No. of Households	638	230
	Population	3962	1408
	Growth Rate @	2.97	3.06
2.	*Dominant Castes (Percentage)*		
	Jat/Gujjar	30	11
	Rajput	14	23
	Brahmins/Banias	10	27
	Schedule Castes and others	46	39
3.	*Infrastructural Facilities*		
	Transportation	Yes	Yes
	Communication	Yes	Yes
	Electrification	Yes	Yes
	Education	(up to secondary)	(up to middle)
	Banking	Yes	No
	Input Distribution Centre	Yes	No
	Marketing	Yes	No

Source: (1) = Census Report, 1991, (2) = Field Survey.

Infrastructural facilities such as transportation and communication, education, banking, agricultural input distribution system and marketing facilities play an important role in the development of groundwater markets in a particular region. Transportation and communication facilities were available in both the villages. In the arid village, Kukanwali, the farmers purchased fertilizer, seed and other inputs from the input distribution centre which was functioning well. Informal marketing practices of agricultural production were prevailing within the villages. The marginal and small farmers benefited from these facilities that were available locally. The other categories of farmers market their farm produce at

Kuchaman city located at a distance of 20 km from the village. Banking facilities were also available within the village. The villagers of Srichandpura village use these facilities that were available at a distance of 5 km outside the village. Both villages were electrified during the early 1980s. After electrification, a great change has been taking place in the agricultural sector in both the selected villages in form of popularization of the water markets. Diesel oil pump sets were replaced by electric pump sets that resulted in a larger use of groundwater for irrigation.

Utilization of Land Resources and Development of Irrigation

In the past two and half decades, the area under irrigation has increased sharply. In Kukanwali village it increased from about 5 to 50 per cent while the increase was from 13 to 45 per cent in Srichandpura village. Fallow land has also increased slightly (*see* Table 15.4). In Srichandpura unlike Kukanwali village. the uncultivated area remained as it was during 1971. In this village, larger proportion of uncultivated land was brought under cultivation because of the increasing demand for food. The farmers keep their holding fallow during rainy season and they allocate it to irrigated crops in *Rabi* season.

Table 15.4: Land use pattern of sample villages

(Area in percentages)

Particulars	*Kukanwali*				*Srichandpura*			
	1971	*1981*	*1991*	*1997*	*1971*	*1981*	*1991*	*1997*
Total Geographical Area (ha.)	1253	1253	1253	1253	438	438	438	438
Uncultivated	6.33	6.89	7.74	8.22	53.65	53.88	26.71	26.01
Cultural Waste	1.55	0.22	0.18	neg.	8.90	7.20	7.53	6.13
Fallow Land	6.78	10.35	6.81	9.34	2.07	1.02	13.71	8.72
Cultivated	85.34	84.52	85.27	82.44	35.38	37.90	52.05	59.14
Irrigated	9.73	23.70	43.45	61.67	36.77	57.23	44.74	76.06
Unirrigated	90.27	76.30	56.55	38.33	63.23	42.77	55.26	23.94

Source: Village Revenue Records.

There was a sharp increase in irrigation facilities in the selected villages during reference period. But this increase was faster in the arid Kukanwali village, where it was recorded as more than six times whereas it just more than doubled in Srichandpura. How did this increase take place and what role have water markets played in this? These questions are explored in the following discussion.

Ownership Pattern of Land and Water Resources

For studies relating to natural resources like groundwater markets, it is essential to understand who owns these resources? In the present section, we deal with the ownership and distribution of resources such as land and sources of water. .

Land Resources

The present section focuses on the distribution of land holding in both the selected villages. In Kukanwali village, about 29 per cent of the marginal and small farmers own only about 7 per cent of the total land while 12 per cent of big farmers owned about 41 per cent of large size farm land. The semi-medium and medium size of farmers are dominant in the village and they constitute about 56 per cent of the total number of holdings, *i.e.* they owned more than half of the total land. In Srichandpura village, the distribution of holdings was in favour of marginal and small size of farmers. They constituted 83 per cent of the total number of holdings and owned about 59 per cent of the total land. One-third of the total land was owned by only 13 per cent of semi-medium and medium size of farmers. There was a very limited proportion of large size of farmers, that is, less than 1 per cent who owned about 7 per cent of the total land. The average size of the holdings was comparatively lower in Srichandpura than that of Kukanwali village (*see* Table 15.5).

Where distribution of irrigated and unirrigated land among different sizes of categories of households is concerned, there was an inverse relationship between the proportion of area irrigated and size of holdings with the exception of the large category in the arid village.

Table 15.5: Farm size classwise distribution of land holding in selected villages

Farm Size Class	*No. of Holdings*	*Distribution of Holding (per cent)*	*Area Owned (Ha)*	*Distribution of Area (per cent)*	*Av. Size of Holding (Ha)*	*Distribution (Per cent)*	
						Irrigated	*Unirri-gated*
			Kukanwali				
Marginal	34	10.86	20.41	1.28	0.6	79.08	20.92
Small	56	17.89	83.76	5.23	1.5	76.79	23.21
Semi-Medium	102	32.59	300.22	18.76	2.94	47.60	52.40
Medium	83	26.52	540.72	33.78	6.51	43.87	56.13
Large	38	12.14	655.52	40.95	17.26	53.87	46.13
Overall	313	100.00	1600.63	100.00	5.11	53.85	49.15
			Srichandpura				
Marginal	124	53.91	73.81	25.19	0.6	82.43	17.57
Small	73	31.74	100.01	34.14	1.37	77.86	22.14
Semi-Medium	24	10.43	62.16	21.22	2.59	52.65	47.35
Medium	7	3.04	35.05	11.96	5.01	45.79	54.21
Large	2	0.87	21.95	7.49	10.98	45.15	54.85
Overall	230	100.00	292.98	100.00	1.27	67.24	32.76

Source: Village Revenue Record.

Sources of Water

An attempt has been made to examine the distribution and ownership pattern of water sources (wells) by using household level data. With certain exceptions, it has been found that there is a positive relationship between the ownership and distribution of wells, and land resources. Where, the nature of well ownership is concerned the figures show that about three-fourths of the total wells were of individual ownership. More than five shareholders owned a very limited proportion of wells. The two shareholders and three to five of them owned wells in equal proportion as 12 per cent. In Srichandpura village, there were wide variations in the ownership pattern of wells. One-third of the total wells here were individually owned and two-thirds were in joint ownership (*see* Table 15.6). Joint ownership existed because the installation of a well was not economically feasible due to the small size of holding. Therefore, the smaller size of farmers invest jointly to install wells to irrigate the limited size of their holding or these were constructed before the division of ancestral property, that is, lands.

The analysis shows that there exist variations in the ownership and distribution of land and water resources. In Kukanwali village, semi-medium and medium whereas marginal and small sizes of farmers in Srichandpura village occupy the prime position in numbers. The size of holding was also substantially lower in Srichandpura as compared to that of Kukanwali village. Largely, the ownership of water sources was individual and joint in Kukanwali and Srichandpura, respectively.

Operation of Groundwater Markets

Water markets function differently under varying conditions. Besides, natural and physical factors such as climatic conditions, soil, quality and quantity of water, depth of water table, etc. the institutional arrangements both formal and informal play an important role in the operation of water markets. How do these institutions determine their operation? How do the various categories of water users behave in the changing situations and environments? An attempt has been made to answer these questions.

Table 15.6: Farm size-wise distribution and nature of well ownership

(Percentage)

Size of Farms	*No. of Wells Owned*	*Nature of Well Ownership*			
		Individual	*Joint*		
			Up to 2	*3 to 5*	*Above 5*
		Kukanwali			
Marginal	1	0	0	0	100
Small	5	80	0	20	0
Semi Medium	13	62	30	8	0
Medium	14	88	0	12	0
Large	0	83	17	0	0
Overall	33	73	12	12	3
		Srichandpura			
Marginal	6	16	17	50	17
Small	7	29	29	29	13
Semi Medium	6	50	0	50	0
Medium	0	0	0	0	0
Large	0	0	0	0	0
Overall	19	32	16	42	11

Source: Field Survey.

Who Participates in Water Markets?

For intensive investigation in groundwater markets it is essential to understand the nature of participation: Who participates in water trading and what factors affect their participation. Past studies (Meinzen-Dick, 1996; Narayanamoorthy, 1994; Shah and Raju, 1988) conducted in different parts of the country show that the well owners who own a limited size holdings have a higher extent of participation than those who owned larger holdings due to the surplus volume of water in their possession after irrigating their own lands. However, this study does not support this hypothesis. In Kukanwali village, only one-third of the total sellers belong to the small and semi-medium categories whereas two-thirds belonged to the medium and large farm size categories. This may be attributed to the following three major reasons for the participation of the large and medium size of households in water trading depending upon situational factors for such as location of well installation, *e.g.* (*i*) The location of lands belonging to the small and marginal (poor) farmers who do not have their own water sources surround the farms that have wells. In such a situation, well owners sell water to them; (*ii*) The seller reduces his own demand and saves water by adopting a less water intensive cropping pattern and sells the surplus water; (*iii*) Availability of water is relatively higher with the larger size of farmers because they have the capacity to chase the water table due to their sound economic conditions (Singh, 2000). The installation of a well and its deepening involves a heavy investment that is often beyond the reach of poor farmers (Moench, 1992; Shah, 1985). Ultimately, well owners having a larger size of holding have the option of selling water to resource poor households. In Kukanwali village, about half the buyers belong to the semi-medium followed by marginal and small size of farm categories, that is about 38 per cent.

In Srichandpura village, unlike Kukanwali the sellers were in equal proportion as marginal, small and semi-medium size of farms (*see* Table 15.7). This may be the major reason for sparing water for sale across the farm size categories. The majority of the buyers belong to the marginal size of farms. In overall terms, the sellers belonged to the medium and large size groups while the buyers were from the marginal and small size of farms. The argument that the resource rich household can extract groundwater in larger quantum holds true in general and of Kukanwali village in particular. Certain well owners having a large size of holding, however, remain out side the water business. Further, an attempt has been made to understand the reasons for participation and non-participation in the trade.

Table 15.7: Farm size-wise distribution of households participating in water business

Category	*Size of Farms*					*Table Sample Size (No.)*
	Marginal	*Small*	*Semi-Medium*	*Medium*	*Large*	
			Kukanwali			
Self-users	0	0	15	77	8	13
Sellers	0	11	21	42	26	19
Buyers	25	13	50	12	0	8
Overall	5	8	25	42	15	40
			Srichandpura			
Self-users	17	50	33	0	0	6
Sellers	33	33	34	0	0	6
Buyers	70	15	15	0	0	13
Overall	48	28	24	0	0	25
			Overall			
Self-users	5	16	21	53	5	19
Sellers	8	16	24	32	20	25
Buyers	52	14	29	5	0	21
Overall	22	15	25	29	9	65

Source: Field Survey.

Factors Affecting Participation and Non-Participation in Water Business

Under what circumstances do farmers participate or not participate in water trading?

In Kukanwali village, about 46 per cent of the self-users reported that they were not in a position to involve themselves in the water business due to a lack of surplus water on account of low discharge. This is due to the fact that a larger area, that is about 60 per cent of the total irrigated area owned by them was allocated to a water intensive crop like wheat and barley (Singh, 2000). In Srichandpura village, the proportion of self-users who did not have surplus water was considerably higher, that is 83 per cent (*see* Table 15.8). The well intensity also influences the water markets. It limits the sale and purchase practices in a particular region. Similarly, 38 and 17 per cent of total self-users in Kukanwali and Srichandpura, respectively reported that they did not have buyers that led to their non-participation in marketing practices. It is important to mention here that non-availability of buyers was not because of a higher water price as in the case of other commodities, but is due to the fact that almost all the neighbours have their own water sources. Transportation of water to far off distances was not in practice due to various reasons, like growing tension among the village communities. In such a situation there is no possibility of water markets.

Table 15.8: Reasons for participation and non-participation in water business

(Percentages)

Particulars	*Kukanwali*	*Srichandpura*	*All*
Self-Users' Reasons for not participating			
• No Surplus Water	46	83	58
• No Buyers	38	17	32
• Water Quality	16	0	10
Sellers' Reasons for Participating			
• Surplus Water	74	83	76
• Profit Earning	26	50	32
• Power Policy	26	17	24
Buyers' Reasons for Participating			
• Owned Land but No Well	88	46	66
• Limited and Fluctuating	12	54	34
• Water Supply			

Source: Field Survey.

In arid regions, quality sets an upper limit. According to some estimates, 73 per cent of India's saline water is in the arid region of Rajasthan (Rathore, 2001). In Kukanwali village, 16 per cent of the total self-users did not participate in marketing practices due to considerations of quality deterioration. If they did so, water quality may deteriorate due to excessive water extraction. Such perceptions were missing in the semi-arid village. This problem is very acute all over the water scarce regions. It varies even in a small fraction of the distance. This village is spread over the larger area and divided into a number of smaller hamlets. The problem relating to water quality also prevails in the village. Keeping in view the future consequences the well owners avoid extracting an excess of water.

Mainly, sellers enter the water markets because they have surplus water after irrigating their own fields. Besides, other factors that encouraged well owners to sell water were profit earning and the existing power pricing policy. The well owners stated that they sold water for the sake of profit earning. They charge a higher water price from their buyers, in order to cover at least the variable expenditure, incurred in the operation of water extraction equipment. In Kukanwali village, more than one-fourth of the sellers and in Srichandpura village half of them sold water for earning a profit. Overall, nearly one-third of the well owners sold water for the sake of profit. The existing power pricing policy was also responsible for encouraging water markets. The well owners try to extract a maximum amount of water against the fixed amount of payment of electricity charges. In Kukanwali village, more than one-fourth and in Srichandpura, 17 per cent of the sellers reported that they extracted water for sale due to the flat-rate power pricing policy.

With a change in the power pricing system, from pro-rate to the flat-rate system, no corresponding change occurred in the water sale prices. The experience of Gujarat shows that with the switching from pro-rate to flat-rate system the decline in water price was nearly three-fourths of the decline in water prices (Shah, 1993a). But in fragile conditions here no change in water price was recorded due to a short supply of water against the fast growing demand for irrigation water. Buyers who do not have their own well had to depend upon other well owners for irrigation water. The proportion of these buyers was considerably higher at 88 per cent in Kukanwali village as compared to 46 per cent in Srichandpura. The fluctuating water supply was also the major cause of the emergence of water markets as found in Srichandpura village.

Development of Groundwater Markets and Externalities

The externalities are non-reimbursed cost or uncharged benefits occurring to people as a result of someone else's action. Typically, we are concerned about externalities that impose costs on others, for instance, extraction of groundwater that results in irreparable loss to society in the long-run. In the present context, how does development of groundwater markets create externalities is explored. Effort has been made to answer this with the help of following illustration (Fig. 15.1).

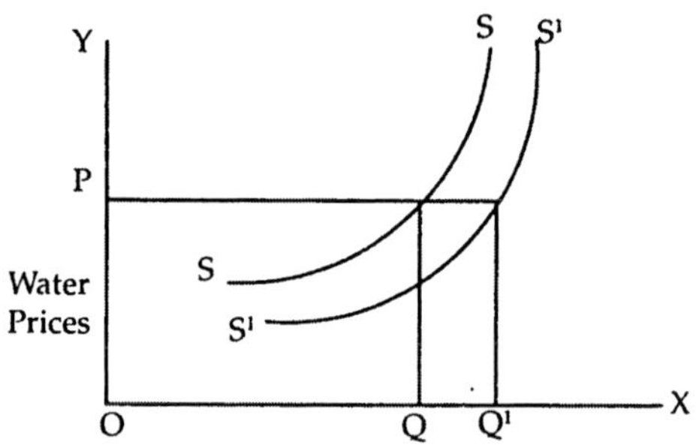

Fig. 15.1: Water Extraction for Sale.

The illustration shows that the seller extracts OQ volume of groundwater at the prevailing OP water price. In similar conditions, the well owner in general and sellers in particular opt for the local options of water saving, as cultivation of less water intensive crops and skipping over the number of irrigation. Thus the seller makes available water for sale to his buyers out of OQ quantity of extracted water. Under the existing price mechanism, the resource rich farmer adopts technological measures to increase the water supply due to his capacity to invest in water extraction equipment. This intervention of the well owner resulted in increase in the supply from SS to S^1S^1 by extracting OQ^1 volume of groundwater at the same OP price level. Now, the seller sells a larger volume of water to his buyers thereby earning more profits than before.

It may be inferred that when the well owner participates in the water business and extracts groundwater at its maximum levels that would result in aquifer depletion. In other words, seller's technological intervention in water extraction creates externalities in the long run.

Terms and Conditions of Water Markets

Broadly, there exists two types of terms and conditions for water trading, namely cash and kind based. Each type has wide variations in its operation across the participants who participate in the water business. With changes in infrastructure and the institutional set-up, the terms and conditions of operation also vary overtime. There are three types of contracts on the basis of which water markets operate (*see* Table 15.9).

(*i*) *Time Based Contract*: Under this contract the seller provides water to the buyer and the former charges on an hourly basis. This condition is prevalent in Srichandpura village. Certain variations in the operation of terms and conditions also exist. If the seller owns an electric motor and electricity charges are paid according to flat-rate system, then the rate is Rs. 30 per hour. In the case of a pro-rate system, the charge is Rs. 5 per unit of electricity consumed. In case, the seller owns and operates diesel oil engine, then the charges are at the rate of Rs. 15 per hour plus fuel charges.

(*ii*) *Crop and Input Sharing Contract*: Under this contract, the seller provides water while the buyer provides land and labour. The paid-out cost of inputs applied and production are shared in equal proportion between the buyer and seller. This condition of water trading prevails in Kukanwali village that represents the arid region. Largely, water markets operate under this condition.

(*iii*) *Crop Sharing Contract*: Under this condition, the buyer provides land and the seller provides water. All the paid out cost of input applied for the crop production is borne by the seller. The seller provides a fixed quantity of agriculture produce, that is, 1.60 quintals per hectare to the landowner. This condition also prevails in Kukanwali village. A limited proportion (12%) of the total buyers and sellers are involved in water transactions under this contract. Because of the small size of the holding and short supply of labour with the landowner on the one hand and resource rich farmer's socio-economic pressure on the poor landowner on the other, the present contract of water markets is chosen.

Table 15.9: Terms and conditions of water trading

Particulars	*Kukanwali*		*Srichandpura*	
	Cash	*Kind*	*Cash*	*Kind*
Water Supply Charges	NA	-Half of the produce # -1.60 qtls./ha. ##	Rs. 30 * Rs. 5 per unit* Rs. 15+ fuel**	NA

NA = denote not applicable.
*Charges for electric pump sets. ** Charges for diesel oil pump sets.
For crop input sharing contract. ## For crop sharing contract.

The types of contract under which the water markets function are determined by the informal village institutions, that is village conventions, and the sellers' priority and motive. The well owners in general and the sellers in particular determine the price level of the water transaction. The changes in price level are correlated with the prices of energy used in water extraction, that is electricity and/or diesel oil from time to time. In water scarce regions, sellers' motive occupies the prime position in the selection of contracts under which water markets work. Always, the seller is intent on getting the maximum benefits from water trading. Under the crop and input sharing contract, the well owner exploits his buyers by using monopsonist power. A crop sharing contract is similar to the reverse land tenancy where land is leased in favour of the rich farmer who has better control over key productive resources. But, exploitation of the resource poor farmers still continues as in the case of land tenancy. Due to the weak bargaining strength of the poor farmer the resource rich enjoys the hegemonistic power to exploit him for his gains in a number of interlinked markets of the rural economy (Murty, 1998).

Functioning of Groundwater Markets

In this section, an attempt has been made to examine the nature of groundwater markets in the selected areas. In Kukanwali village, the average command area of each well was 6.40 hectares of land, which was substantially higher than in Srichandpura village, that is 3.62 hectare. This is because in the arid village the farmers use the sprinkler irrigation and more-over they grow less water consuming crops like mustard and gram in a larger proportion. In semi-arid villages, the farmers use flow irrigation. Thus, the water-saving technology and approach adopted in the arid village has helped to cover a larger area under each well (NABARD, 1989). The well owners used nearly two-thirds of the total extraction of water to irrigate their own land and the rest of the of water was sold to their buyers. In Srichandpura village the situation was completely reversed for here water was used to irrigate the buyers' land in a larger proportion (*see* Table 15.10). It may be due to the fact that the well owners with a smaller size of holding have substantial surplus water that is available for sale.

Table 15.10: Functioning of groundwater markets in sample villages

Particulars	*Kukanwali*	*Srichandpura*	*Overall*
Area Irrigated by Sellers (Ha.)	6.40	3.62	5.53
Proportion of land irrigated by			
Sellers	64.69	38.40	62.93
Buyers	35.31	61.60	37.07
Annual Income Received from sale of Water (Rs.)	17368	9237	15417
Annual Operational & maintenance (Rs.)	17518	7608	15388
Expenditure Recovered by sale of Water (per cent)	99.14	121.41	100.19
Average Investment Made for Sale of Water (Rs.)	12723	6500	11556

Source: Field Survey.

There is considerable difference in the income received by the sellers from sale of water. It may be attributed to the variations in the terms and conditions of water markets across the selected villages. In Kukanwali village, sellers received the water charges in kind that were substantially higher than that of cash payments. The water charges were considerable higher than the cost of electricity used for extraction of groundwater. There was considerable difference in the operational expenditure incurred in water extraction in both the selected villages. It was more than double per well in the arid village. One of the reasons was the capacity of the pumpsets used for withdrawal of groundwater. In Srichandpura village, the pumpsets were installed up to a capacity of 10 HP capacity. Pumpsets of larger capacity, that is 15 HP and above have to pay higher amount of electricity charges as fixed by the State Electricity Board (Singh, 2000).

Largely, in the prevailing water market conditions the sellers were able to meet all operational expenditure including annual charges paid for electricity/diesel consumption, maintenance charges and interest paid on the investment made for the sale of water. In Kukanwali and Srichandpura villages, an average area of 2.26 and 2.23 hectare was irrigated through marketing practices respectively. It shows that there is slight difference in the area on which the sellers provided irrigation facilities after meeting their own water requirement. In Kukanwali village, the sellers were able to recover the operational and maintenance (O&M) expenditure to a larger extent, that is, about 99 per cent while in Srichandpura village, not only was the O&M expenditure met from the sale of water but the sellers could also obtain profits from the water business. In overall terms, the marketing practices were profitable to the sellers because these recovered all the O&M expenditure. The irrigation expenditure incurred on the crop production at well owners' farm is transferred to their buyers. It may be inferred that the water markets work on the principle of profit maximization. Sellers were not ready to compromise on the terms and conditions of water markets. Thus the buyers were compelled to participate in water trading on the existing terms and conditions due to the limited livelihood options available to them. If they resisted the prevailing conditions, than they would be opted out from the marketing practices. Because of short-term gains the farmers have to participate in water trading. In case the buyers opted for remaining outside the water trade then they would have to migrate elsewhere to earn their livelihood which is what generally happened in the water scarce area of Rajasthan. Thus, the monopsony power of the sellers dominate the water markets scenario in scarcity conditions where the sellers act in a concerted manner. On the other hand, further development of groundwater markets was dependent upon the accessibility to infrastructural and institutional facilities like transportation and communication, institutional credit and agricultural marketing (Singh, 1998). In the context of the availability of basic facilities, the farmers always prefer short-term gains in the water trade which, has unfavourable consequences in the long run.

Institutions in Groundwater Markets

Broadly, social customs, shared beliefs, taboos, national ideologies and their operationalization, bureaucratic and policy environment, market and non-market exchange system, structure of property rights and rules for changing them — all these fall in the arena of institutional economics (Shah, 1993a). Institutions may be formal or informal. Formal institutions refer to rules as coded in law by the governing structures of the legislature and judiciary. Informal institutions include conventions and code of behaviour (North, 1990). Both formal and informal institutions play an important role in managing natural resources. In changing conditions, informal institutions collapse or become ineffective or change their structure. Here, we concentrate on the role of institutions in the development of groundwater markets. In the present context, an attempt has been made to identify both the formal and informal institutions and their impact on the working of the water markets (Table 15.11).

Formal Structure of Institutions

As mentioned, the formal institutions are constitutionally recognized. The Asian experience shows that there are no such institutions that may influence water markets directly. Water rights are associated with land ownership. Under the Transfer of Property Act and Land Acquisition Act, groundwater extraction rights cannot be separated from land ownership and the individual interested in the water rights must have an ownership interest in the land (Singh, 1991). The fact that a farmer owns a plot of land endows him with private access to and privilege in the use of groundwater. Hence groundwater rights are not clearly defined. In other words, the term 'Land Lords' coincides with 'Water Lords'. Similarly, in the absence of clearly defined water rights, water markets cannot be constitutionally recognized.

Public policies are interments that influence the working of institutions in either way. The institutional credit subsidized inputs, power pricing policy, provision of agricultural marketing facilities as well as technological interventions can also be treated as formal institutions. Experience shows that not only do these institutions play an important role in increasing the demand for irrigation water but that they also contribute considerably to the emergence of water markets. The analysis shows that these institutions have an indirect relationship with water markets while there are direct linkage between them and groundwater development.

Informal Structure of Institutions

In the farming community, an informal arrangement has been recognized for groundwater use by the haves and the have-nots. This arrangement that can be known as the water market has not been legally recognized. The farming community fixes the terms and conditions generally followed by the buyers and sellers itself. The prevailing village conventions, socio-economic and environmental factors play an important role in the emergence and operation of groundwater markets in a particular region. Therefore, the functioning of water markets is area specific. Similarly, they are also treated as informal water markets. The international experience shows that formal markets are found in North and South America, whereas informal markets are more prevalent in South Asia (Easter *etal.*, 1999; Thoban, 1997).

The analysis reveals that the informal institutions in the form of village traditions, socio-economic conditions of the households, social relations and behaviour among the water users excercise a stronghold in the operation of water markets. On the other hand, the nature of the water right has the direct implication on water markets. Public policies are one of the important interments that influence directly the workings of institutions relating to water.

Table 15.11: Status of institutions and their implications on groundwater markets

Institutions	*Kukanwali*	*Srichandpura*	*Implications*
Formal			
1. Property Rights	Undefined	Undefined	Direct (unrecognized)
2. Extraction of groundwater	Restricted	Restricted	Indirect
3. Power Supply and Pricing Policy	Higher	Lower (at minimum)	Indirect
4. Institutional Credit	Within village	Outside village	Indirect
5. Subsidized Inputs	Within village	Outside village	Indirect
6. Technological intervention	Adopted by well owners	Partially adopted by the well owners	Indirect
7. Marketing Facilities	Within village	Outside village	Indirect
8. Size of Holding	Medium (overall)	Small (overall)	Direct
9. Well Ownership	Individual	Joint	Unclear
Informal			
10. Village Conventions	Dominant	Dominant	Direct
11. Price determination	Water users	Water users	Direct
12. Sellers' Priority	Dominant	Dominant	Direct
13. Sellers' Motive	Profit maximization	Profit maximisation	Direct
14. Social Relations	Dominant	Dominant	Direct
15. Mode of Payments	Kind	Cash	Direct

Source: Based on Field and Literature Survey.

Conclusions and Policy Implications

The analysis brings out some interesting aspects. The issues that emerge from our in-depth analysis provide meaningful insights that have wider policy implications. The main conclusions that emerge from the study are as below:

(*i*) Water markets mitigate the inequalities in the accessibility to groundwater resources. The individuals who do not have their own well have access to groundwater irrigation through water markets. Thus, the resource poor farmers can allocate a large proportion of total cropped area to irrigated crops and make the best use of their limited size of holding only because of the emergence of the groundwater markets. No doubt, water markets play an important role in minimizing the inequalities among the resource poor and rich farmers in the short run. But, in long run they may result in an adverse impact. The faster and excessive use of groundwater can cause an increase in the inequity among the farming community in the long run. Therefore, urgent and serious attention is needed to the emerging adverse implications in long run.

(*ii*) Water markets work on the principle of profit maximization. The sellers sell water to their buyers with the motive of gaining profits. They adopt strategies at the farm level to make available groundwater for sale. They continue their efforts for water augmentation, which helps to meet the increasing demand for groundwater for their own and buyers use. The sellers further, are not ready to compromise with the prevailing terms and conditions of the water markets which presently are in their favour. The buyers have to purchase water accordingly otherwise they will be deprived of access to groundwater irrigation. Therefore, it may be concluded that in water scarce regions, water markets work under monoposony conditions, where sellers act in a concerted manner.

(*iii*) The terms and conditions of groundwater markets that is, kind and cash also work differently. The survey of literature indicates that in water scarce regions water prices remain substantially higher than those of water-endowed regions. The experience shows that in these regions buyers use groundwater in a sustainable manner due to the higher water price. But, this argument absolutely refutes the reality, because, when the payments for groundwater are made in kind as fixed share of quantity produced, the efforts of sellers continue to make more water available to his buyer with a view to increase agricultural production so that they may get the maximum crop share as a water price.

(*iv*) Various institutional provisions have been made to check over-exploitation of groundwater. This includes the withdrawal of institutional support for digging new wells and a restriction on electric connections for irrigation purposes. Some, other restrictions are proposed on groundwater extraction (Singh, 1998) particularly in the problematic areas (Rajasthan, 1997). These measures have proved to be ineffective to regulate the extraction of groundwater. Often, such measures are taken when the situation reaches an alarming stage. In other words, when the potential zone is converted into a dark zone or an over-exploited position, only then are the proposed restrictions supposed to be exercised. If some precautionary measures are undertaken then we may be in a position to escape from reaching the critical stage. In such a situation, water markets play an undesirable role in taking the situation from bad to worse.

(*v*) The study indicates that the existing institutional set-up, both formal and informal, behaves in an adverse manner. Under the existing property rights groundwater is considered as an open resource in which farmers make a private investment thinking that they have absolute rights to groundwater beneath their land. Therefore, users treat it as private property. This results in unchecked extraction of groundwater. The unrestricted access makes the situation more complicated by converting Common Property Resources (CPRs) into Open Access Resources (OARs). Informal institutional arrangements also add their undesirable role in groundwater depletion.

Policy Implications

Some of the specific and important policy implications that could be derived from the main conclusions are as follows:

- In order to make the institutions more effective, there is an urgent need to define water rights. Groundwater should be treated as a common property resource in a real sense.
- To encourage and/or discourage the groundwater markets in particular conditions; an integral approach at various levels is needed in general and for formal institutions in particular.
- Community-based action is required for the efficient use of water resources in scarcity conditions through making the effective working of informal institutions. In this regard, the model of the Pani-Panchayat in Maharashtra (Pangare and Lokur, 1996) can be experimented upon in other scarce regions, where the community determines the use pattern of scarce resources. Under the Pani-Panchayat system, the community decides the cropping pattern keeping in view the availability of water. During scarcity of water, no farmer is allowed to grow water intensive crops on his fields. In case any individual violates the provisions made by the panchayat, the farmer is punished as per the norms.

References

Aron, Janin (2002). "Growth and Institutions: A Review of the Evidences" *The World Bank Research Observer*, Vol. 15, No. 1.

Barah, B.C. *et al.* (1993). "Economics of Water Sharing in Drought Prone Areas (Part I)," Department of Economics, University of Hyderabad (memeo).

Bhatia, Bela (1992). "Lush Fields and Parched Throats: Political Economy of Groundwater in Gujarat", *Economic and Political Weekly*. Vol. 27, No. 51-52.

Dhawan, B.D. (1995). *Groundwater Depletion, Land Degradation and Irrigation Agriculture in India*, Commonwealth Publishers: New Delhi.

Dubash, Navroz K. (2000). "Ecologically and Socially Embedded Exchange: Gujarat Model of Markets", *Economic and Political Weekly*, Vol. 35, No. 14.

Easter, K. William *et al.* (1999). "Formal and Informal Markets for Water: Institutions, Performance and Constraints", *The World Bank Observer*, Vol. 14, No. 1.

Government of Rajasthan (1997). "Draft Rajasthan Groundwater Bill," Jaipur.

Janakarajan, S. (1993). "Economic and Social Implications of Groundwater Irrigation: Some Evidences from South India", *Journal of Agricultural Economics*, Vol. 48, No. 1.

Janakarajan, S. (1997). "Consequences of Aquifer Over-exploitation: Prosperity and Depletion", *Review of Development and Change*, Vol. 11. No. 1.

Jodha, N.S. (1995). *Sustainable Development in Fragile Environment*, Centre for Environment Education, Ahmedabad.

Joshi, B.H. (1999). *Development of Agricultural Economics Through Water Management*, Ray Publishing House: Jaipur.

Kerr, John M. *et al.* (1997). *Natural Resource Economics: Theory and Application in India*, Oxford & IBH: New Delhi.

Meinzen-Dick, Ruth (1996). *Groundwater Markets in Pakistan: Participation and Productivity*, International Food Policy Research Institute, Washington, D.C.

Moench, Marcus (1992). "Chasing the Watertable: Equity and Sustainability in Groundwater Management", *Economic and Political Weeklys*, Vol. 27, No. 51-2.

Moench, Marcus (1994). "Approach to Groundwater Management: To Control or Enable?" *Economic and Political Weekly*, Vol. 29, No. 51-52.

Murty, C.S. (1998). *Land Lease Market in Agriculture*, Commonwealth Publisher: New Delhi.

NABARD (1989). *Sprinkler Irrigation in Semi-Arid Areas of Rajasthan: An Ex-Post Evaluation Study*, Bombay.

Nagaraj, N. and Chanderkant, M.G. (1997). "Intra and Inter-Generational Equity Effects of Irrigation Well Failure: Farmers in Hard Areas of India", *Economic and Political Weekly*, Vol. 32, No. 13.

Narayanamoorthy, A. (1994). "Who Sells More and Who Sells Less on Deep Bore-well, Water Business in Pudukkottai District", *Journal of India Water Resource Society*, Vol. 14, No. 1-4.

North, Douglas (1990). *Institution, Institutional Change and Economic Performance*, Cambridge University Press: Cambridge.

Pangare, Ganesh and Lokur, Vasudha (1996). *Good Society: The Pani Panchayat Model of Sustainable Development*, Indian National Trust for Art and Cultural Heritage, New Delhi.

Rathore, M.S. (2001). "Groundwater Drought in Rajasthan: A Man Made Disaster", paper presented in fourth *International Conference on Rajasthan in New Millennium*, December 28-30.

Saleth, R. Maria (1996). *Water Institutions in India: Economic, Law and Policy*, Commonwealth Publishers: New Delhi.

Satyasai, K.J.S. *et al.* (1997). "Terms of Transactions in Groundwater Markets: A Study in Anantapur District of Andhra Pradesh", *Indian Journal of Agricultural Economics*, Vol. 52, No. 4.

Shah, Tushaar (1985). Transforming Groundwater Markets into Powerful Instrument of Small Farmers Development: Lessons from Punjab, Uttar Pradesh and Gujarat, Institute of Rural Management, Anand, (memeo).

Shah, Tushaar (1992). "Sustainable Development of Groundwater Resources: Lessons from Junagadh District", *Economic and Political Weekly*, Vol. 27, No. 10-11.

Shah, Tushaar (1993a). "Institutional Foundations of Natural Resources Management", Working Paper No. 45, Institute of Rural Management, Anand (IRMA).

Shah, Tushaar (1993). *Groundwater Markets and Irrigation Development: Political Economy and Practical Policy*, Oxford University Press: Bombay.

Shah, Tushaar and Raju, K, Vengama (1988). "Groundwater Markets and Small Farmer Development", *Economic and Political Weekly*, Vol. 23, No. 13.

Singh, Chhatrapati (1991). *Water Rights and Principles of Water Resource Management*, Indian Law Institute, NM Tripathi Pvt. Ltd.: Bombay.

Singh, Dalbir (1998). "Managing Groundwater Resources for Sustainable Agricultural Development in Rajasthan: An Institutional Approach," Working Paper No. 106, Institute of Development Studies, Jaipur.

Singh, Dalbir (1999). "Emerging Issues and Challenges in Groundwater Markets: An Overview", in Ramesh Chand and V.P.S. Arora (ed.) *Agricultural Inputs and Output Market Reforms in India*, Advance Publishing Concept, New Delhi.

Singh, Dalbir (2000). "Groundwater Markets and Institutional Mechanism in Fragile Environments", Institute of Development Studies, Jaipur (memeo).

Thobani, Mateen (1997). "Formal Water Markets: Why, When, and How to Introduce Tradable Water Rights", *The World Bank Research Observer*, Vol. 12, No. 2.

Winpenny, James (1994). *Managing Water as an Economic Resource*, Routledge: London.

16

Irrigation Reforms, Institutions and Livelihoods

Opportunities and Challenges Case of Andhra Pradesh, South India

BALA RAJU NIKKU

I. Introduction

The state of Andhra Pradesh is the fifth largest state in the country with a total population of 72.7 millions.[1] This state can be divided into three main regions, namely: Coastal Andhra consisting nine districts, Telangana region comprising ten districts which are relatively drier than the other two regions, and the Rayalaseema region which represents four districts. In the state the irrigation sector reform programme was initiated by issuing a "White Paper" in the month June 1996. The paper highlighted the facts, performance and the investments made in the irrigation sector till date. It also recommended the need for a reform programme due to an increasing gap between the irrigation potential created and utilized, system decay and inability of the state to invest more budgets in the sector. Reforms in the irrigation sector are seen as vital to both the system health and economic development of the state. Consequently, Andhra Pradesh Farmer's Management of Irrigation Systems (APFMIS) Act was legislated unanimously in the assembly in March 1997, followed by the enactment of elections for 10,292 Water Users Associations (WUAs) which were conducted and office bearers for the associations were elected in June 1997. Simultaneously a 'Water Charges Review Committee' was constituted in the month of December 1997. The new irrigation sector policy was articulated in March-May 1998. After the formation of WUAs and Distributory Committees (DCs) trainings were conducted for the Presidents of WUAs and Presidents of DCs at the district headquarters and WALAMTARI, Hyderabad. The funding was met from the irrigation component of AP Economic restructuring project, financed by World Bank. Irrigation reforms in the state were among the first in a series of reforms launched by the *Telugudesam* Government.

The Irrigation Sector Profile

Andhra Pradesh is richly endowed with water resources and is appropriately called a river state and the rice bowl of the nation. Large rivers such as the Krishna, Godavari, and Pennar and several minor rivers flow across the state. Together these rivers drain 70 per cent of its land area. The water potential of Andhra Pradesh is estimated to be 7.78 million hectare metre (ha m) or 2,746 TMC (thousand million cubic feet). The ultimate irrigation potential from all sources is estimated to be 9.50 m ha. It includes 7.30 m ha from surface water and 2.20 m ha from groundwater (Raju, 2002).

As irrigation has been the primary source of agricultural growth in the state, it has always been accorded top priority. Traditionally, the irrigation sector has been allocated with sufficient plan funds next to the power sector. Over the years, the state has been able to create an irrigation potential of 4.84 million hectares out of which 2.2 million ha is contributed by tank irrigation. There are 15 major irrigation projects, 75 medium irrigation projects, and approximately 12,264 tanks in operation in the state.[2]

Plan Expenditure for Irrigation Sector

During the implementation of the First Five-Year Plan (1951-6) the total amount that was spent on major and medium irrigation and minor irrigation was 374.7 millions and 35.2 millions, respectively. The IP created during that period was 0.079 m ha under major and medium and 0.025 m ha under minor irrigation. By the time the Seventh Plan was implemented, the budget allocations were as high as 13,064.0 millions for the major and medium and 1314.0 millions for the medium irrigation projects. The potential created during the same plan period was 0.089 m ha under the major and medium and 0.067m ha under minor irrigation schemes. The total irrigation potential both under major and medium and minor irrigation in the state was increased to 4.96 m ha at the end of the Eighth Plan period (1992-7).[3]

Reforms in Irrigation Section

'Reform' can be defined as improvement in a system, an organization, a law, etc., by making changes to improve the behaviour. In other words, a change that is made to a social system, an organization, etc., in order to improve or correct it. Reforms in irrigation system management are visualized as necessary to address the issue of food availability in the future by the governments in power. Two major kinds of "reform" efforts on the part of the state are visible —"reform" of existing systems through state intervention (often accompanied by new construction), and "reform" of existing systems by the shift of responsibilities from the state to the "farmers" (Hunt, 2001). In most countries irrigation reforms policies been conceived by the government in power often in collaboration with or at the initiation of multi or bilateral funding agencies like the World Bank. These reforms can broadly be categorized as technical and institutional in nature.

The institutional reforms on the technical front are being organzied by conducting training for the Water User Association (WUA) leaders and irrigation staff in order to build the capacity and skills of the users. The formation of the competent authority from the irrigation department to provide the technical inputs and the WUA as a corporate body implementing the works with inputs from the state can be understood as a process of both technical and institutional aspects of the reform. In addition, the provisions in the act, namely "recall" and "social audit" to bring about the transparency and decision-making strengthens the institutional functioning. The main thrust of the reforms has been the transfer of management of major, medium and minor irrigation systems from the state management to the Water User Associations' (WUAs). The Government of AP has issued more than 100 government orders and memos since 1997 to create these institutions and to make them functional.

The Approach

The "Model of AP irrigation reforms" is unique especially in the scale of its implementation and operation. The approach followed was a "Bang Bang"[4] approach unlike "gradualist approaches" in implementation of the reform programmes. There are 10,292 associations which were formed and made functional with a uniform legal back up. All the irrigated area (except in some tribal pockets) in the state was brought under the management of WUAs.

The state is committed to the implementation of the PIM approach throughout the state uniformly. It has developed a *Vision 2020* document and an *Irrigation sector policy paper*[5] to guide the sectoral development. With assistance from the World Bank-assisted Andhra Pradesh Economic Restructuring Project (APERP) project which has been used initially to fund the minimum rehabilitation and for O&M works to be taken by the newly-formed associations. A clear legal framework Andhra Pradesh Farmer's Management of Irrigation Systems (APFMIS Act of 1997) extended to the whole state has further legitimized the reform process. The newly-formed WUAs are the local level organizations that are empowered with fee collection, operation, maintenance, water management, distribution and conflict-resolution. The driving force behind the reforms has been the political will and the recognition of the need to reduce the state expenditure for the irrigation sector and to protect the system decay by involving users. The Act further emphasizes, "scientific and systematic development and maintenance of irrigation infrastructure is considered best possible through farmers' organizations". The Act also empowers, "such farmers' organizations have to be given an effective role in the management and maintenance of the irrigation system for effective and reliable supply and distribution of water" (APFMIS Act, p.1).

Evolution and Features of Reform Policy

The linkages developed over a period of time with the World Bank by Chandrababu Naidu, the present Chief Minister of Andhra Pradesh, have facilitated the reform process not only in the irrigation sector but in other sectors also. The reform environment in the state and the leadership that showed considerable political will to implement reforms in the state might have facilitated the financial borrowings from the World Bank in the name of Andhra Pradesh Economic Restructuring Project (APERP). The APERP Programme aims to rehabilitate 2.45 million ha under various irrigation projects in the state. The minimum rehabilitation is envisaged at an amount of Rs.1350/ha. The Irrigation Component (IC) of the Andhra Pradesh Economic Restructuring Project (APERP-IC) provides support totalling 12,986 million to the Government of Andhra Pradesh's irrigation sector reform programme. For the development of the WUAs, a $141 million rupee budget had been spent initially according to the government reports.

The main components under APERP-IC[6] are: (*a*) Irrigation Performance Improvement which consists of: (*i*) minimum rehabilitation of major and medium irrigation schemes (2.45 million ha) with a fund allocation of Rs. 4,549 millions, and (*ii*) recurrent maintenance on major, medium and minor schemes (3.7 million ha) to ensure their sustainable functioning with a total allocated budget of Rs. 6,518 million. (*b*) Scheme Improvement/Modernization and Farmer Turnover (SIFT) by implementing pilot projects to test more intensive modernization on selective projects with a funding outlay of 1,202 millions rupees (*c*); Agricultural Intensification programme by intensification of agricultural extension services to Farmer Organizations (FOs), training and skills development of Agriculture Department staff, dissemination of information and productivity enhancement for increasing the irrigated agriculture productivity and water use efficiency, agriculture technologies, and on-farm demonstrations. The total budget allocated under this programme is Rs.113 millions. (*d*) Institutional Development of FOs and government departments by making newly formed FOs fully functional and improving management capabilities through training for farmers, government staff and NGOs. The budget allocated is Rs. 603 million. The expected outcome of these measures would include the full recovery of water charges, complete financing of irrigation O&M, an improved performance of irrigation schemes, and finally increased crop production.

Andhra Pradesh Model of Irrigation Reforms

Early arguments in favour of Irrigation Management Transfer (IMT) to water-user organizations were based on experiments that were carried out in the Philippines and Turkey. Successful examples of collective organizations much before the Water Users Associations can be seen in farmer managed irrigation systems from Nepal & Tubewell Cooperatives in Gujarat, India. However, these institutions do not represent the issues involved with the large scale irrigation system management. In large systems individuals cannot mobilize capital investments required to carry out O&M and to operate the system on a self-financing basis. Where as irrigation reforms in the state of Andhra Pradesh aimed to achieve production, efficiency and equity in the irrigation sector by transferring irrigation management to the WUAs.

The earlier research evidence shows that IMT is relatively easy and successful to implement where the irrigation systems are central to the agricultural development of the state. The average farm size should be large enough for a significant proportion of the command area. The investments made to irrigate a plot should be less when compared to the gross value of the total agricultural production out of the plot. These are some of the pre institutional arrangements that have prevailed in Mexico, USA, New Zealand, Colombia[7] and Turkey from which the other countries could learn while implementing IMT in their own countries.

II. Policy Translation and Institutional Arrangements

In this section I would like to present the initial evidences that are drawn from an ongoing research study. The field studies were conducted between 2000-2002. The findings are based on the study of one single irrigation project and the WUA falling under a branch canal. The analysis focuses on the implementation of the reform policy and its implications at the local level on the system management and livelihoods of the users in a command area.

Defining a Water User

According to the APFMIS Act "water user" means and includes any individual or corporate body or a society using water for agriculture, domestic, non domestic, power, commercial, industrial or any other purpose from a government or the corporation source of irrigation.

The definition can be interpreted, as water users are the individuals and organizations that are legally exploiting and using water for irrigation and other purposes. The unique features of the act are that, it recognizes the registered tenants also as water users. It is a known fact that none of the landowners implement or practice the tenancy act due to the fear of losing land rights. It was observed that only oral agreements do exist. There are no tenants technically or legally observed within the WUAs and hence they cannot participate in the voting for the WUA. This lead to the inactive or non-participation of the tenants and other small farmers in WUA decision-making process.

The "definition" of water users can also be interpreted or understood as someone using water for his/her own benefit or for some one else. A user can benefit from a service provided by an organization or an individual. Hence the WUA as an association of farmers who enjoy land rights in a declared command can also be seen as a user. The WUA was visualized as a corporate body in the act and was empowered to make transactions to sell water to the other users for non-agricultural purposes on a payment basis. The evidence from the study show that no single WUA could make use of the provisions of the act rather they remained as bodies that depend on the external resources.

Rights and Responsibilities of Water Users

The institutional rules that individuals use to order specific relationships with one another that govern the operation and maintenance of irrigation systems are greatly influenced by the physical and social environment (Yoder *et al.*,1988). Rights to the resource are often constructed by the participation of the users. These rights are transferable, negotiable and often compulsory in order to have an access to the common resource. The issue of water rights is a very new issue and the relevance of the rights issue in irrigation reform can be understood at three levels. They are at the level of the WUA and the individual farmers within it. The second is at the intra-system level, within the community of irrigators. The third is at the inter-system or regional level (Mollinga, 2001).

The Andhra Pradesh Farmer's Management of Irrigation Systems (APFMIS) Act provides seven basic rights and fifteen responsibilities that all the user members of the WUA are expected to follow. Some of the important rights are: to receive water as per the specified quota for use and to have the freedom for growing any crop, other than those prohibited by law, adjusting the areas within the water allocated. The act also authorizes the user to sell or transfer the water share to any other water user within the operational area of WUA. The member can suggest improvements/modifications in water deliveries and to receive equitable benefits from the activities of the organization. A year after the formation of the WUAs, a land mark provision was included in the act by recognizing not only the tenants but "any person who is in lawful possession and '*enjoyment of the land*' (italics my own) can claim membership and shall pay the water charges and the fees as may be described" in which case the WUA shall not refuse the membership of such person for the purposes of said act.[8] When it comes to responsibilities, the user is expected to adhere to the water delivery schedules and not to tamper with the system by breaching, cross bundling, damaging the structure in the minor of field channel. The user is also expected to avoid misuse and wastage of water and to conserve water by making judicious use of the irrigation supplies. The user is also expected to respect the easement rights and other customary practices in vogue in the system.

The field evidence shows that the members are not aware of their rights and responsibilities. A study conducted by GIDR–IWMI[9] in the year 2000 reported that about 52 per cent of the members were not even aware of their membership in a WUA. Unawareness regarding the WUA was more prominent especially among farmers with less than one hectare of operational land holding and also among tenants, than other categories of farmers. The reasons for the lack of awareness can be attributed to the fact that the reform programme is state-initiated and implemented from the top but not demanded by the users. The high level of unawareness in the initial years of the implementation of the reform programme suggests a large gap between the paper declaration of the APFMIS Act in which all irrigators are members of WUA by having a piece of land in the command area. Five years after the WUA formation though the awareness level has increased yet the participation of the members in the WUA activities has not occurred. Observations about the type of leaders elected or nominated for the WUA should take into account the low levels of interest or lack of understanding about the functions and powers of a WUA. It is however reported by the users that the next elections that are to be held would be held in a highly competitive political environment. The reasons reported are due to the increased awareness of the functions and powers of a WUA and perception of WUA as a source of funds.

The Leadership Pattern

The upper caste, village elite, big farmers and active party workers represent the majority section of the leadership of the WUA. A Majority of them have a direct stake in the water availability and distribution in order to have better access to irrigate their own fields. The leaders who are elected to the WUA reported various reasons such as they came to know about the WUA formation through the newspapers and their senior leaders in the party asked them to file a nomination. A majority of the present WUA leaders have political linkages and have served political posts. Some of the WUA leaders reported that this was an opportunity to serve farmers and this also saw the WUAs as the new institutions around water through which they could strengthen their political career. Some of the WUA leaders were nominated by the local politicians, that is, mostly by the MLAs in the jurisdiction in which the WUA command falls in. The WUA representatives in the study reveal that a majority of them belong to the ruling Telugudesam Party, followed by the Congress and the Communist Party. The reform programme in the state claims to be 'a political or non-political' but there is clear evidence to show the political linkages of the elected WUA representatives. In the NSP sub-division, Madhira it was evident from the discussions with the WUA leaders that out of nine WUAs four presidents belong to the TDP, two belong to the Congress and two reported that they belonged to the Communist Party.

Irrigation System Management: Role of WUA

The Act suggests about fourteen responsibilities of a Farmers' Organizations (FOs).

It provides the functional and administrative autonomy to the FOs to prepare their own operational plan and water distribution based on the water allocation, in order to serve the command. However, the envisaged objectives in the Act are not achievable by the farmers organizations. The study shows that the WUA leadership had only involved itself in taking up minimum rehabilitation (MR) and operation and maintenance (O&M) works when compared to the other functions.

The Repairs and Rehabilitation

The main reason reported for the inefficiency of the irrigation systems is due to the system decay. The repairs to the systems and canal network; and rehabilitation of the structures under the supervision of the WUAs are an area of focus in the reform programme. It is an effort towards system up-gradation to bring back the original design and efficiency of the irrigation systems. There is a genuine need for repairs and rehabilitation, however, the original design of the irrigation schemes are no more relevant to the present cropping patterns as is claimed by the irrigation staff. They also reported that it wouldn't be possible because the whole canal network would need to be redesigned which would involve much funding and administrative clearances. However, the WUA should be empowered to effect minor changes to the canals with the approval of the department. In the original design certain areas were meant for a particular crop according to the soil and rainfall pattern of the area. But over time due to failure in the irrigation supplies the present situation on the ground had changed.

Apart from changes in the cropping and rainfall patterns the infrastructure is tampered with and in many places it is in appropriate and obsolete. The present 'infrastructure intervention' is an opportunity to the department as well to the WUAs to reshape the design according to the present needs. Whereas the practice suggests that the older system pattern without any modifications is being followed which results in wastage of money that is invested into the system repairs.

When the canals were designed at the major and minor head reaches, irrigated dry crops were suggested. The tail reaches of the canals were localized for the wet cropping pattern. Today this pattern is the reverse of the original cropping plan. Institutional decentralization coupled with the hydraulic decentralization is presently required. Rethinking of suitable designs, re-localization of areas and suitable cropping pattern is essential to achieve the reform objectives.

The minimum rehabilitation programme was executed through the elected WUA and DCs are authorized to carry out the works on majors and branch canal while the project committee which is yet to form can execute works on the main canal. The works have to be carried out by the agency but not by the individuals or by the WUA representatives. The rule is to eliminate the middlemen or the contractor, who used to be the main people to execute the works. A "mobilization advance" was made available for farmers to start the works. Subsequent payments will be made after the completion of work by the WUA. Peoples' contribution amounting to fifteen per cent of the total value of the work is mandatory to receive any assistance from the department. According to the government statements 49,000 works have been taken up at a cost of Rs. 4 billion ($94 million) by the end of 1999-2000.

The visible impact of the reform programme has been that the minimum rehabilitation works and the O&M works have been taken throughout the irrigated command area by all the WUAs in each year after their formation since 1997. The works have lead to floe water to most of the tail end areas. Acknowledging the positive developments of the programme, it was also evident that the participation from the middle and marginal farmers is lacking. The presidents in the name of the WUA (agency) had carried out most of the works and hence there are no profits shown in the accounts of the WUA. In the first year, *i.e.* 1998-9 funds were given on a per acre basis. Per acre Rs. 150 to 200 were spent to carry out the rehabilitation and O&M works. Works such as earth removing, silt work, bunds strengthening, and jungle clearance had been carried out and had created many working days. But later the presidents had started using machines that replaced labour and adversely affected the livelihood of the landless.

Let us take the experience of one WUA, No. 169. The WUA for the year 1998-9 had spent six lakh rupees under MR and O&M budgets. The total budget was spent on earthwork and other manual labour. It had created eleven thousands of working days according to a departmental calculation in that area.[10] Whereas in the following years these works were replaced by constructing structures and hence 90 per cent of the budget was spent on materials rather than on labour. Government has also allowed the use of machines to do the earthen work that once again curtailed human labour, which is the source of their livelihood. It was also observed that in the year 2001-02, WUA presidents did not show any interest in carring out the works because the government instead of paying a 40 per cent advance asked for a fifteen per cent contribution to be deposited. The grants are also much lesse than the grants released in the first year. The same WUA No. 169 had received only Rs. 1,60,000 under the heads of MR and O&M grants. From this year onwards the WUAs will be receiving their share of the revenue collected in the form of cess. Many WUAs have not yet received their full share because the payment to the WUAs was dependent on the payments of the cess made by the farmers. The WUA receives its share of cess from more than one mandal revenue office because the command of the WUA has fallen into more than one mandal jurisdiction. Often this lead to serious delays in depositing the WUA share and as a result the WUAs could not take up the works in time.[11] The collection of the cess however remains with the revenue department, though it was planned to hand this over to the WUA.

Bridging the Gap Ayacut: A Myth

There are different definitions to define what is "gap ayacut". According to one definition it can be defined as the gap between the irrigation potential created and utilized. Whereas in this case there is more than one definition to define the gap. Discussions with the irrigation staff were held to understand their view in defining the gap. Some of them had shared that they are refering to the gap bridged before and after APERP. At the macro level the state leadership claims that there had been a significant achievement by bridging the gap ayacut in the first two years of the implementation of the reforms. More than seven lakh acres were brought under irrigation by bridging the gap ayacut was the claim by the political leaders in the government. One can assume that if additional areas are brought under irrigation facilitates, there will be development of agriculture and hence increase in the economic activities in that area resulting in an enhanced base for the construction of new livelihoods. The present study attempted to locate the newly-irrigated areas so that the results of the reform can be studied. Interviews with the farmers and the irrigation department officials at the lower level confirm that there has been an increase in the area that received irrigation water in the first two years. This was a result of the repairs like cleaning of canals and weed removal. Some section of the irrigation staff were of the opinion that these are not the newly irrigated areas but the areas that were never officially reported and hence there is an increase in the figures regarding the gap bridged. To find out what was the state level experience I looked at the secondary sources available and published by the Bureau of Economics for the same years.

Evidence from Secondary Analysis

Looking at the state aggregate data on the source of irrigation and the extent of contribution by different sources of major irrigation, *i.e.* canal irrigation reported as the same in the years 1998-9 and 1999-2000 with 16.34 lakh ha in the state.

Table 16.1: Source of irrigation.

Year	*Canals*	*Tanks*	*Dug wells*	*Tubewells*	*Others*	*Total (lakh.ha)*
1994-5	16.06	6.92	8.81	5.99	1.81	39.59
1996-7	16.29	8.44	9.87	7.41	1.94	43.95
1997-8	15.37	5.62	9.03	7.74	1.68	39.44
1998-9	16.34	8.10	9.81	9.16	1.98	45.39
1999-2000	16.34	6.52	9.00	10.00	1.98	43.84

Source: Economic Survey, 2000-2001, Government of Andhra Pradesh, Planning Department, A.P Secretariat, Hyderabad, p. 152.

At the same time the total irrigated area by all sources of irrigation shows a decrease from 45.39 lakh ha to 43.84 lakh ha showing an increase in the gap. If this is correct how does one understand the state governments claim of bridging the gap of seven lakh acres.

Evidence from Filed Level Data

Kalluru Division of NSP (Left canal) consists a total of 1,53,000 acres of irrigated ayacut. Out of the total, Madhira Branch Canal (MBC) represents a total ayacut of 97,285 acres. The MBC is about 34 kms length and then becomes Nidanapuram major with 20 kms length. The total ayacut of MBC falls in the jurisdiction of Kalluru sub-division *i.e.* a total of nine WUAs, Tiruvuru sub-division with two WUAs command, and Madhira sub-division with a total of 50,360 acres of localized ayacut representing nine WUAs (Table 16.2).

Table 16.2: WUA wise ayacut development (1998 to 2001) Madhira sub-division

(In hectare)

WUA	*Local-ized Aya-cut*	*Aya-cut Ha*	*Total*	*Deve+ illegal 1998-1999*	*Grand 1999-2000*	*Total*	*Deve+ illegal*	*Grand 2000-01*	*Total*	*Deve+ illegal*
		Wet	*ID*	*Total ha*	*Wet*	*ID*	*Total ha*	*Wet*	*ID*	*Total ha*
177	2107	972	1556	2528(+)	992	1600	2592(+)	1036	1532	2568 (+)
178	2943	304	2818	3122(+)	273	2794	3067(–)	329	2662	2991 (–0
179	1547	87	1540	1627(+)	162	1382	1544 (–)	4	1530	1534 (–)
180	2020	449	1612	2061(+)	537	1686	2223(+)	424	1617	2041(–)
181	1985	40	1769	1809(–)	27	1595	1622 (–)	24	1873	1897 (+)
182	2040	473	1280	1753(–)	569	1257	1826 (+)	612	1336	1948 (+)
183	2815	438	1904	2342(–)	517	2034	2551 (+)	289	2230	2519 (–)
184	2536	13	2305	2318(–)	0	1774	1774 (–)	829	1480	2309 (+)
185	2460	112	2098	2210(–)	0	1638	1638(–)	402	1749	2151(+)
Total	20453	2888	16882	1977(–)	3077	15760	18837 (–)	3949	16,009	19,958+

Source: Compiled from WUA and Sub-division Records.

The total localized ayacut in the sub-division (9WUAs) has been 20,453 ha whereas the total ayacut development in the year 1998-9 is only 19,970 ha (483 ha less than localized) whereas in the following year 1999-2000, the total ayacut development reported was 18, 837 ha only, again 1,616 ha less than the localized. In the year 2000-01 the total ayacut irrigated was reported as 19,958 ha.

In the three consecutive years there has been development

in the irrigated ayacut in the sub-division. However, the total localized ayacut is yet to be fully irrigated. Irrigation development has been taking place outside the localized command area. However I could not get enough evidence of why there has been development in the year 1999-2000 when compared to the previous year and it has shown an increase in the year 2000-2001.

The data analysis suggests that there are discrepancies in reporting the data and also that different practices and methods are followed in reporting the data. I tried to compare these statistics with the revenue department statistics. While doing the comparison I faced the problem of a "common unit". The irrigation department records are prepared "canal-wise" that may cross more than one village, whereas the revenue department's records are revenue villages-wise. It can be concluded that the data is being constructed, reported and used in order to meet the targets and priorities of participating departments. There is an urgent need for reconciliation of command, simplification of reporting and common reporting methods by all the participating departments.

Water Rights and Distribution

One of the objective of the reform policy is to provide irrigation water to the tail enders. They had been deprived of the irrigation water to which they are entitled legally. However, it was observed that water regulation and distribution is an outcome of many activities that are often governed by socio-economic, technical and political factors. The state government by means of an act, tried to bring equal access to water by the users irrespective of head and tail reaches. Access to water is derived by the prevailing "rights" system. In many places, the right to water is linked with the land ownership. However, in some other places the water right is negotiated quite independently of the land ownership. In this case, through the original investments made in the construction of the canal and by means of participation in the cleaning and maintenance of the canals every year, users maintain their rights over the water.

In Andhra Pradesh also, there is evidences of local institutions like the *Kudimaramattu, Kalva pedda systems* under which canal cleaning and repairs are be organized. The farmer members contribute their labour and hence establish their right. These local institutions do not function especially in the major canal systems because the government used to control them and invest in their maintenance. Even after the formation of the WUAs the repairs and rehabilitation of the canals was carried out by funds provided by the state and no user contribution was encouraged. Moreover the requirement of fifteen per cent contribution of the total value of the work to be shared by the users, was not net. The local government and agencies have a strong tendency to ignore these local systems of rights. In the case of Andhra Pradesh, the state had transferred the management activities of the irrigation systems but the right to structures and the canal assets still rest with the state. Hence, the WUAs are treated as implementing agencies that are not managing the resources including the infrastructure. In the absence of these rights the WUAs are technically and socially not well equipped to manage the irrigation systems efficiently.

Dynamics of Water Supply

To be able to supply water to the plot of the last farmer at the tail end, there is a need for coordination between the irrigation staff, the WUA and the farmers. This is lacking even after the formation of the WUAs. The WUAs are not able to govern the behaviour and attitudes of its members. No strict punishments and reward systems are in place. The Luskar (the link between the irrigation department and farmers) play an important role in deciding the access to water by a user. There is ample evidence that there is a tendency that the users had to *'satisfy'* the Luskar who is the last in the irrigation hierarchy responsible for the water distribution. (I have not come across any women Luskars who involved in water distribution to get more water or a favour from him.)

To correct these imbalances, the water distribution is supposed to be done by the WUA representatives with the help of the irrigation staff. But in practice the irrigation staff plays a dominant role because the WUA does not have any personnel to supervise or distribute and at the same time it does not have the tools to operate the structures. Many of the WUA presidents interviewed had shared that they do not have any 'say' in the water distribution and supply decisions. However, they can make representations if they are in need of water or request the irrigation departments higher-level staff to consider their requests.

One of the main objectives of the reform is that the WUAs will take up the water distribution and supply in their respective jurisdictions. There will be lesser losses and wastage of water which can be used more efficiently. However, the intentions of the policy could not be translated into action totally. The coordination among all the WUAs on a single or a major branch canal is not evident. The reason for this lies in that each WUA president wanted more water for his or her area as this meant more power. This as a strategy of satisfying their own WUA members will ensure re-election as a president for another term. Another reason was absence of a proper water distribution and operational plan at the WUA level as well as at the department level. In the absence of such a plan the water distribution has become an activity of 'push and pull'. If there is a political pressure for water from downstream, the water will be released to that area irrespective of the need of other areas. Many of the WUA presidents are also active political party workers and try to exert pressure through their party leaders for more water. The party leaders seize these opportunities to establish their authority and capitalize on the farmer's vote bank.

However, there are changes in terms of water selling by private people which had existed before the creation of the WUAs. In the past whenever a water crisis occurred especially in the tail reaches, the farmers raised money to contribute to the expenses of a group of people who voluntarily involved themselves to negotiate with the department and if necessary use pressure with the upstream farmers to bring water to the tail reaches. The money collected, often spent on bribes to the irrigation staff, food, and liquor and travel expenses of the group. These practices were no longer current with the initiation of the WUAs. The WUA president who is available within the village is now consulted whenever there are canal water shortages.

The main observation is that farmers subverted the rules and regulations of water distribution and tried to secure more water to their plots. The causes for this behaviour was stated as scarcity of water and unreliable supply. Especially in the event of an existing scarcity situation with no reliability on water availability, deviation from the rules and tampering with the water structures was reported as high. To address the issue there should have been the formulation of a plan by the local institutions to face the water scarcity and unreliable supplies. In the absence of such initiatives by the WUA, the water distribution is usually left to the decision of the department staff. This had reinforced the faith of the users in the department's ability only to solve the issue. In this process the WUA as a local institution lost the confidence of its members.

Hence water distribution and supply has become a function of interactions between the technical, human and institutional factors.

Technical factors:	The design of the canal, the control structures, the flow measurement devices, the location of the plot, etc.
Human factors:	The farmer's behaviour and water usage practices, rent seeking behaviour, political linkages.
Institutional Factors	The rules, regulations, punishments, and rewards.

Factors and combinations that govern the access to water :

- The location of the plot and its proximity to the type of the structure
- Water levels maintained and supply over a time period
- Type of crop
- Type of cropping pattern in the upper reaches
- Risk taking capacity of the farmer
- Technical skills of the department staff
- Existing water distribution practices, rules, sanctions, and punishments

- Irrigation staff's decisions and the role of the Luskar in distributing the water
- Coordination between the WUA and ID and among the WUAs and the Irrigation staff
- Farmer's political linkages and their pressure mechanisms
- Individual farmer caste, community, and economic and political affiliations
- Owning agricultural implements
- Local political aspirations of the political representatives
- Departmental decisions that come from the top level
- Policy matters and involvement of bureaucracy

This list can be as long as one can divide and sub-divide the entries. This is what makes water control and distribution a more complex phenomenon. The present water distribution practices show clearly that in the absence of sharing rules that head reach farmers continue to appropriate water on a first come first basis. The WUA could not play an effective role in the equitable distribution of water. Further evidence has to be collected, but if this becomes a valid observation one had to question the role the WUA and the objectives of IMT as a tool to address the inequitable water distribution.

Joint Azamaish

The method of 'Joint Azamaish' was introduced and made mandatory by the state. Representatives from the revenue, irrigation, agriculture departments and the WUA representatives participate in this exercise. The representatives will walk to make a survey and report the developed ayacut and the type of the crop and the yields. In the past, the revenue department with the help of its village-level employees collected the cess from the farmers on a village to village basis. The agriculture department conducts its experiments and announces the average yield on an area basis. The irrigation department maintains its own records about the area under irrigation. There used to be a difference in the statistics that are reported by the revenue and irrigation departments. To address these gaps in recording the method of Joint Azamaish has been introduced.

BOX: 16.1. WHAT GOVERNS THE WATER SUPPLY AND ACCESS TO WATER?

Kalluru irrigation division represents four distributory committees (DCs) (the second tier in the PIM model). There are three DCs on the MBC. When I had approached one of the DC presidents to know more about the watei distribution in MBC, he had shared that the other two DC presidents were not willing to listen to suggestions made by him regarding water allocations and distribution. He is at the head reach of the MBC. There was a meeting organized by the irrigation department in the first week of November 2001 which the Superintendent Engineer also had attended. Alternative plans like water distribution based on the "proportionate sharing" was proposed to solve the water scarcity issue, the actual designed 700-900 cusecs. Hence, the water will also be divided for each major on the MBC proportionately. So that even a tail major would receive its share of water was one suggestion. Once the water is in the major it is the WUA president and the TC members who distribute the available water to their members. This suggestion seemed viable and equitable in distribution. However, the technical problem raised by the staff was that if there is proportional sharing many of the off-takes (OTs) will not function if the designed discharge is less than the original design.

Another alternative suggestion was to operate the canal majors at the "full discharge level" in a rotation method hence all the major canals could function. To this suggestion the tail reach WUA presidents wanted that their majors should receive the water first and accordingly the rotation should follow to the head reach which the head reach WUA presidents did not agree with. The suggestion from the WUA presidents from Madhira Sub-division (tail end of MBC) is that first supply water to the down stream majors (Nidanapuram major) because the cropping pattern is mostly irrigated dry and it does not need much water and they are in high need of water during November to save their standing crops. The head reach farmer representatives did not agree saying that, 'they cannot stop their farmers' from using water that is available in the canal and flowing down stream. The meeting ended without any concrete plan and what was observed later is that the department took decisions and retained its technical advantage over a WUA.

This case of water distribution and supply in a branch canal can be applied to the entire system. The lack of coordination between the Water Users Associations left the department with no option but. This case also suggests that social pressures influence such decisions more than the technical feasibilities. How can we understand the complex water distribution in terms of the collective action principles? Collective action is visible when all the farmers come together to represent their need and to demand more water to their particular canal. Once the water (resource) is available in the canal, each member wants to have a greater share or first preference for his or her respective plots. It proves that "rational choices collectively" become "irrational individually".

The practices and methods adapted by the departments involved and the different data sets in use show that the data is being used to meet their own targets and needs by the participating departments. Added to this the government order empowers the WUA to participate in this exercise with a clear mandate of getting the process right. The WUA receives its share of financial commitment for future repairs and maintenance out of the total cess collected from a WUA jurisdiction. The present rate of water cess had been increased to two hundred rupees from sixty rupees for wet paddy crop per acre in season. Out of the cess collected fifty per cent will be deposited with the state to meet the administrative costs. The remaining fifty per cent is divided among the WUA, DC, PC and Gram Panchayat in a 50 : 20 : 20 : 10 ratio.

The Practice

The government wanted to handover the collection of the cess to WUAs, thereby relieving the revenue departments of this activity. As a policy tool the joint azamaish seems a good strategy to check the existing gaps. However, in the implementation there seems to be some practical problems. First, the revenue department does not want to lose the power and income, which it derives from dealing with the land. There has been a change within the revenue department with formation of village secretariat and appointment of a village secretary. Some of the functions or powers of a Mandal Revenue Officer were transferred to the Mandal Development Officer (MDO) by bringing these newly-formed village secretaries under his/her the control. Since the water cess collection is linked with the land ownership and titles, the newly-appointed villages secretaries are not able to grapple the intricacies of the revenue collection.

Many of the WUA presidents are of the opinion that they do not want to take the responsibility of collecting the cess. They see that the collection of cess means that one needs to deal with the farmers and to be strict about it. This act of the president may lead the farmers not to vote for the same (WUA president) in the next elections if he/she wants to recontest and elect. There are also arguments questioning whether the WUA can manage these functions without any biases. Presently, the cess collection is still with the village secretaries and the WUA is supposed to help the office in the collection of the cess by motivating the farmer members. There is no such evidence that the WUAs is actually involved in mobilizing the farmers. The cess collection is also affected because of the failure of the crop and hence poor percentages of collections were recorded. This in turn affected the share of the revenue that each WUA had received. The amounts were as low as seven thousand rupees in some cases. As claimed by a WUA president with the money available he could not even attempt to clean the canals. The president also stated that the users did not contribute their labour in canal cleaning. This showes how the user's perceptions and attitude have changed regarding the voluntary contribution of labour after the introduction of the WUAs and the works carried out by the WUAs with the help of the assured funding. Hence, there is a need to find strategies to mobilize the user's participation and alternative sources of income by the WUAs that is crucial for their own sustainability.

III. Reforms: An Opportunity to Create Livelihoods

The concept of livelihood is rapidly gaining acceptance as a valuable means of understanding the factors, that influence people's lives and well-being (Ellis, 1998). Interventions like irrigation reforms can provide work opportunities to the users if the user family was given an access to these irrigation resources, through which the user family can participate in the activities and earn some income. Pursuing this approach further will lead to interventions in land (re)distribution, access to market, credit facilities, education and health. The research question addressed in the paper is whether the present model of PIM programme can be successful in the state? The small and marginal farmers representing their small land holding pattern are cultivating a large portion of the irrigated area. The participation of these farmers is the key to the success of the programme. In theory the small and marginal farmers will not show interest in PIM/IMT if they need to invest more time and effort by attending meetings and other organizational requirements. If the net gains or benefits are low by being WUA members when compared to their time spent in the WUA activities the participation will also be less.

Conflicting Priorities: Agricultures vs. Electricity

When it comes to the priority in using the irrigation water, the state government has protected its priorities than the farmers needs, claims the DC President Mr Ranga Reddy Yachavarapu. He argues that the state had given priority to the production of electricity than the reserve water for the irrigation. According to him water scarcity for irrigation had been created. He shows the evidence that the total water available in Nagarjunasagar is supposed to meet the requirements of both the right canal and left canal command users. Whereas the State is favouring Krishna Delta farmers by releasing water from the dam into the river in the name of electricity production. The data, which is provided by the irrigation department, show that the total water released into the NSP right canal (the daily discharges from 1-5-2000 to 30-4-2001) is 182.203 TMCs. Whereas into the left canal is 167.923 and into the Krishna Delta (river) is 240.039 TMC. The previous year data also confirms the same disparity (from 1-5-1999 to 30-4-2000), that is, into the Right canal 202.509 TMC, to the left canal 176.287 TMC and into Delta through the river 400.672 TMCs water releases made by the department. In the absence of a project level committee the government has been deciding the closure and opening of the reservoirs and the allocations of the water at the system level. One of the reasons for the government not to hold elections for the project committees was that in case of the PC the State cannot impose its own decisions directly. The DC president, questions the state machinery that which government order that authorizes the releases that are made into the river, in the name of electricity. He claims that though the NSP was constructed as a multipurpose project, the prime concern is to protect the farmer's interests and agricultural purposes and not the production of electricity. He had filed a public interest litigation (PIL) case against the state and satirically puts that the *state can buy power (electricity) but not water.* He also made it a point, that there has to be a GO released about the water releases and closing dates and government has to stick to the commitments made by itself. He argues "If the state fails to protect the interest of the farmers, *i.e.* supplying irrigable water in time, in sufficient quantities which has also made a right under the APFMIS Act, the state had to pay the compensation to the farmers".

Production Thinking vs. Livelihood Thinking

In irrigation discussions, especially in large canal irrigation systems the main focus is on how to make this system management efficient, productive and hence financially viable. The main reason that governments invest in irrigation development projects is to ensure development of agriculture and hence economic development. Here, the main question one needs to ask and one that is often neglected is whose economic development and at whose cost? The discussions on livelihoods of especially, small, marginal, tenant farmers and share croppers and landless, has never been the focus of the irrigation efficiency discussions. This view considers the production achieved from a unit of land and water (production thinking). Further the overall objective of the irrigation development is also to alleviate poverty traditionally, and irrigation is considered as a powerful tool to address poverty. Often we are caught in calculating the rate of returns and production figures rather than seeking answers for equity questions. Privatization or market mechanisms are suggested as the mechanisms for the improvement of the system management. There were less rather no efforts made to understand in detail the livelihood patterns of the members in an irrigated command area and how their livelihoods were constructed and influenced by the irrigation system management.

Conclusions

- There is a need to establish strong linkages between the WUAs and the department. The formation of Project Committees is yet to be done by the state. Without the formation of project committees, the decision-making rests with the irrigation department at a higher level and hence one cannot claim the present system as a participatory irrigation management (PIM) programme.
- The formation of Project Committees is an urgent need because it can forge linkages right from the WUAs, to DCs. A true partnership can be established between the farmer organizations at all levels.
- The irrigation department needs to assess its new roles as a "technical and consultancy body" instead of as a provider as in the past. There is a need for attitudinal changes not only at the higher levels but also at the last link in the irrigation hierarchy.
- IMT might face problems while organizing small and marginal farmers because they need to spend most of their time earning a livelihood from different sources. Hence if IMT can facilitate economic activities suitable to all types of farmer groups then it will receive the participation of more farmer members. Any irrigation reform programme needs to aim at meeting the livelihood needs of the marginal, small and tenant farmers and landless families. Broad-based multi-sectoral institutional reform is required to improve the access to credit and agricultural implements and to substantiate agricultural income and the land tenure of these communities.
- A successful IMT model should meet the full-time farming needs of the users in order to help them to earn a substantial proportion of their livelihoods from irrigated farming. Agricultural production is linked to the reliable supply of water at particular intervals. Whereas supply factors are dependent on many other elements of the system management. If

the supplies are good the farmers pay the water cess. Hence, the agricultural production is directly linked with better supplies and willingness-to-pay the water fees by the users. Therefore, the reform programme should not only aim at the production aspects but needs to focus on the economic capacity building of the users.

The paper has tried to present the initial evidences from the field with an aim to broaden the understanding of implication of reform policy and institutional functioning at the local level. The high political will and support that the reform programme enjoyed at the beginning of the programme has been diluted. As a result, the institutions that are created at the local level are struggling to survive. The reform approach to institutional change needs to be further tested. The "AP model of irrigation reforms" offers an understanding of how these "models" evolve and are being constructed. The strengths and limitation of these models need to be studied before they are blindly applied in other places. Understanding the local realities and aspirations is necessary before "effective and just" reforms can be designed and implemented.

Notes

1. According to the 2001 census data.
2. Government of Andhra Pradesh, 1999. *Vision 2020*. Hyderabad. Whereas the tanks are in addition to the tanks vested under the control of the *Gram Panchayat* numbering 60,000, mainly used for drinking water, animals use, fisheries and small-scale wet irrigation.
3. The figures do not include expenditure on CADA up to the end of the Seventh Plan. The figures are adopted from the *Eighth Five-Year Plan (1990-95)*, AP-F&P (PLG) Department, Volume 11, and October 1990.
4. Andhra Pradesh reform programme began in 1997 with the formation of 10,292 WUAs covering 4.8 million ha of irrigated command in the state.
5. The Irrigation Policy articulates the state's vision for the irrigation sector and sets out the reform agenda. Published in 1999. The *Vision 2020* document states the vision of the state in 14 key areas identified

as growth engines, and is the basic document to the government's planning process.

6. More details can be downloaded from www.apirrigation.com.
7. The Mexico Irrigation reform programme began in 1991 and 1998, 3.0 million ha of irrigation were transferred to farmer-managed irrigation districts. The Programme in Turkey began in 1993 and by 1997, 1.3 million ha were transferred. For detailed studies, see Ramirez and Vargas, 1999 and see Svendsen and Nott, 1997.
8. The important amendment, right to vote or membership to a tenant member in a command area was declared in the Andhra Pradesh Farmers' Management of Irrigation Systems (Amendment) Act, 1998 which came into force with effect from 11th September, 1998.
9. For more details please refer Parthasarathy (2001, 2002).
10. As per the department rate per person 2.5 cubic metres earthwork can be done and per cubic metre rupees 21.50 will be paid.
11. One revenue mandal is a group of 20-30 villages. It is the second level of district administration headed by the Mandal Revenue Officer. At the Panchayat level the cess will be collected and debited to the Mandal Office, which in turn debit the share of the WUA in the form of DD drawn in favour of WUA.

References

Ellis, Frank (1998). "Household strategies and Rural Livelihood Diversification", *The Journal of Development Studies*, Vol. 35, Institute of Development Studies, University of Sussex, London.

Government of Andhra Pradesh (1997). *The Andhra Pradesh Farmers Management of Irrigation Systems Act (Act 11 of 1997): Acts and Rules*, Department of Irrigation and Command Area Development, GoAP, Hyderabad.

Hunt, C. Robert (2001). "Commentary: Irrigation Management and 'reform', *International Journal of Water*, Inderscience Enterprises Ltd., France.

Mollinga, P.P. (2001). *"Power in Motion: A Critical Assessment of Canal Irrigation Reform, with a Focus on India"*, Working paper/Monograph series No1, Indian Network on Participatory Irrigation Management, New Delhi.

Raju, K.V. (2002). "Participatory Irrigation Management in Andhra Pradesh: The Way Forward", in *Users in Water Management: The Andhra Model and its Replicability in India* (ed) by Hooja Rakesh, Ganesh Pangare and K.V. Raju, Rawat Publications: Jaipur and New Delhi.

Parthasarathy, R. (2002). "Comparing Gujarat with Andhra Pradesh: Reforms in Irrigation Management and People's Participation", in

Users in Water Management: The Andhra Model and its Replicability in India (ed) by Hooja Rakesh, Ganesh Pangare and K.V. Raju, Rawat Publications, Jaipur and New Delhi.

Parthasarathy, R. and Harish Joshi *et al.* (2001). "Access to Water and Equity dimensions of Irrigation Management Transfer : A comparison of Gujarat and Andhra Pradesh, India". Report submitted to the International Water Management Institute, Colombo, Sri Lanka, Gujarat Institute of Development Research, Ahmedabad, India.

Yoder, R. and Martin Edward (1988). "Water Rights and Equity Issues: A Case from Nepal" in *Searching for Equity: Conceptions of Justice and Equity in Peasant Irrigation*, Boelens Rand Davilia, G. (ed) Van Grocum Publishers: Assen.

SECTION-IV

AQUATIC ECOSYSTEMS AND SERVICES

17

Aquatic Ecosystems, Ecosystem Services and Ecological Economics

BRIJ GOPAL

As water is one of the most valuable resources on which the very survival of humans depends, it has often determined the progress and fate of civilizations, kingdoms and societies. Today, it has a direct bearing on every sector of the modern economy as its availability regulates agriculture, industry and energy. However, one readily forgets that there would be no life without water. We not only need water for ourselves and our activities but also for the millions of living organisms — from microscopic bacteria to giant trees and mammals — which directly or indirectly contribute to our existence. And, it is not only the living and human world that depends on water, but our physical non-living environment as well. This dependence is best echoed in the couplet by poet Rahim, "*Rahiman pani rakhiye, pani bin sab sun; pani gaye na ubre, moti, manas, chun*". Here, *moti* (pearl) clearly symbolizes the biodiversity and *chun* (flour) symbolizes the non-living environment whereas *manas* (human) refers to humans and their culture. Thus, water is not merely a resource or commodity but is a provider of most other resources. Further, it is a major force that, through the hydrological cycle, unites together all living and non-living components of the Earth — atmosphere, geosphere, hydrosphere and biosphere — into a self-sustaining, dynamic and resilient living system.

Aquatic Ecosystems

Water is the predominant component of the environment in several parts of the earth's landscape. Environments such as springs, streams, rivers, lakes, ponds, tanks, and marshes, and so on are not just storages of water but are habitats to a very large diversity of organisms. These habitats differ greatly in their hydrological, physical, chemical and biological characteristics, and are accordingly recognized as distinct ecosystems. Some of the more important characteristics are described below briefly.

The Physico-chemical Environment

The aquatic environment imposes several limitations on the survival and growth of organisms. The availability of oxygen and carbon dioxide is severely restricted due to their limited solubility. Saturation of soils with water eliminates the air from interstitial spaces and hence, the oxygen for roots and soil organisms. High density of water, relative to that of air, results in a steep increase in pressure on the organisms, and the light passing through water changes rapidly in its spectral composition as well as intensity. The presence of suspended organic or mineral matter further reduces light penetration as well as oxygen availability. Flow, particularly high velocity and turbulence, causes further constraints. Equally great stress is caused by periodic flooding and drying, and the greater is the amplitude of flooding, the greater is the stress.

Freshwater Biota

Numerous plants and animals, ranging from microscopic algae to large trees and from protozoa to mammals, possess various morphological, anatomical, and physiological adaptations that allow them to survive and grow in water. They differ greatly in their response to hydrological regimes which include gradients of depth, duration, frequency, amplitude and timing of flooding of land, as well as flow velocity. The responses vary also with the stage in the lifecycle. Seed germination, seedling establishment, vegetative growth and reproduction of various aquatic plants occurs under different water regimes. Seeds of many species require a wetting-drying cycle, and in most aquatic plants seeds germinate on moist soils or in shallow water when light is available. Seedlings of most emergent and woody species often do not tolerate flooding though their adults may survive deep flooding or even complete submergence for varying periods. Accordingly, plants exhibit a large variation in their spatial and temporal distribution in different aquatic ecosystems. Trees, shrubs and large rooted aquatic plants are confined to relatively shallow standing waters, riparian areas of slow flowing streams, and banks of fast-flowing rivers. Further zonation occurs along the hydrological gradient. Free-floating plants occur mostly in standing waters though they may form significant stands along sheltered margins of rivers and streams. Rarely, submerged species anchored to bottom substrata also occur in swiftly flowing streams.

Many organisms are planktonic (freely drifting within surface layers of water). They include numerous algae, crustaceans, and larval stages of several groups of organisms (including fish). Phyto- and zooplankton occur in all habitats but maximum development of their populations occurs in standing waters. A variety of invertebrates occur on the bottom sediments of shallow and slow flowing waters. A majority of insects lay eggs in waterlogged or shallow flooded habitats where their larval stage develop. Fish are common throughout the river as they depend on different habitats, particularly for feeding and breeding. Fish, as also amphibia and reptiles, breed in shallow, standing waters though their adults may grow and feed in deep and running waters. Some fish can swim against the current in fast flowing streams also. Some fishes can breathe in air and hence, survive in oxygen deficient waters. Numerous species of birds (waterfowl) inhabit shallow and deepwater bodies, feeding, breeding and/or nesting there. Other aquatic fauna include amphibians (frogs), reptiles (turtles, crocodiles, and even snakes), and even mammals (e.g., rhinoceros, beaver, otter, dolphin). Many other animals do not reside permanently in water but are adapted to utilize aquatic habitats periodically for various purposes.

Interactions between Biota and Abiotic Environment

Plants and animals inhabiting the aquatic ecosystems interact variously resulting in different communities. These interactions include competition, herbivory, carnivory and detritivory. Interactions occur also between other abiotic components such as hydrology and nutrients, light gases and temperature. The outcome of these interactions affect the biota while the biota themselves influence the physical, chemical and hydrological environment. The increased availability of nutrients results in greater primary production (eutrophication) and a decline in species richness because a few species that can efficiently utilize the nutrients outcompete others. Also, the increased amounts of decomposing organic matter cause oxygen depletion, and the anoxic conditions trigger a series of chemical changes such as denitrification, and reduction of sulphates and carbon compounds to hydrogen sulphide and methane, respectively. Further, the plants provide niches for colonization by a diversity of organisms (periphyton, epiphytes, parasites etc.) which may in turn serve as food of animals such as molluscs and fish. The community functions are also directly governed by hydrology. In deep water systems, primary production is predominantly by phytoplankton and hence, energy transfer occurs mainly through the grazing pathway in which zooplankton and fish play a major role. In many aquatic ecosystems (wetlands) that are dominated by macrophytes, large herbivores (e.g., rhinos, water buffaloes) remove a greater part of the annual production. In other aquatic ecosystems, primary production is converted to detritus which may decompose and mineralize completely, or may be used by detritivores which in turn support a food chain, or may accumulate as peat.

Plants influence the hydrology and topographic features of aquatic systems. Plants offer resistance to flow, thereby reducing its velocity. They also reduce the capacity of the waterbody to hold or carry water. This favours settling of suspended particulates and infiltration of water through soil. The rise in water levels causes flooding of neighbouring areas. The roots and rhizomes help in increasing the hydraulic conductivity of the soil, and hence, promote infiltration. Aquatic vegetation may enhance or reduce the loss of water through evapo-transpiration. Besides aiding and abetting siltation, plants may also contribute to the process by producing large amounts of organic matter. The topographic changes consequently alter the hydrological regimes that may trigger changes in biotic communities. Several animals also have considerable impacts on hydrology. For example, beavers fell and drag trees to create dams across streams, causing major changes in stream hydrology. Some animals such as rodents and otters make burrows or tunnels on lakeshores and riverbanks, leading to erosion and flooding. Larger animals such as buffaloes and rhinos affect hydrology by grazing and wallowing in water.

Influence of Biota on Water Quality

Plants have a major influence on nutrient cycles and water quality. They may act as sources or sinks of nutrients in the aquatic ecosystem. Plants that remove large amounts of nutrients from the sediments return them to water after their death and decay. Plants may also take up nutrients from the water column and accumulate them in their tissues, thereby help improve the water quality. Similarly, most plants take up and accumulate in different plant organs varying amounts of toxic substances (such as heavy metals and pesticides). Some plants exhibit luxury consumption, that is, the uptake of nutrients increases manifold with their increased availability. Transfer and biomagnification of nutrients and toxic substances along the food chain also influence the water quality.

Several other biological processes affect water quality. Precipitation of calcium and phosphorus from water is associated with the photosynthetic activity of many submerged plants. Active transport of oxygen from leaves to the roots, and its diffusion into the soil, cause precipitation of iron and manganese on the root surfaces. Decomposition of organic matter under anoxic conditions leads to the loss of nitrogen and sulphur through reduction reactions.

Interactions with Watershed

Since Forbes (1887) described lakes as microcosms, it was a general belief that lakes are closed systems. Similarly, for more than a century, rivers whose studies focused on flows, sediment loads, channel forms and temporal dynamics, were considered as unidirectional systems which change, from their headwaters to mouths, in their flow, channel form, biotic communities, and fluxes of materials and energy from catchments to oceans.

However, it was also recognized long ago that freshwater ecosystems do not function in isolation from their surrounding landscape. Through their particular hydrological attributes, they actively interact with and depend upon other systems in their watershed. They receive nutrients and sediments along with the water. Remains of dead organisms, particularly plant litter, are often a significant source of energy and nutrients for aquatic organisms. Many terrestrial animals, including humans, utilize the aquatic resources. These interactions, discussed in detail, among others, by Loucks (1990), are most intense and important at the interface between land and water which is often referred to as an ecotone (Holland, 1988; Naiman and Décamps, 1990; Naiman *et al*., 1993). Both lakes and rivers are now seen as an integral component of their watershed. A river ecosystem is no longer a simple linear feature delimited by the bed and banks of the main channel, but a three-dimensional system dependent upon longitudinal, lateral and vertical transfers of material, energy and biota (Petts and Amoros, 1996). Differences in the structure and functioning of various aquatic ecosystems arise from differences in their evolutionary history, geology, climate, watershed characteristics (geomorphology and biota in particular), anthropogenic impacts, and size, besides hydrology which is influenced by all these factors. Details are available in several recent books (Mitsch and Gosselink, 1993; Keddy, 2000; Wetzel, 2000; Calow and Petts, 1992, 1994).

Eco-system Services

The phrase "eco-system services" refers to a wide range of conditions and processes through which natural ecosystems, and the species that are part of them, help sustain and fulfil human life (Daily *et al.*, 1997). These services maintain biodiversity and production of goods whose harvest and trade represent the familiar parts of the human economy. Besides goods, the eco-system services support life in several other ways. All ecosystems include plants that convert atmospheric carbon dioxide to organic matter (biomass) through photosynthesis. This organic matter is the food (source of energy) of humans and all other animals. Timber, fuel, fibre, thatch, chemicals, medicines, and many other products result from the process of photosynthesis. The plants absorb nutrients and other chemical substances and pass them on to the animals with food. While providing these goods, ecosystems provide another service — the reduction in carbon dioxide concentration in the atmosphere as well as the supply of oxygen. The fossil fuels (coal and petroleum) that we use today are the goods provided by the ecosystems existing long before the arrival of humans on the earth. Those ecosystems also serviced the humans by reducing the CO^2 levels in the atmosphere. Plants check erosion and loss of the top layer of fertile soil. The micro-organisms and various soil animals help maintain soil fertility by decomposing the organic matter.

There are several services that are provided specifically by the aquatic systems. For example, the rivers have for millennia provided "treatment" of human wastes and various forms of organic matter. Floodplain wetland eco-systems have controlled floods and treated wastes passing through them from upland terrestrial systems. Wetlands recharge aquifers and thus provide safe groundwater while mitigating floods. Mangrove wetlands provide protection against cyclonic storms in coastal areas. All aquatic ecosystems help moderate the microclimate of the areas surrounding them. Many eco-system processes in wetlands help in the treatment of wastewaters, and now people are taking advantage of this service by constructing wetlands specifically for this purpose. Aquatic ecosystems of all kinds provide beauty to the landscape and enhance its aesthetic value. Recreation (swimming, surfing, boating, angling, river rafting, etc.) is another service provided by aquatic ecosystems. The service provided by a swimming pool is not the same as that by a natural lake or sea.

Ecological Economics

Water in rivers, lakes and wetlands is generally treated as a resource or commodity, and various water bodies are at best considered to be multiple resource systems. It is readily forgotten that the very existence of aquatic biological resources is dependent upon the availability, quantity and quality (in terms of physical and chemical characteristics) of water, and in turn, the biological resources also exert influence on the water, particularly its quality. Not only water and the biological resources cannot be considered separately, but the aquatic ecosystems interact also with their watershed. Without their respective watersheds, the water bodies have no existence. Further, all watershed-based human activities exert demand on water and/or contribute to the degradation of water quality (influx of silt, nutrients and other pollutants from land with surface or subsurface runoff) and affect the amount of water in the system through reduced flows and increasing siltation of the water body.

Economists tend to focus on water as a commodity that could be priced. Further, they consider allocation of the water resources to various uses such as agriculture, industry, domestic supplies, etc. Recent concern about biodiversity and environment has led to the discussion about allocation of a small portion of water for 'nature'. One needs to realize that the services such as fish production cannot be provided by rivers with negligible or small flows which do not allow interaction with the floodplains that are their breeding and feeding grounds. Similarly, a channelized river or a small body of water cannot provide the major ecosystem service of "treatment of wastes". Disruption of flows by diversion and withdrawal of water from rivers, drainage of wetlands, conversion of mangroves to paddy fields and similar impacts on aquatic ecosystems should be evaluated not only in terms of the immediate gains but also in terms of both short- and long-term losses of eco-system services many of which cannot be readily substituted.

I have recently discussed the management of aquatic ecosystems under different property regimes (Gopal 2002). I would like to reiterate that the aquatic resources are multiple use systems where emphasis on one or the other use adversely affects other uses. The functions and values of aquatic resources depend upon the catchment basins, and hence, the management of the aquatic resources must consider areas beyond their physical boundaries, *i.e.* in a watershed or river basin perspective.

References

Calow, P. and G.E. Petts. (Eds). (1992). *Rivers Handbook*, Vol. 1. Blackwells: Oxford, p. 526.

—— (Eds). (1994). *Rivers Handbook*, Vol. 2. Blackwells: Oxford, p. 523.

Daily, G.C., Alexander, S., Ehrlich, P.R., Goulder, L., Lubchenco, J., Matson, P.A., Mooney, H.A., Postel, S., Schneider, S.H., Tilman, D. and Woodwell, G.M. (1997). "Ecosystem services: Benefits supplied to human societies by natural ecosystems. Issues in *Ecology* 2." Ecological Society of America, Washington, D.C.

Forbes, S.A. (1887). "The lake as a microcosm". *Bull. Peoria Illinois Sci. Assoc.* Reprinted in: *Bull. Ill. Nat. Hist. Surv.* 15 (1925): 537-50.

Gopal, B. (2002). "Management issues of aquatic resources including wetlands". In Marothia, D.K. (Ed) *Institutionalizing Common Pool Resources*. Concept Publishing Co.: New Delhi, pp. 674-88.

Holland, M.M. (1988). SCOPE/MAB technical consultations on landscape boundaries. Report of a SCOPE-MAB Workshop on ecotones. *Biology International*, Special Issue 17: 47-106.

Keddy, P.A. (2000). *Wetland Ecology: Principles and Conservation*. Cambridge University Press : Cambridge.

Loucks, O. (1990). Land-water interactions. In Patten, B.C. *et al.* (Eds) *Wetlands and Shallow Continental Water Bodies*, Vol. 1: 243-58, SPB Academic Publishing: The Hague.

Mitsch, W.J. and J.G. Gosselink. (1993). *Wetlands*. Van Nostrand-Reinhold: New York.

Naiman, R.J. and H. Décamps (Eds) (1990). *The Ecology and Management of Aquatic-Terrestrial Ecotones. MAB series* 4. Unesco, Paris, and Parthenon Publishing Group: Carnforth, U.K.

Naiman, R.J., Decamps, H. & Pollock, M. (1993). "The role of riparian corridors in maintaining regional biodiversity", *Ecological Applications* 3: 209-12.

Petts, G.E. and C. Amoros, (1996). *Fluvial Hydrosystems*. Chapman & Hall: London, p. 322.

Wetzel, R.G. (2000). *Limnology*. Third edition. John Wiley: New York.

18

Valuing the Hydrological Impact of Changing Land-use

A Case of Yamuna Floodplain Wetland Ecosystems, Delhi

PUSHPAM KUMAR, AMIT LOVE,
RADHEY SHYAM SHARMA, C.R. BABU

Introduction

Wetlands are natural capital assets that provide a variety of goods and services to society. These goods and services are generated through the ecological functions performed by them. Hydrological functions especially, the groundwater recharge performed by floodplains is of prime importance as it feeds into the fresh groundwater reserve of the area. This water reserve is critical for society especially in the water scarce environment of the semi-arid tropics. Conversion of wetland areas for alternative developmental purposes leads to the disruption of these ecological functions and their subsequent benefits.

Pressure for conversion of wetlands for developmental purposes is very high especially in case of urban riparian wetlands. This is, particularly true, in the case of the floodplains of the Yamuna river corridor in Delhi. Floodplains are approximately 6.5 per cent of the total area of Delhi. Of the total stretch of the floodplains present in Delhi, the twenty-five kilometres stretch extending from Wazirabad to Okhla is perhaps the most threatened ecosystem of Delhi because of the immense anthropogenic pressures on this riparian habitat.

Delhi being an ever-expanding metropolitan area needs greater land for habitation and township development. And the floodplain becomes an easy target for conversion by the city planners. However, the conversion of the area impairs the ecological functions and the corresponding benefits which go unacknowledged and unaccounted for by the urban planners. The damage inflicted on the functions of floodplain wetland ecosystem though hidden and social in nature do not appear in the cost-benefit analysis of the conversion strategy and hence the appraisal of the wetland conversion plan is bound to be imperfect. Of the many ecological functions of the floodplains, the hydrological function of recharging of groundwater is the most prominent.

In this study groundwater recharge has been evaluated by applying the production function approach. However, the production function approach has been used through the agriculture production function in the floodplains of the Yamuna River in the corridors of Delhi. A number of development options including, the partial channelization of the river channel have been suggested for this flood area by various development agencies. To highlight the impact of change in landuse on the groundwater recharge potential of the floodplains, groundwater recharge occurring through the floodplains annually was estimated using the water balance method.

The Study Area

The twenty-five kilometre stretch of floodplains of Yamuna river corridor extending from Wazirabad barrage to Okhla barrage (Fig. 18.1) was taken into consideration for an assessment of the extent of groundwater recharge.

The area spreads over an area of approximately 3250 ha. Three barrages are present on the river, *viz.* (*i*) Wazirabad barrage, (*ii*) Yamuna barrage, and (*iii*) Okhla barrage.

These barrages regulate the flow of the river for a major part of the year except during the floods after the city experiences seasonal rains on "monsoon". The horizontal extent of the study area is bounded on both sides between the embankments which have been constructed to restrict the spread of floodwaters inside the city during the monsoons. The floodplains are predominantly dry areas except during the monsoons when seasonal flooding of the river Yamuna occurs. The local people living on the floodplains use this vast stretch of land for practicing seasonal agriculture.

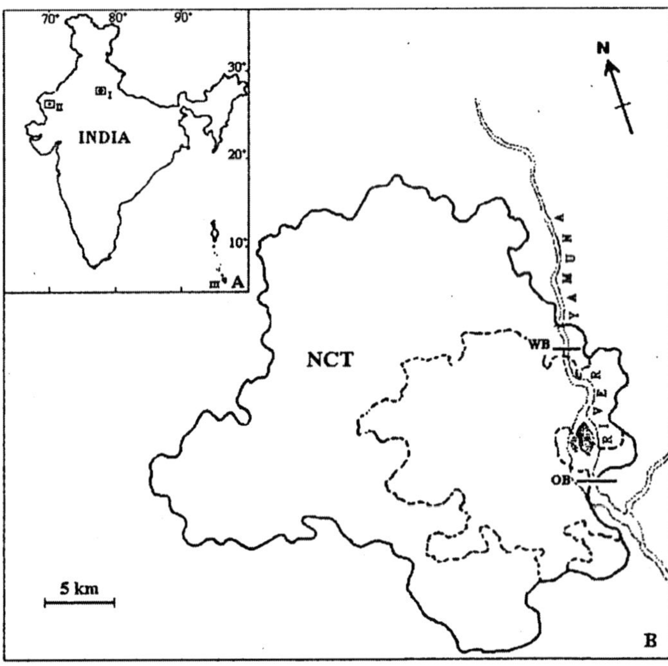

Fig. 18.1: Map showing the Location of Study Area A: Map of India Showing Location of Delhi Region; B: Map of National Capital Territory of Delhi Showing the Stretch of Yamuna River Corridor from Wazirabad barrage (WB) to Okhla barrage (OB).

Methodological Framework for Estimation of Groundwater Recharge through the Floodplains

Groundwater recharge to the aquifer through the floodplains cannot be measured directly, it can only be inferred by indirect methods. Therefore, to estimate the groundwater recharge from the floodplains to the shallow aquifers present in the study area, the water balance method (Brassington, 1993) was employed. This method has been widely used by hydrologists in studies related to applied water resources planning.

The water balance of a system is represented by the following equation:

Inputs = Outputs + Change in soil moisture.

Inputs and outputs are the inflows and outflows of the water into and from the study area, respectively and soil moisture storage includes soil retention and groundwater recharge.

Water Balance of the Study Area

Water balance was calculated based upon certain assumptions which are: (*i*) aquifers are disposed evenly in the floodplains and behave as a single unit; (*ii*) infiltration and percolation rates are similar in the whole area; and (*iii*) rate of withdrawal of water is the same in all the areas.

To estimate groundwater recharge through the floodplains, major inputs and outputs of surface water to the study area were identified and quantified. Schematic representation of the water balance of the study area is shown in Fig. 18.2.

The major sources of surface water entering the study area (inputs) are:

(*i*) Water released from the Wazirabad Barrage (WB);
(*ii*) Run off generated from Delhi area (RO); and
(*iii*) Sewage out falling into the river (S).

The major sources of outflows of surface water from the study area are:

(*i*) Water released from Okhla barrage and Agra canal (WOA);
(*ii*) Water taken for Indraprastha and Rajghat thermal power plants (WTPP); and
(*iii*) Evapo-transpirational losses of Delhi area (E).

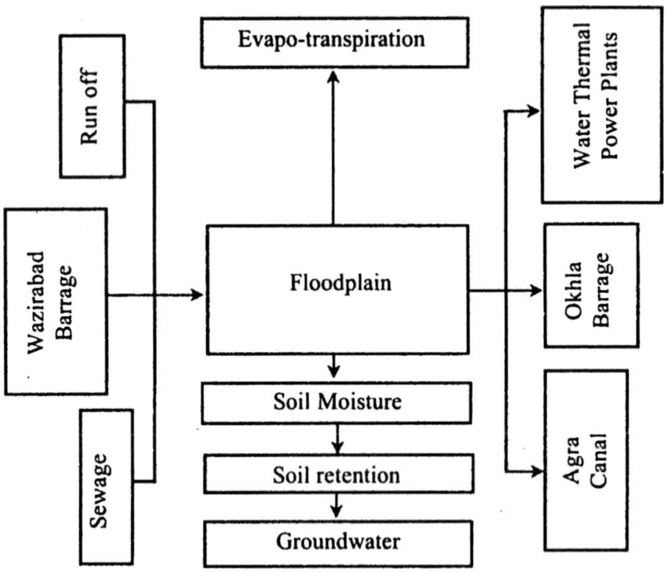

Fig. 18.2: Schematic Representation of Water Balance of the Study Area.

Based upon the major inputs and outputs of surface water to the area the following water balance equation was formulated:

$$WB+RO+S=WOA+WTPP \pm SM$$

where, SM represents the change in soil moisture.

Major Inputs and Outputs of Surface Water

Estimates of the volume of water released from Wazirabad Barrage, Okhla Barrage and Agra Canal; and water extracted by I.P. and Rajghat thermal power plants were obtained from the Flood and Irrigation Department of Delhi, Uttar Pradesh, Haryana and the amount of sewage out-falling in the river was obtained from the Central Pollution Control Board. Whereas, the volume of runoff generated and evapo-transpirational losses of Delhi were computed especially for this study, the methods used for which are described briefly below:

(*i*) *Runoff generated from Delhi*: Rainfall runoff generated from Delhi is an important source of surface water during the rainy season. Volume of runoff generated was calculated using the average value of runoff coefficients of different blocks of Delhi.

(*ii*) *Evapo-transpiration*: Evapo-transpiration is the loss of water that occurs by the process of evaporation from various free water surfaces and transpirational losses effected by vegetation. For the calculation of evapo-transpirational losses from the Delhi region the Thornwaite method (Thornwaite, C.W., 1948) was used. In this method the average monthly air temperature is used as the primary variable for an estimation of potential evapo-transpiration (PET) from an area. PET values were subjected to latitudinal correction taking into account the geographic position of Delhi. PET values provide the upper limit of evapo-transpiration, which occurs from an area. Actual evapo-transpiration (AET) values are of more relevance for various hydrologic calculations. AET was taken at 40 per cent of the PET values (Mutreja, 1986).

The term "soil moisture" in the water balance equation implies the water that enters the soil. It is calculated as the residual term of the water balance equation in the present study. After entering the soil the water distributes itself as soil moisture retention and as the water that goes for recharging of the aquifers. It may be noted that the dynamics of the water in the soil is influenced by the physical characteristics of the soil. Therefore, particle size distribution, an important physical characteristic of the soil, influences its many physical and hydraulic properties. Particle size distribution of the soils of the study area was estimated by using the wet sieving method and it was found to be the characteristic sandy loam type of soil.

Utilizing the information on soil type of the study area other parameters related to hydraulic properties of the soil were calculated. The groundwater recharged to the aquifer was calculated by taking into account the difference between the total volume of water entering the soil and the field capacity of the soil of the study area.

Analysis of the Water Balance

For analysis, the units of different inputs and outputs were converted into a uniform unit of cubic metres per month. The total inflow to the study areas was calculated on a monthly basis using estimates of all the major sources of inputs mentioned earlier. Similarly, the total outflow of surface water was calculated on a monthly basis using estimates of all the major outputs described previously. Inputs and outputs of the study area follow a similar trend around the year with low surface water flow in the river for the most part of the year followed by a characteristic peak (Fig. 18.3) when the volume of water in the river and the floodplain increases during the monsoon.

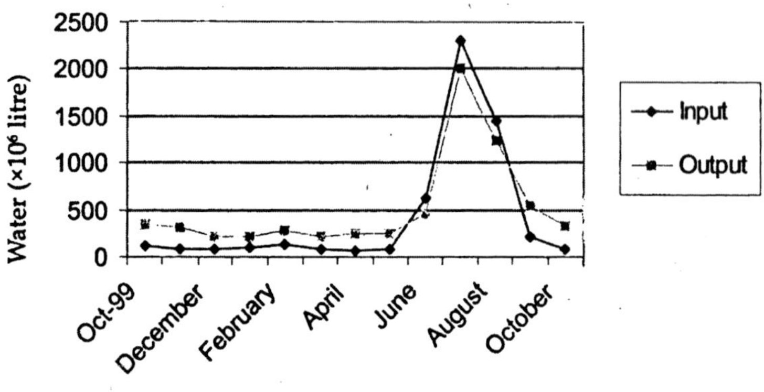

Fig. 18.3: Water Inputs and Outputs Profile of the Study Area.

As can be seen from Fig. 18.3, there was a sudden rise in the flow of water into the study area in the month of June, whereas, a maximum flow of 2.3×10^9 mcm occurs in the month of July. After attaining a maximum there is a decline in the river flow so as to reach the normal flow around October. In the three months of the monsoon period approximately 80.71 per cent of the total surface water that enters into the study area in a year, flows through the river and the floodplains resulting in floods in the study area. A majority of the groundwater recharge occurring through the floodplains takes place during the seasonal floods.

Interestingly, the soil moisture store of the study area showed a negative balance for nine months (Fig. 18.4) in a year implying that outputs from the study area are more than the inputs. Such a negative balance can result in a reduction of the groundwater levels of the area. During the monsoon season, soil moisture is positive (Fig. 18.4) indicating a possible increase in the groundwater reserve of the area. During this period a large amount of water (Table 18.1) is available in the study area, and this is more than sufficient to saturate the floodplain aquifers.

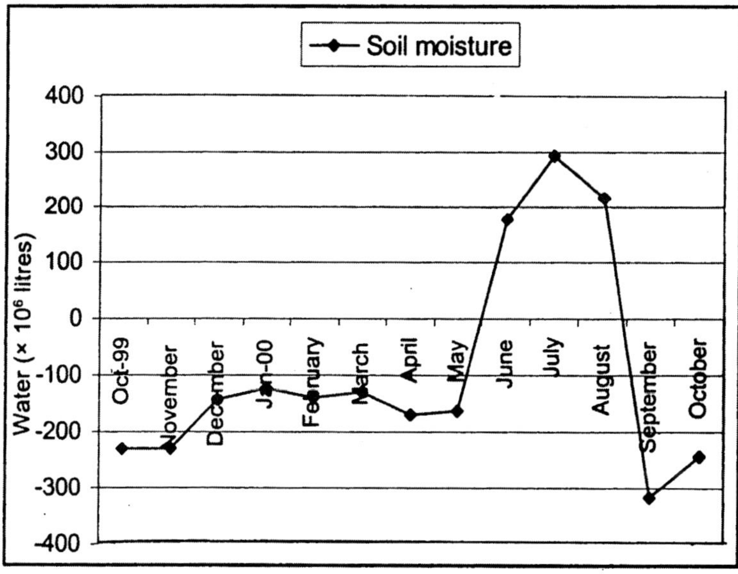

Fig. 18.4: Variations in the Soil Moisture in the Study Area from October 1999-October 2000.

The volume of water that can enter into the subsurface water system of the study area depends upon the maximum water table elevation during the lean season in the floodplains and the porosity of the soils. Approximately, 2.34×10^7 kl of water enters the groundwater reserve after the monsoons resulting in an increase in the ground water level from 3m to 2.28 m between pre- and post-monsoon seasons.

Table 18.1: Water balance for the period October, 1999 to October, 2000

Month	*Inputs*	*Outputs*	*Soil moisture*
October 1999	117787038	349210640	-231423602
November	77968800	307221211	-229252411
December	76466273	219695603	-143229330
January 2000	91230929	215958443	-124727514
February	137530401	275995439	-138465038
March	86931896	217576003	-130644107
April	71754062	243090612	-171336550
May	88034634	251189203	-163154569
June	630409547	453917440	176492107
July	$2.3*10^9$	2010000000	295000000
August	$1.4*10^9$	1239000000	218000000
September	219853161	536917440	-317064279
October	81973920	324154511	-242180591

A critical analysis of the water balance of the study area showed that excess water was available in the soil moisture store even after complete recharging of the floodplain aquifers. This can be explained only if there are certain other sinks available beside the floodplain aquifers. The presence of possible sinks can be explained as: (*i*) floodplain aquifers have hydrologic connections with the aquifers of the neighbouring city areas which led to lateral flow of groundwater to the connected aquifers in the city. This is evident from the change in groundwater levels in the city areas during post-monsoon seasons; and (*ii*) the water balance model uses a simplified hydrologic system where soil moisture is the only water store. Therefore, the volume of water that is present above the surface of the floodplains during the floods and the water present in seasonal pools after the floods recede is also shown as soil moisture in the water balance equation.

The water balance of the study area has provided key insights into the general pattern of hydrological regimes of the area. Some of the salient findings of this study on the water balance of the floodplain are: (*i*) sandy aquifers present in the active floodplain area are saturated after the monsoons; (*ii*) annually 2.34×10^7 kl water is recharged into the aquifer leading to an increment of 0.72m in the water table; (*iii*) aquifers present in the city areas are also recharged due to lateral migration of groundwater from the floodplain aquifers to the connected aquifers in the city; and (*iv*) groundwater recharged from the floodplains is an invaluable fresh water reserve which can be used for providing drinking water to the people of Delhi and for practicing dry season agriculture.

The availability of water can help agriculture or it can be used to extract water for household consumption in Delhi. Both of these benefits are not exclusive of each other. However, we derive the economic estimates of water through the agricultural production function. For the purpose of economic valuation, only groundwater recharged into the aquifers in the study area was taken into consideration.

This recharged water helps agricultural activities in the floodplain where farmers incur a lower cost on irrigation (lower cost of pumping of the water) and hence the following section focuses on the valuation of recharged water through the agricultural production function.

Table 18.2: Groundwater recharge

Area of the demarcated study zone:	3.25×10^7 m
Water recharged into the available aquifer space:	4.09×10^7kl/annum
Water reaching into the groundwater reserve:	2.34×10^7kl/annum
Depth of water table during pre-monsoon period:	3m
Increase in the mean height of the groundwater table:	0.72m
Depth of water table during post-monsoon period:	2.28m

Methodological Framework for Valuation of Water Recharge

Under this framework, the incremental change in the water table is studied in its impact on the cost of irrigation through the pumping costs of the farmers in the region. This cost avoided or that incurred gives a measure of the welfare of farmers and it is this welfare which we are measuring here.

Assuming that farmers produce l = 1...*n* crops, irrigated by groundwater. Let y_i be the aggregate output of the *i*th crop produced by the farmers. The production of y_t requires a water input W_i, abstracted through shallow tubewells, and $j = 1$ of other variable inputs (*e.g.* fertilizers, seed, labour), which are denoted as $x_i, \ldots, x_j$ or in vector form as X_j. Because of the relationship between recharge and the level of water in the aquifer it is assumed that the amount of water available to the farmer for extraction is dependent on the groundwater level, *R*. The aggregate production function for crop *i* can be expressed as:

$$y_i = y_i\,[x_{i1} \ldots x_{ij}, W_i\,(R)] \text{ for all } i \quad (1)$$

and the associated costs of producing y_i are:

$$C_i = C_x X_j + cw\,(R)\,W_i \text{ for all } i, \quad (2)$$

where C_j constitutes the minimum costs associated with producing y_i during a single growing season, c_w is the cost of pumping water and C_x is a vector of $c_{xi} \ldots c_{xj}$ strictly positive, input prices associated with the variable inputs $x_{i1} \ldots x_{ij}$. We also assume that c_w is an increasing function of the groundwater level, *R*, to allow for the possibility of increased pumping costs from greater depths, *i.e.*, $c'w > 0$, $c''\, w > 0$. We first assume that there exists an inverse demand curve for the aggregate crop output, y_i:

$$P_i = P_i\,(y_i) \text{ for all } i, \quad (3)$$

where P_i is the market price for y_i, and all other marketed inputs prices are assumed to be constant.

Denoting S_i as the social welfare arising from producing y_i, S_i is measured as the area under the demand curve (3) less the cost of the inputs used in production:

$$\begin{aligned} S_i &= S_i[x_{i1} \ldots x_{ij}, W_i(R)] \\ &= \int_0^{y} P_i(u)du - C_x X_j - c_w(R) W_i \text{ for all } i, j \end{aligned} \quad (4)$$

To maximize (4) we find the optimal values of input x_{ij} and water input W_i through setting the following first order conditions to zero:

$$\frac{\partial S_i}{\partial x_{ij}} = P_i(y_i)\frac{\partial y_i}{\partial x_{ij}} - c_{xj} = 0 \text{ for all } i, j\,. \quad (5)$$

Equations (5) and (6) are the standard optimality conditions indicating that the socially efficient level of input use occurs where the value of the marginal product of each input equals its price. We assume that each farmer is a price-taker, so that this welfare optimum is also the competitive equilibrium,

$$\frac{\partial S_i}{\partial W_i} = P_i(y_i)\frac{\partial y_i}{\partial W_i} - c_w(R) = 0 \text{ for all } i \tag{6}$$

The first order conditions in Eqs. (5) and (6) can be used to define optimal input demand functions for all other inputs as $x_{ij}^* = x_{ij}^* [c_{xj}, c_w (R),]$ and for water as $W_i^* = W_i^* [c_{xj}, c_w (R)]$. In turn, the optimal production and welfare functions are defined as $y_i^* = y_i^* [x_j^*, ..., x_j^*, W_j^* (R)$ and $S_i^* = S_i^* (x_{ij}^* ..., W_j^* (R); c_w (R)]$.[1]

From the above relationships, we would like to solve explicitly for the effects of a change in groundwater levels, R, due to a fall in recharge rates on social welfare (*i.e.*, the farmers) in the region for our purpose. Assuming that all other inputs are held constant at their optimal levels, and that all input and output prices (with the exception of c_w) are unchanged, it follows from the envelope theorem that:

$$\frac{dS_i}{dR} = [P_i(y_i^*)\frac{\partial y_i}{\partial W_i} - c_w][(\frac{\partial W_i}{\partial c_w}\frac{\partial c_w}{\partial R} + \frac{\partial W_i}{\partial R}) - W_i^*(\frac{\partial c_w}{\partial R})] \tag{7}$$

The net welfare change is, therefore, the effect of a change in groundwater levels on the value of the marginal product of water in production, less the per unit change in cost of water input. Given the Equation (7), as long as per unit pumping costs are not prohibitively high, one would expect an increase in groundwater levels (to a point that leads to a welfare benefit, or at least to maintain the initial welfare levels, whereas a decrease in groundwater levels would result in a welfare loss, either due to increased pumping costs and/or change in productivity.

If we now assume that all farmers face the same production

---

[1] Denotes optimally chosen quantities.

and cost relationships (1) and (2) for each crop *i* and are price takers, then it is possible to derive the aggregate welfare effects of a non-marginal change in groundwater levels. Let there be 1...*k* farmers producing y_{ik} output of crop *i* and using w_{ik} water inputs. It follows that by integrating Eqn. (7) over R_0 (old level) to R_1 (new level) and aggregating across all *K* farmers yields the welfare change for the farmers in the region. The welfare effects of a marginal change in groundwater levels on the aggregate output of crop *i* will be given as:

$$\Delta S_i = \sum_{k=1}^{k} \frac{\Delta S_{ik}}{dR} = \sum_{k=1}^{k} \int_{R_0}^{R_1} \left[P_i(y_i^*) \frac{\partial y_{ik}}{\partial W_{ik}} - cw_k \right]$$

$$x \left[\left(\frac{\partial W_{ik}}{\partial c\, w_k} \frac{\partial c\, w_k}{\partial R} + \frac{\partial W_{ik}}{\partial R} \right) \right] - W_{ik}^* \left(\frac{\partial c\, w_k}{\partial R} \right) dR \qquad (8)$$

Implementing the above welfare measure in Eqn. (8) requires knowledge of the production function for each crop, as well as how the equilibrium output and inputs change with *R*. Alternatively, we could measure the aggregate welfare effects directly from changes in social welfare, S_i, in Eq. (4) above. This would imply:

$$\Delta S_i = (S_{R_1}) - (S_{R_0}) = \int_0^{y_1} P_i(y_i^*) dy$$

$$- C_x X_j^* - c_w(R_1) W_j^*(R_1) - \int_0^{y_1} P_i(y_i^*) dy$$

$$- C_x X_j^* + c_w(R_0) W_j^*(R_0) \text{ for all } i, j, \qquad (9)$$

where y_0 is the initial output level and y_1 is the final output level. To use Eq. (9) as a welfare measure we would also need to estimate production functions for each crop and calculate optimal level of inputs and outputs. We return to these welfare measures in the next section where, using the information from estimated production functions, we use both measures to calculate welfare change for our sample of wheat and vegetable farmers in the floodplain.

Estimating Production Functions for Wheat and Vegetables in the Floodplains of Yamuna

In the production functions estimated below, we assume that output (y) depends on land (L), labour (B), seeds (S), fertilizer (F) and water inputs (W). The farmers in the Yamuna floodplains area mainly grow wheat and vegetables (Turnip, Radish, Cabbage, Tomato and Onion). Accordingly, the crops have been divided into six groups because of the different nature of water requirement, fertilizer application, and other factors. Wheat is grown in October-April and vegetables are grown throughout the year. Estimation of production relationships for different crops that is wheat and vegetables has been done.

Linear and log-linear functional forms have been considered for wheat and vegetable production. The linear form assumes constant marginal products and excludes any interaction between the inputs. Although the lack of interaction terms is restrictive, we observe in the literature that linear relationships are likely, particularly for wheat production and with low levels of inputs. The log-linear form assumes constant input elasticities and variable marginal products. The coefficients estimated by using this form represent output elasticities of individual variables and the sum of these elasticities indicate the nature of returns to scale. The production function has been taken as the generalized Cobb-Douglas production function which are written as:

$$Y_i = \alpha L_i^{\beta_1} B_i^{\beta_2} F_{1i}^{\beta 3} F_{2i}^{\beta 4} W_i^{\beta_5}$$
$$i = 1 \ldots\ldots 6 \text{ crops.} \quad (10)$$

The estimated linear and log-linear production functions for wheat are:

$$Y = \alpha + \beta 1 \ 1\text{n L} + \beta 2 \ 1\text{n} \ \beta 3 \ 1\text{n F1} = \beta 4 \ 1\text{n F2} + \beta \ 1\text{n W} + \mu, \quad (11)$$

and μ is the random disturbance associated with the production function.

The production function for different vegetables have been estimated separately.

Table 18.3: List of variables

Variable	*Definition*
Y	Output (kg)
L	Land (ha)
B	Labour (man-hour)
F_1	Chemical Fertiliser (DAP + urea in kg)
F_2	Dung (kg)
W	Water (I)
LY	LN (Y)
LL	LN (Land)
LB	LN (Labour)
LF_1	LN (Fertiliser)
LF_2	Dung
LW	LN (Water)

Valuing the Groundwater Recharge Function

Hydrological evidence for the relationship between flood extent and recharge to village wells show that there is a fluctuation with flood extent and mean water depth of the shallow aquifer. The effect of planned channelization/embankment will have an impact on the producer welfare within the wetlands through changes in flood extent therefore groundwater recharge.

We have found an increment in the mean depth of the water table in the floodplain area from 3 m to 2.28 m, on average after flooding every year. The expected change in welfare associated with this reduction in recharge is sought to be calculated. This exogenous change affects the farmers decision-making process during the farming season, *i.e.* after decisions on other inputs have already been taken since the effect of the reduced recharge will not be felt until after the dry season agriculture has started.

The welfare change measure for non-marginal changes in *R* (level of naturally recharged groundwater) is given by Eq. (8). This welfare change measure is used together with the results of the production function estimates to calculate welfare changes for individual farmers. We also assume that farmers in the Yamuna floodplain area are price takers and hence face a "horizontal" demand function, *i.e.* $P_i(y_i) = P_i$.

From Eq. (8) we see that the effect of R on welfare is felt through a change in water input due to increased costs and/or a change in water availability. This second effect will occur only if a change in recharge were to cause a decline in groundwater levels below 6 m. This is unlikely to happen within a single season and we do not therefore consider this aspect in calculating welfare change. Instead we consider the effect of changing pumping costs on water input and use the production function estimated earlier for the purpose of estimating welfare changes. However, in order to do so, we need to calculate $(\partial W_i/\partial c_w)=0$, the marginal change in water demand due to a marginal change in the cost of pumping. Holding all other inputs constant and noting that only water input will vary, we use the log linear production functions together with the optimality conditions in Eqs. (5) and (6) to solve for W_i as:

$$W^*_i = \left(\frac{C_w}{P_i\alpha\beta_5 L^{\beta 1} B^{\beta_2} F_1^{\beta_3} F_2^{\beta_4}}\right)^{1/(\beta_5-1)}, \tag{12}$$

where L, B, S and F_1 and F_2 are all the other inputs in the specified production function (for crop i) with estimated parameters β_L, β_B, β_s and β_F.[1]

$$\frac{\partial W_i}{\partial c_w} = \frac{1}{\beta_5-1}\left(\frac{C_w}{P_i\alpha\beta_5 L^{\beta_1} B^{\beta_2} F_1^{\beta 3} F_2^{\beta 4}}\right)^{(2-\beta_5)/(\beta_5-1)}$$

$$x\left(\frac{1}{P_i\alpha\beta_5 L^{\beta_1} B^{\beta_2} F_1^{\beta_3} F_2^{\beta_4}}\right) \tag{13}$$

[1] For the vegetable production function, the variable S (seeds/seedlings) is not included and is therefore not included in the estimation of W_i either

This is calculated for each farmer, using the estimated values for the relevant parameters and constant terms and the market price of the crop.

We now calculate welfare change due to a drop in groundwater levels to 3m, for individual farmers, using the welfare measures in Eq. (8) or Eq. (9). However, we have used the expression (9) to derive the value of change in the welfare.

Empirical Estimation

For this, first of all production for all the crops have been estimated. A generalized Cobb-Douglas model has been adopted for estimation purposes. To generate the data for the estimation, a survey of five major vegetables—tomato, turnip, onion, radish, cabbage and wheat (almost the entire crops grown) are conducted in the floodplain. The survey was designed for each crop in the month of April-May (pre-monsoon) and August- September (post-monsoon) of the years 2000 and 2001, around 15 per cent of total farmers in the floodplain were surveyed and the questions were asked primarily on inputs (water, land, labour, fertilizers, seeds, etc.) and the physical production for different crops. The focus of question was crop-wise and hence the farmers (sometimes) were repeatedly questioned regarding each crop knowing that they follow a multiple cropping system in the area. The log-linear estimate has been preferred to the linear one because of its better results.

From Table 18.5 it is clear that welfare change which is in fact the revenue earned on account of recharged water or cost saved due to the recharging facilities provided by the floodplain wetland read as –ve in the case of tomato, radish and onion, this means that for these crops water is not optimally used. For other crops, the values of the welfare change are +ve. Overall the level of welfare change is +ve to the order of more than Rs. 5 lakh for the cultivated area of approximately 772 ha in the floodplains. It should be noted here that this value is of the recharge function alone.

Table 18.4: Results for the crop production function: Dependent variable: Crops output

Explanatory variable	*Wheat*	*Onion*	*Tomato*	*Turnip*	*Radish*	*Cabbage*
Ln (DUD)	0.68	0.39	0.734	0.627	0.696	-0.66
	(8.40)	(1.59)	(5.636)	(5.665)	(6.100)	(-2.31)
Ln (W)	-1.10	0.36	0.421	0.133	0.252	0.17
	(-4.89)	(2.38)	(1.607)	(0.418)	(0.893)	(0.36)
Ln (MH)	-3.96	0.12	0.610	0.484	0.627	0.78
	(-4.96)	(0.47)	(1.593)	(1.416)	(1.608)	(1.43)
Constant	40.77	2.01	4.455	0.052	-1.894	8.82
	(5.78)	(0.76)	(3.357)	(0.015)	(-0.521)	(1.67)
R^2	0.837	0.707	0.638	0.673	0.628	0.285
Adjusted R^2	0.805	0.617	0.598	0.611	0.582	0.172
Durbin Watsan Stat.	2.089	1.428	1.500	1.432	1.693	1.329
F-Statistics	25.79	7.848	15.904	10.981	13.559	2.521
Observations	19	18	21	20	28	23

Table 18.5: Welfare change

Sl. No.	*Crop*	*Welfare Change for All Farmers (Rs./Ha)*	*Welfare Change per farmer (Rs./Ha)*	*Land under cultivation for each crop (Ha)*	*Total Welfare Change (Rs.)*
1.	Wheat	133092.8	7828.98	88.90	695996.94
2.	Cabbage	73001.16	3318.23	186.44	618651.73
3.	Onion	-45025.1	-2401.40	161.00	-386625.40
4.	Radish	-20105.3	-718.04	256.32	-184049.29
5.	Tomato	-29012.4	-1450.62	194.40	-282000.52
6.	Turnip	7497.202	394.58	187.46	73969.76
	Total				535943.22

Conclusions

The floodplains of the Yamuna river play a crucial role in the urban ecosystem of Delhi particularly with respect to groundwater recharge due to their characteristic position in the landscape of Delhi. A large amount of fresh water also enters the groundwater system of Delhi from the floodplains due to lateral migration of groundwater from the floodplain aquifers to the connected aquifers in the city. Any shift in the land use of the floodplains affecting the groundwater recharge potential, will result in a lowering of the water levels not only in the floodplains but also in the neighbouring city areas. This reduction in the groundwater recharge will affect the fresh ground stocks of the area thus having a profound impact on the water availability scenario in some parts of the city. The magnitude of recharge from the floodplains to the city areas and delimitation of the area of influence of recharge requires detailed hydrological investigations. In the absence of such studies the long term effects of land-use change on hydrological regimes in Delhi is open to speculation. Reduction in the groundwater levels due to change in land-use will lead to a technological shift from shallow bore wells which are presently used for practicing dry season agriculture in the study area to deep bore wells.

References

Acharya, G. (1998). "Hydrological-economic linkages in water resource management". Ph.D. dissertation, University of York (Unpublished).

Acharya, G. & E.B. Barbier (2000). "Valuing groundwater recharge through agricultural production in the Hadejia-Nguru wetlands in northern Nigeria", *Agricultural Economics* 22: 247-59.

Brassington, R. (1993). *Field Hydrogeology*. John Wiley & Sons: NY.

Ellis, Gregory M., and Anthony C. Fisher, (1987). 'Valuing the Environment as Input'. *Journal of Environmental Management*, 25: 149-56.

Mutreja, K.N. (1986). *Applied Hydrogeology*, Tata McGraw Hill Co.: New Delhi.

Thornthwaite, C.W. (1948). "An approach towards a rational classification of climate". *Geographical Review* 38: 55-94.

SECTION-V

DIRECTION FOR THE FUTURE

19

Emerging Institutions for Rural Development

YOGINDER K. ALAGH

Introduction

Somehow a mindset has been created that globalization and corporatization are the only sustainable ways of developing rural India. I, genuinely, believe that Indian agriculture is peasant based. Markets are a very powerful way of supporting atomistic peasant-based institutions. But community organizations and limited and focused cooperative forms of institutions are critical at the present phase of development. However these institution do not figure in the central policy debates, because these are set in notion by think tanks with little roots in India while the debate on sustainable development policy with alternative mindsets does not seem to have a market in official policy-making institutions, abroad and at home. But that does not mean that the alternative theories do not explain developments or have no application possibilities. In the world of ideas, turnarounds anyway take place in ways we do not understand. I show in this lecture that a number of arguments used in the dominant thinking are irrelevant. Community-based institutions with a sharp focus will have to be central to the next phase of rural development policies in India. Profitability rates have gone down in the agricultural sector in the period of liberalization. Instead of critically examining the causes, the argument still continues, on the old lines. There is no thinking by finance, planning and central bank policy makers on meshing strategic developmental policy rules with market reform and so the faster the deterioration, the more shrill the repetition. Now the trade debate and its meshing with sustainable development possibilities is also getting priority. India's Aggregate Measure of Support is probably no longer negative, but to some there is no change in the argument. Fortunately responding to some of us our international negotiators have taken a more sensible position. But the changes needed at home are still not sharply defined. I have argued in my Gandhi Smruti lectures that this centring of the debate with arguments outside India, is at best insufficient and at worse immoral (Alagh in Gangrade and Radhakrishna, 2000). But my purpose here is to argue that alternative possibilities are not only possible, but even global experience suggests that there are better ways of organizing rural India, so that it can unleash its own creative energies.

The Problems

We explore three critical areas. First, how can communities and individuals in fragile regions and at risk, turn around? In other words, how can sustainable development strategies be operationalized. The institutional question here is the socio-economic rules and incentives and disincentives, so that organizations and structures are in a position to foster the replicability of success stories. Second what are the policy and macro implications of supporting these strategies. Here the knowledge base is thinner, but the timing is appropriate. The nineties was the period of 'globalization'. After the East Asian crisis, the Washington Consensus is being superimposed with this (is giving up to?), in many countries, (See Jomo, K.S., 2000 for the growth and distribution impacts of the East Asian crisis), although the new strand of reasoning has little impact on India. In fact until 1999/2000, even the Economic Survey had little discussion of the East Asian meltdown. Development strategies are by definition mobilizational in the sense of resource raising, but do we have the wisdom to develop systems which will genuinely help those who help themselves, or will, resources, technology and organization support always and only, be a part of commerce, the urge of dominance and double speak. Here, even the blueprints are vague. At least the rules of structural adjustment were clear — "abolish controls, reduce government, reduce tariffs and never have special policies for employment or development of backward areas." The rules of helping those who help themselves are not known. The power of modern technology and decentralized markets is known. But how do we use this power for areas where markets and resource backups are weak, is not quite clear. What are some of the areas where development concerns will need structures at the regional, national and global level? Third and finally, can partnerships be talked about if the language of the other is not known. What are the requirements of participation and transparency? How can they be made a part of functioning systems? This is a question of alternate thinking styles at which India was very good at one stage, but alas no more. However, the least we should do is to borrow sensibly from experience elsewhere.

Sustainable Livelihood for the Marginalized

Consistent agricultural growth over a period of time has led to a decline in rural poverty levels in the areas where such growth takes place. The so-called Green Revolution areas are a case in point. Agricultural growth of 3.5 per cent annual plus for two decades or so invariably led to a critical elimination of hunger and significant declines in poverty levels. But such growth took place in areas with good soils and assured rainfall or irrigation supplies. FAO studies showed that the elasticity of cropping intensively w.r.t. irrigation was around 0 3 and so assured water supply was land augmenting and of land productivity w.r.t. irrigation was above 0.5 (*see* Agriculture Towards 2010). As the Japanese economist Y. Hayami, has shown (1981) this kind of growth raised the demand for labour, employment went up and poverty levels declined. The model of atomistic peasant agriculture worked here. The benefits of the state-supported agricultural research could reach the farmer, provided land rights were established. Input and output disposal markets, worked since irrigation technology and market support were very much, a part of this strategy, it worked in selected areas.

Planning work in India, for example, worked around it (*see* Alagh, 1988, 1989, 1990 and Planning Commission, 1989). In the early eighties the critics called it the favoured crop, favoured region model. Another critic described the planning of this strategy as limited and linear thinking. The problem, however, was in areas where these initial favourable conditions did not exist. They were bypassed in the growth process. In India, in the first phase of the Green Revolution, in a-fifth of the districts growth was negative and in another two-fifths it couldn't keep up with population growth levels. In Sahelian Africa, many countries in the rest of Africa and in some countries in Latin America, the situation was worse and continued to remain so. The prime issue of institutional development and policy is to reverse this process. With all the advances made in understanding of both the organization of agriculture, technology and resource management, precedence of mass poverty and hunger, contrasts with the claims of universal progress.

An interesting aspect of these problems is its relationship with environmental problems. These are "fragile eco-regions", the arid and semi-arid regions described in the FAO-UNESCO agro-economic atlas of the World (*see* Alexandratos, 1993). These are the hill slopes with declining tree cover and rainfall causing soil erosion; the coastal areas with mangroves disappearing; the saline lands and the problematic soils. These are, as in the plateaux and arid regions in India, areas in which through time, communities had established a balance between carrying capacity and human need. There was poverty, but also time honoured practices of sustaining the fragile resource base with activities, technologies and customs which had evolved through centuries of experimentation and adaption. In the last century, dramatic reductions in mortality and resource demands from outside had rudely shaken the carrying capacity balance of such areas.

In the preparation for the Rio Conference and particularly at the Hague Symposium and Declaration, some of us had argued that the problems of the global commons cannot be meaningfully negotiated without "addressing this problem". Sea warming, global warming and the biodiversity questions, cannot be addressed through global negotiations, if local communities are out of balance with their land and water resources. This was articulated in The Hague Declaration of which I was a signatory and for which I had worked and was drafted by Mahabub-ul Haq and J. Pronk (1992). These ideas were ultimately reflected in the Rio Declaration and Agenda 21.

Very little organized work is available on successful models under these conditions. In the late eighties, in India an attempt was made to build up a set of best practice cases, which had worked. I had summarized them in work done for starting an agro-climatic policy exercise by the Planning Commission and had written for WIDER (Alagh, 1991). The cases had some common characteristics. The economic rates of return to investment were high — 18 per cent plus on the investment made. Substantial food and energy deficits of the rural communities studied were met. The technology consisted of a land and water development followed by the introduction of appropriate "cropping" sequences. On the hill slope it was watershed development, contour building, gully plugging, work along the ridge contours. In coastal areas, it was acquifer management. In saline waterlogged soils, soil amendments and drainage. Vegetation cover was a part of the strategy. Appropriate tree cover for consolidating soil and either tree crops or the "recommended" crops, followed the land and water development strategy. Generally, a low yielding cereal was substituted by a two-crop sequence or tree cover either of which helped to consolidate the soil further.

From the replicability angle there were some interesting features. Leadership at the community level was important in all these cases. Generally, the leadership group had some science background. The leaders were from NGOs or social work organizations, retired army personnel, civil servants and in one case a farmer. The technology for the land and water development part was generally available in the institutions in the region, although some adaptions were made locally, for example, in the saline waterlogged soil reclamation project. In each of the cases, the major work was at the community level. Individual landholders had to cooperate for well-defined purposes. In fragile agro-ecological regimes limited cooperation is a precondition for land and water development strategies to succeed. If one farmer stays out, the contour bunds of the others will be washed out in the next monsoon. The atomistic model alone cannot work here.

The economics of these efforts led to interesting questions. While these projects had high internal rates of returns on the investments made, they ran into financial losses. Generally, markets were weak in fragile regions, output prices were lower than border prices and input prices like soil amendments or water pumps higher. Also in the initial phases, land productivity levels are lower and improve as the effort proceeds with the organic composition of the soil.

Sometimes low value productivity crops are needed to improve soil composition. While generating employment or improving access to food and energy, such activities need initial subsidies. Ignacy Sachs suggested that these be funded as "front up costs", in his summary paper for the Hague Conference. (Maurice Strong described the "two seminal meetings on environment and development ... Founex in 1971 and the Hague in 1991.") In this paper prepared for the Rio Conference and extensively reprinted, Sachs asks for:

> "a welcome shift from crop-oriented policies to production systems-oriented ones, with special emphasis for the needs of small farmers". Sachs pointed out that "Alagh (1991) gives many examples of watershed development projects with a short payback period. The techniques for such projects are well known and their impact at the level of the community would be very favourable. Yet they need public funding for the front up costs. Alagh argues in favour of an agro-climatic planning in terms of alternative agricultural and farming systems in order to overcome the shortcomings of a favoured crop-region approach."

This was integrated in Agenda 21 and the Rio Conference. The effort by community-level agencies has now reached such a stage that in countries like India the approach is no longer at a pilot level, but forms the beginning of a movement. The literature on case studies that I had initiated has been extensively enlarged and a table of eight cases I had presented has been replicated for over fifty efforts (K. Chopra and G. Kadekodi, 1993). More important, the largest NGO in the land and water field supported by the EEC (Sadguru in Panchmahals) and another group supported by a German programme (WOTER in Ahmednagar) had put under tree cover 1,24,316 acres, as verified by independent evaluations. This is just a little less than the tree cover lost in India in the years 1990-95. The EEC supported group, the advisory committee of which I chair, has recently completed a training programme for Sahelian tribal leaders supported by NORAD and the Aga Khan Foundation.

The requirements however are not in hundreds of thousand, but millions of hectares. Government support for such programmes in India has just about kept constant. International support has gone down. This was a period of "economic reform." Why do we find it difficult to help those who help themselves even in the core areas of local and global concern? As preparations of the Kyoto and Johannesburg meetings show, in relation to Rio targets, and more recently, the Planning Commission in relation to a review of the current plan, we are well behind in the interrelated areas of land, water and energy.

Local, National and Global Rules

The problem of imposing a hard budget constraint at the local level and helping those who help themselves, is a difficult one to address. Another way of settling the problem, is to harness the great vitality of decentralized markets in replicating widespread rural growth, within the core areas of local and global concern.The need is to harness these decentralized markets with institutions and organizations which foster limited and well focused areas of community and cooperative action that will achieve widespread rural growth. In fact in the current phase even if you think of the problematic seriously, you risk being called Luddite. (For the ludicrous nature of the policy binds we get into, see Y.K. Alagh, 2001.) When I was India's Energy Minister, forward looking local leaders had introduced schemes where if a village or industrial estate promised to pay the price of electricity, facilities would be provided to them urgently. I asked that national and international loan money should be accessed to such areas, since they had met the preconditions of "reform". I was told I was "islanding" reform. In spite of the professional literature on Alagh's insidious "islanding" proposals I would suggest that islanding is a desirable feature, because if a community is willing to pay the price of reform, it should not be asked to wait for the reform as designed by a global or national institution, for the system as a whole.

There is interestingly a lot of literature by political scientists, which shows that the mass of people in poor countries was not interested in the Washington Consensus kind of economic reform. Many of these studies have a very incomplete understanding of the reform processes in poor countries. Islanding would mean that we would relate reform to communities and also force ourselves to work out the implications for higher order systems of such issues. This is important because the Comprehensive Development Framework should not mean setting the reform objectives from the metropolis, and bringing in safety networks later, but a genuine attempt in participative growth. For developing the structure or incentive and disincentive systems for this kind of growth, we can begin with a taxonomy of complimentarities of policy rules at different levels of policy making, like no level can spend more resources than they have access to. But resources, which are short or binding constraints at national or global levels, are elastic at the local levels. However, their mobilization requires policy changes at higher levels. For example, it is easy to buy a tax free bond of the New York civic bodies, but very little attention has been paid to markets for municipal paper in developing countries and the fiscal reform that has to precede them (*see* Vaidya, 1999, for a description of an exceptional effort in Ahmedabad and the difficulties faced). Watershed development, for settled agriculture alternating with tree crops, reclamation of saline lands, farmer run lower level irrigation systems, aquifer management in difficult situations, like coastal aquifers, tribal irrigation cooperatives, tank irrigation have all been reported as success stories and studied. The question is that of replicability on a larger scale. We try to set out some rules, which if applied in functioning policies may reverse the tide:

1. The success stories are community- and leadership-based, with leadership coming from diverse sources: a progressive farmer, an NGO, a local army retired person, a "concerned" civil servant, a scientist working in the field. The leaders either had a science background or knew enough to adapt from a nearby science institution. The organization structure was neither purely private ownership, nor fully community or social control. The leadership invariably argued for aggressively functioning markets and land ownership was private and agricultural operations at the household level. However, there was for land or water management, limited and well defined cooperation. This could be drainage, soil shaping, contour management, improvement and management of lower level canals, desilting of tanks, raising embankments, fish culture, market development, controlled grazing and so on. They estimated the land and water development costs, the labour component, "outside finance", the output in terms of food requirements met, energy requirements met and fodder supplies. There were estimates of "economic rates of return on the investment", *i.e.* at accounting border prices, with a shadow wage rate 25 per cent higher than the market rate. Financial rates of return at market prices were also estimated. These studies showed high economic rates

of return, 18 per cent plus, making them very productive investments;

2. There have to be well identified shelfs of a large number of small projects on land, water and other infrastructure projects available for financing (*see* Alagh, FAO/RAP, 2002);
3. Financial institutions have to design structures such that community collateral is possible for viable projects. Self-help financing groups are only one such group. Land and water development groups, local infrastructure projects, in road or communication sectors, productionizing products developed in R&D institutions, training for production with improved techniques, market development schemes developed by local and community groups would be other examples;
4. Lending through a weather or project cycle would be necessary. NABARD had started a scheme of this kind in 1991 as a part of an agro-economic regionalization strategy started by the author, gave it up in 1993 and is again starting it now (*see* Kapur Committee, Reserve Bank of India, 2000 for details);
5. Developing policy "champions" for sorting out administrative, financial and procedural issues at local, regional and national levels, when problems arise with these kind of development strategies. It is reasonably certain that problems are going to arise in development experiments which are off the beaten track. The question then is, is there somebody in the policy decision making structure who will sort out the problem. ADB reports in a detailed study of farmer managed irrigation systems, that the failure cases were those where such support did not exist. Failure here is defined as performance levels in water delivery lower than by government agencies.

There is by now considerable acceptance of these needs in the global debates. For example, one of the more serious preparatory meetings for the World Summit on Sustainable Development at Johannesburg was the Expert Thematic Round Table on Promoting Sustainable Development in A Globalizing World, February, 2002. This Expert group squarely addressed the issues we have been pleading for as listed above. The former Swedish P.M. Ulsten, the present author and the former Environment Minister of New Zealand were the monitors for the different sections of the discussion. Amongst others, the following proposals for action emerged from the discussion of the Round Table:

- *Improve* investment processes in developing countries and countries with economies in transition to facilitate access to credit lines as well as to preferential terms of financing and of providing funds for collateral support systems and sharing of investment risk. In this context, provide securities for local institutions involved in infrastructure development and specific knowledge-based activities to support sustainable economic growth, through, for example, creation of collaterals, interest differentials and trading of financial papers. These processes should be targeted, amongst others, to artisan and producer groups linked with local and global markets, local government agencies providing social and economic infrastructure, and farming and rural communities.
- *Improve* coordination among international financial institutions and redirect funds to sustainable development projects.
- *Develop* new or *strengthen* existing mechanisms such as the Clean Development Mechanism (CDM), to finance or re-finance community projects in rural areas aimed at land and water development, agricultural diversification and agro-processing, development of infrastructure, trade, and rural energy supply.
- *Use* debt swap mechanisms to finance sustainable development projects including projects for environmental conservation.

- *Create* fairer trading systems through, for example, initiating consumer-producer partnerships for the production and consumption of goods and services that have positive social and environmental impacts and developing production and product standards for improving the competitiveness of those goods and services.
- *Strengthen* international support to developing countries for efforts of sustainable agricultural practices, while the global agricultural markets are being reformed.
- *Study* for the purpose of replication, existing models for providing access of rural communities to ICTs in order to enhance the level of information in rural communities on productions, crops, markets, prices and technologies as well as in support of medical services and education.
- *Promote* South-South cooperation and partnership aimed at bridging the so-called digital and technological divides, thereby drawing upon lessons learnt by some developing countries in creating opportunities for participation in global markets as well as capacities to benefit from new technologies.
- *Establish* a reliable knowledge and data base on global consumption to assist decision-makers in designing sustainable consumption policies and enhance understanding. (WSSD Secretariat at www.johannesburgsummit.org).

The real issue for India is whether it should be a part of this debate or should it pursue its agenda of a corporate agriculture.

Rural Diversification

Some experts say that India is not changing and is stagnating. They say that the country was changing fast in the eighties and is not doing so now. It is true that the rate of change is slower in the nineties, but the historical process of change is there and until recently, the economy was also growing fast. For example, it is no longer true that 74 per cent of the work force depends on agriculture. In fact even in 1961,while 75.9 per cent of the workforce was dependent on agriculture, only 73.5 per cent was dependent on crop production. This has now gone down to 54.3 per cent, according to the National Sample Survey's 55th Round. However, we keep on saying two-thirds, sometimes three-fourths of our people depend on agriculture. There may be some change in the census numbers when they come, but that will not change the big picture. While only 54 per cent workers are engaged in crops, those in livestock, dairying, poultry and so on has doubled, from 2 per cent to 4 per cent of the workforce. However, we keep on saying two thirds, sometimes three fourths of our people depend on agriculture. As people are better off, they eat more eggs, drink more milk and eat vegetable, fruit and cheese. I showed in my Lal Bahadur Shastri Lectures (Alagh, 1992), a decade ago that as India grows faster grains grow a little more than 2 per cent annually, edible oil, sugar, cotton and milk at around 5 per cent annually and poultry and fish at 6/7 per cent plus. This has happened in the nineties and by now this diversification of the food basket is well known. Also that it takes place at all income levels, apart from the lowest decile. However, some trends which had started in the late eighties and nineties have now deaccelerated as the following statistics show:

Table 19.1: Trends in workforce per sector

Sector	*Per cent of work force depending on sector*		
	1961	1993/94	1999/2000
Crop Production	73.5	58.1	54.3
Livestock	2.0	4.5	4.1
Forestry and Fishing	0.5	0.7	0.5

Source: NSSO and Census for 1961.

First, the per cent of workers depending on "logging, forestry and fishing" was 0.5 per cent of the work force in 1961 and is still stuck at 0.5 per cent in 1999/2000. Workers in forestry have gone down, those in fishing have gone up and their combined total remains the same. Thereby hangs a tale. Demand for forestry products is going up, like everything else. In fact we are avid readers of newspapers and so the demand for newsprint is exploding, also wood and all the other nice things that tree crops produce, like, curry patta, resins, medicinal plants and many other wonders are actively sought after. In the late eighties, Pant Saheb was the chairman of the PM's Advisory Board on Energy and convinced the country very correctly that even in an import control regime, forests should not be cut and wood should be freely imported. This continues, totally unneccesarily for all forest products.

The ban on logging is absolutely necessary, but the forests should be nurtured to produce everything else. This is not happening and we are becoming one of the largest importers of forest products, while many other countries are large exporters, say of medicinal plants, organic chemicals of forest origin and so on. These are all on OGL. Import duties in many cases are 10 to 15 per cent. The first thing to do is to clean up the tax system, so that forest produce, particularly what is called minor forest produce is not discriminated against. Apple trees get protection at 60 per cent, some trees at 40 per cent, but most at 10 per cent. No one knows why? The second thing is to do a market reform and let producer and joint forestry management committees enter into contracts with those who can sell. The role of cooperatives is clear. The third thing is to clean up the financing and technology support systems (*see* Lele, Saxena, Alagh and Mitra, 2000). Today a tribal who grows a tree on his own land, can't get financial accommodation for the sixth year, if he has seen the tree through five years with his own savings. Trees are not collateral. In fact trees are a part of our religion and then the National Forest Policy, so they dare not be put a value on! If we get into the business of nurturing our forests, the sector will lead the next agricultural revolution. Poultry and dairying grew at 5 to 7 per cent annual. Forest produce demand is growing at that rate. But production growth is zero. All because of perverse policies. Now that we have Chattisgarh and the North-East is planning a resource-based revival I hope we will get our act together. The outlines of the required policies are clear as we have seen. We have both the technology and the market at home and abroad.

Fish is a strange story. It was growing fast at around 7 per cent annual. Local demand was growing and exports were doing well. The western region in the country was doing well in inland fishing and in the deep sea, where Maharashtra and Gujarat held sway. And then in the mid-nineties it all stopped. Major projects like the FFDA were completed and no new strategies were worked out. The growth of fishermen's cooperatives stopped. We do know that East Asia was our dynamic market and when they melted down one part of the fast growing market crashed out. Forty per cent of our market growing between 8 to 10 per cent stopped growing, but that should be only half the story. What happened to the other half? The inability to tap local demand was a failure. The success stories earlier were not replicated. As dairying, fish, poultry and forest produce, becomes more important, knowledge about replicating success must be implemented. We must find out the answers for continued growth soon.

I have deliberatively talked of sectors where cooperative or well-defined organizations with community-based organization have suceeded on a replicable scale but where growth is not seen subsequently, because I believe that reviving these branches of the economy is possible in a business like manner. These problems are complex, but it is not that modern information and knowledge systems are incompatible with them. Major think tanks working on the character of the Neo-Fordist Technological Revolution, like the FAST Group of the EEC, the flexible industrial specialization networks and others have emphasized that it is strongly compatible with networking and decentralization. As the Science and Technology Minister of India, I had convened for UNESCO a Prepcom meeting at Bangalore for the World Science Conference at Budapest which strongly reiterated that spread of technology was an institutional and not just a technology issue. These issues need to be addressed when the largest clock manufacturer in the world located at Morbi in Saurashtra wants to relocate in China, as also the Surat diamond cutters (*see* Saumya Kanti Mitra, 2001).

Serious research during the last decade and a half has shown fairly conclusively that the tremendous oppor-tunities that are available with the new technology requires groups and systems which can manage its interdisciplinary nature, since applications cut across areas like biotechnology, communications and computerization. If the preconditions are available it spreads very fast, both through space and sectors of the economy and society. But if the infrastructure is not there, both physical and human, vast areas will be left out, including some in the developed world. There is also the need for a quick response. As Ricardo Petrella of the EEC's FAST Group pointed out each generation of innovations is building on the corpses of earlier ones. State and parastatal agencies find it difficult to perform in this framework. Small flexible groups responding to need, work. Again the need for partnerships, of community initiatives to back those who work is obvious. I argued with the Finance Minister that in this budget this should be one of his central tasks so that the Ajanta' stay home.

Much the same kind of approach is necessary in diverse fields like education, health, urban problems, and even in the debate on industrial restructuring policies. The question of synchronising and phasing reform has again gathered urgency after the East Asian meltdown as the recent work of experts like J. Stiglitz and R. Recupero has shown, but this debate was there all along. Ten years ago, in a WIDER monograph I had shown that in a partially liberalizing economy, efficient firms can start making losses, because the industries supplying inputs to them get more effective protection. This was the analytical base of the level playing field argument of synchronising reform in interconnected industries and phasing it, to avoid negative protection to efficient producers, Lance Taylor, in his *Rocky Road to Reform* described this as a Phased Pricing Strategy. (Taylor, 1994) Similar approaches are now again being suggested for the transition to sustainable development based on market principles in sectors like energy, agro-based and material-based industries, which have important environmental consequences.

The Planning Commission has shown in its mid-term review of the Plan that the agricultural growth rate has gone down in the nineties in a decisive manner as compared to the earlier decades. This should have led to some introspection. Instead, we are being told to do more of the same, namely, let the state and community organizations withdraw. Let the process of globalization be complete. But public investment in agriculture has gone down in a big way. Exchange control is no longer there. Public sector input supplying agencies have withered away, in areas like seeds, pesticides, credit and fertilizers. Controls on trade, investment and output are much lesser. Why is output growth less, then, in spite of excellent weather, and why is the agricultural investment rate going down ?

A basic feature of the agricultural economy, that no one is talking about is the fact that according to the detailed cost of cultivation studies analysed by the Commission on Agricultural Costs and Prices, the relative increase in input prices is much higher than in the past and is not compensated by productivity increases (Table 19.2). Since output prices are not rising, fast enough, agricultural profitability rates are going down (*see* CACP, 2000). Now this is a fairly serious matter. In the nineties, when the role of the private sector expands very considerably in supply of seeds, agrochemicals and credit, input prices go up in a significant manner and in a relative sense. Of course, with this structure growth would go down. Also as the Planning Commission points out, agricultural growth in backward regions has also gone down, because of a collapse of public investment. So we have both diminishing returns and cost push declines in profitability, in a period of substantial privatisation of agriculture. While the constraints on diversification in relation to domestic demand and exports can be explained by the structure of price support in India, the rise in input prices and falling profitability can only be explained by the inefficiency in the working of the market. The case for reform in the state-sponsored pricing system to encourage diversification of Indian agriculture is very clear. So is the need to remove bureaucratic controls on commodity movements in India and for exports. But the effectiveness of large companies in supplying cheaper inputs in village markets of small size needs study. There presence is necessary. Is it sufficient?

Table 19.2: Relative Price Movements in Indian Agriculture

Year	*Price Index*	
	Intermediate Consumption Purchased	*All Commodities Sold*
1989-90	401.4	393.5
1997-98	920.1	785.4
Percentage Increase in 97-98 over 89-90	229	199

Source: CACP, 2000, Table 4.2, p.302.

At a more basic level, the argument that since the aggregate measure of support to Indian agriculture was negative, trade liberalization would lead to great gains to agriculture, is now revised by the Bretton Woods institutions. As compared to the Purcell-Gulati estimates of negative AMS of 28 per cent, the findings for the nineties by the Purcell-Blairel studies is that the AMS is close to zero. In my arguments for the food security meeting organized recently by the Lal Bahadur Shastri Academy, I showed that we are not prepared for the WTO negotiations with our old mindset, for as J. Bhagwati has argued the WTO negotiations are conducted with World Bank-IMF data bases (*see* Alagh, 2001). In fact the argument now is that given IMF definitions of government expenditure, Indian subsidies to agriculture are around 30 per cent as compared to around a third of that level in Asian countries. The intellectual base of the present debate is indeed weak. Sharad Joshi has criticized this reasoning as Alagh Economics (Sharad Joshi, 2000). Unfortunately for us, he does not even understand the argument, unfortunately, because he was supposed to prepare us for facing the impact of the WTO agreements. The estimates, he gives as Commerce Minister, for the 1988-90 period, are all Purcell-Gulati estimates of the World Bank. The Bank has changed them since Purcell now works with Blairel, but we continue with our old swansong. It is also strange for Joshi to criticize me for ignoring the work of Tyagi, since it was effectively used for agricultural pricing only, when I was Chairman, APC. But anyway the WTO doesn't go by Joshi or by my use of APC estimates. In fact my critiques of the earlier Bank/WTO estimates were based on Tyagi's work. The fact that some of our debates don't even understand this is sad. Fortunately, our negotiators at Doha did understand that neglect of protecting policy initiatives in the future which would emerge if the old taxing agriculture arguments were carried into the future would be dangerous for India. Now that we have negotiated a vantage position for ourselves for which full credit is due to Murasoli Maran and his team, the important issue is to prepare both for the negotiations in detail and for the actual trade, that would follow. As he has said, we gave up a little in the environment area, for advantages in agriculture. I would in fact put it more positively. To use an expression by Gary Sampson, who has written extensively on the WTO, we have introduced the concept of development in the WTO discussions at an operative level, we should, therefore, now be able to steer the environment and trade discussions to our advantage. This should most certainly be true for the agriculture sector and lends urgency to the arguments in this talk.

When in July the details of the discussion by the US and the EU were coming out in driblets, it was clear that the pressure would be on us to reduce support to agriculture, when the negotiations begin seriously, which for us meant post- Doha. Economists, bureucrats and others who said this would not happen, should look at the Chinese experience. China has the same arguments. They do not accept the estimates of subsidies that the WTO uses, since their budgets use different concepts from those used, for example by the IMF for government expenditure. The Chinese State Planning Commission has made this clear, as late as November, but China has accepted the negotiating position that subsidies to agriculture, currently estimated for WTO purposes at 10 per cent of agricultural GDP, will be brought down to 8.5 per cent, although the pressure was to reduce them to 5 per cent. In India'a case the clauses on food security, livelihood security and sustainable rural development will enable her to save that part of the policy system, which is worth saving. This is a big gain.

A number of our policies are not worth saving and that is our problem, but we have the space and the time in which we can reform ourself and that is important. In fact if we are serious, next year's budget should be for the year of the WTO. Subsidies to crops which are out of sync with demand both at home and abroad cannot be through procurement because you build up unwanted commodity mountains that way.

The real policy options then are to build markets, develop information systems of the economic opportunities available, provide financial institutions which work and provide finances for communication, processing, standardization, quality upgradation and trade. Diversification is in any case not possible if the farmer does not have access to improved energy and water supplies and becomes much easier if the land development technologies are accessible. It is well known that he pays for such facilities when the service is good. Notice the queues when reliable energy is made available at a price. All of this means a great initiative to release the Indian farmers energies by improving his access to inputs and services. It means market-driven input supplies. This means that a number of organizations which cannot cover the last mile have to close down and this becomes difficult when they live on government subsidies and many others who succeed need to have access at least to bank finance. Cooperatives of the type we are pleading for have to be put in the driver's seat. The Rural Infrastructure Fund was meant to do this. But somewhere along the line it became a vehicle to finance the official irrigation plan. We have to revive it or develop a new vehicle to power the Indian farmer and his organizations to diversify, trade and improve his income. All of that will now be WTO compatible. After Doha, the main challenge is at home. The road map abroad is no longer hazy. Again we have in the nineties agreed to very low tariff bounds for forest products, fruits, vegetables, dairy products, juices, medicinal plants and all the products from which the farm sector can make money. This is tragic and needs immediate reversal at the policy level.

The role of sponsoring policies for alternative distribution channels so that there is genuine competition in input supplies is not foreseen in Indian policy reform at all. Cooperatives, non-profit organizations and partnerships between private sector, coops, NGOs or local governments is not foreseen at all. A Committee that I chaired presented a draft law for cooperatives to set up producer companies with corporate alliances (*see* Alagh, 2000). It saved the cooperative principle by providing the one share one vote basis, but as V. Kurien and the Independent Cooperative Initiative declared in the Anand Declaration in December 2000, the legislation must be passed. It has now been tabled on the floor of Parliament by the Law Minister, Arun Jaitly and we hope the Parliamentary Standing Committee under Pranab Mukherjee will support it.

There is a similar vast literature on irrigation performance and management systems. The struggle for defining rules which permit local initiative without open ended subsidies is being recognized. More important is that now an awareness exists that a set of disincentives, will have to be evolved in other words rules which discourage perverse behaviour (*see* references, below for many case studies).

There is an interesting literature which is now emerging which empirically questions the proposition that decentralization and self-management automatically increase performance levels. There can be a decline in government expenditures on the water sector. More important there can be lack of a policy focus which monitors performance and makes an effort to anticipate and develop systemic solutions to emerging problems. Empirical studies therefore show mixed results. The available results are of an extremely tentative nature needing systematic empirical study.

The literature on capacities and designs leads to interesting questions. On the one hand it is argued that there has been an excessive emphasis on building physical capacities. Traditional management systems in particular it is argued are being swamped. On the other hand it is argued that physical capacities have fallen short. The argument that the link between techno-managerial arrangements and local performance goals must be understood before measuring performance or proposing interventions probably contains an important clue (Ambler, 1991, p.11). Alagh and Buch have shown that in planning distribution systems in developing countries that will require hydraulic techniques developed elsewhere, for example, full supply hydraulic distribution systems operating in the developed world, needs a lot of innovative systems work. In the California Aqueduct or the Canal de Provence, there were very few farmers as compared to the examples they studied. The behaviour of large farming populations then has to be studied and built into capacity and design parameters. Alagh and Buch (1995) show the use of acreage response studies and water allocation and distribution systems working on conjunctive use principles with the operation authority monitoring and if necessary intervening. The Mekong Basin Indicative Plan asks for such work to be intensified. They refer to the Alagh-Buch model as "an excellant example of how agricultural econometric models are used in water resource planning to help design and manage large irrigation systems" (Mekong River Commission, 1996). This is an area of high priority work for project design and policies.

While generally a recognition of water rights will lead to smaller dams, the need for basin and interbasin transfer of water will remain. For example the stipulation in the Mekong agreement to cooperate on the maintenance of flows, "To enable the acceptable natural reverse flow of the Tonle Sap to take place during the wet season; " (Agreement on the Cooperation for the Sustainable Development of the Mekong River Basin, Art. 6), shows a sensitivity to water rights not seen earlier. These sensitivities will have important design and structural implications and work of an interdisciplinary nature not seen earlier, when the economist, the social anthropologist and the social worker were seen as add-ons after project design rather than a part of the engineering design team. Replication of these examples will need systematic work, training and manuals.

There are recent examples of difficult cultural issues being grossly mishandled by those who would know better. The important issue is that while socio-economic rehabilitation policies have been given priority, the underlying premises of the debates on sustainable development and rehabilitation, have a strong developed country focus and very little effort is made to adjust institutions and the terms of debate to the structures of poor countries where the developments take place. This is a part of the larger failure to place reform debates in specific socio-cultural settings and therefore the lack of enduring change in institutions, including those in the water sector (Alagh, 1999).

On Limited Wants

A basic issue that has to be accepted in this century is that in spite of great technological progress of the kind that Gallopin and Colombo and others have outlined, the lifestyle of the rich is simply not feasible for the world as a whole. Now that China and India are growing faster, this question which was raised at the Hague meeting by Gammini Correa, cannot be avoided for long.

The notion that wants should be limited as a moral question is there in many religions. Yet most practical people find it unacceptable. In fact, the powerful counter factual is to the effect that given limited resources if everybody maximizes his pleasure, the world will be the happiest possible place. In the early years of Independence, apart from morality, there was the compulsion that savings rates were low and economic requirements were very pressing. It is now being argued that the concept of forced savings which is an outcome which is the other side of the coin of limited wants, is both unnecessary and inefficient, in the sense that it would misallocate resources. From this premise it would follow that we should bravely go in for opening up India's consumer goods and service industries for example, for international investment and competition.

The problem I suggest lies in another direction. Wants can be created today in a manner such that was never possible earlier. Modern communication is both an industry and a consumer product. It works with a devastating power and logic and profit seeking that goes behind itself. I am not against modern technology and modern communication. I think it is silly to keep on sticking to outdated technologies. In fact I support the Department of Space when it says that it wants to leap frog and make its own facilities for the beaming of the latest messages on data products, communication and entertainment. I am proud of the fact that we have a Satellite Policy Document, which I have always whole-heartedly supported. Regulation, however, as our Satcom Policy shows is important for both cultural and economic reasons and is being followed across the globe since nations do tend to defend their cultural identities. The only prerequisite is that it must be transparent and fair.

The more serious questions on unrestricted pursuit of consumerism and utilitarian pleasure as universal objective arises from the fact that in a world of finite resources this is simply an unfeasible objective. Even if technical change continues at the pace at which it is continuing today it is extremely likely that by the time the Chinese and Indian per capita levels cross around a quarter of the current levels in the United States, the world will destroy itself. As a Senior Adviser to the Secretary General of the Rio Conference, I was given the privilege of being "one of 40 leading thinkers from all over the World" to prepare. *The Hague Report on Sustainable Development: From Concept to Action*, which formed part of the preparations for the UN Conference on Environment and Development, by Mr. Maurice Strong. These "leaders" had the following to say:

> "Sustainable development is concerned with models of material consumption which are replicable, while respecting cultural diversity. Such models recognise the limitations of environmental space. They do not treat ecospace as a free good, to be plundered at the free will of any nation or any individual. They put a price on this space, reflecting its relatively scarcity today and tomorrow. These models of development treat ecological space as any other scarce asset and are concerned with policies of sensible asset management."

> "At the global level, sustainable models of development require no less than a new global ethics, a clear understanding that the World cannot be made safe without the willing cooperation of everyone. Concern for common survival must also lead to policies for a more equitable order, based on fundamental global reform."
>
> "Out of such an unequal World our planet simply cannot emerge without some basic reforms. Nor can shared responsibility be created for the health of the Global Commons without some measure of shared global responsibility. Global sustainability without global justice will always remain an elusive goal."

The UNCED Conference itself accepted this doctrine in the Rio Declaration in the following words:

> "*Principle 8*. To achieve sustainable development and a higher quality of life for all people, States should reduce and eliminate unsustainable patterns of production and consumption and promote appropriate demographic policies."

I would suggest that India can afford to neglect this compulsion at considerable risk to its future. A more pressing concern is the fact that the Planning Commission has brought out that not only public, but private savings are now falling and other things remaining the same this means a lower growth rate. The dream merchants seem to have sold the good life too soon.

Conclusion

The word coop as a political scientist of the new emerging systems has argued appears very often now. There are cooperative political arrangements at different levels of government, cooperation with non-governmental agencies, cooperative companies and cooperative arrangements with companies. Ismail Seragaldin has argued that developing countries will have to enter into contract arrangements for health and agriculture with the big genome company in the private sector emerging at the beginning of this century. It seems to be recognised the world over that strategic alliances can be worked out such that limited and focused degrees of community participation become the norm in the reform process.

Decentralized market principles are easily grafted on this mode. The falling profit rates in Indian agriculture need fresh thinking. In a period when there has been large scale withdrawal of the state from input supplies, the rising input costs per unit, require fresh design of institutions which supply the farm sector. In addition to corporate enterprises, these will be cooperative and community-based organizations. I believe that a growing and diversified agriculture holds the key to Indian development. I hope we will champion it and the reform it needs. I have tried to give the main outlines. It is doable, provided the will is there.

A massive programme for land and water development has to be the centre piece. It must be led by a public initiative, with a very large financing component. Newer institutional forms like cooperatives, NGOs, non-profit companies and others must be designed. The new technologies in cereals like the bajra which can be stored, paddy, including the hybrids must get off the research success stage. Shorter duration cereals will pave the way for diversification to oilseeds, pulses, fruits and vegetables. Markets and their development and processing are the key. Strategic alliances must be encouraged.

Removal of restrictions on trade will help agriculture and farmer-based organizations. The net gains from trade may be small at the present stage, but are definitely worth internalizing. Government policies towards newer strategic organizations like the Alagh Committee's proposals on corporate structures for cooperatives must be accepted, rather than protecting the *sarkari* coops as the Independent Cooperative Initiative has argued.

The organizational and financial rules to make these

arrangements and institutions effective will need considerable innovativeness. Problems of community collateral, recognition of hard budget constraints and yet flexibility at the local level, needs policy coordination of a high order. Experience is that *mantras* of one or the other kind do not succeed. Hands on policy systems which monitor initial reform systems, anticipate and solve problems as they emerge (and one can be sure that problems would emerge) are the preconditions of success as the ADBs Vision of Asian agriculture estimates through a review of best practice cases (ADB, 2000). India needs to develop such a perspective.

(This is a revised version of the author's Gadgil Memorial Lecture.)

References

Alagh, Y.K. (1988). Guidelines for Agroclimatic Planning: A Draft for Discussion, *Journal of Land Development*.

———, (1991). *Indian Development Planning and Policy.* WIDER Studies in Development Economics, Vikas: Helsinki and Delhi.

———, (1991). Sustainable Development, From Concept to Action: Techniques for Planners, UNCED.

———, (1994). *Planning and Policies for Indian Agricultural Research,* 25th Lal Bahadur Shastri Lecture, reprinted in ICAR, Landmarks in Indian Agriculture.

———, *Next Phase of Agroclimatic Research,* in ARPU, Vol. 2 of same title, Concept: New Delhi.

———, (2000). Water and Food Security in South Asia, Invited Plenary Lecture at World Water Forum and Ministerial Meeting, Hague.

———, (2000). *Gandhi and the Indian Economy in the 21st Century,* in Gangrade and Radhakrishna, (ed.) *Gandhi and the Changing Facets of India,* Gandhi Smruti, pp. 44-62.

———, (2001). Development and Governance, in Hans van Ginkel and Ramesh Thakur, *UNU Millennium Series,* pp. 73-84.

———, (2001). Report of the High Level Committee on Legislation for Corporatization of Cooperatives, New Delhi, Ministry of Company Affairs.

———, (2001). Working Out Power Economics, *Indian Express,* Jan.16.

———, (2002). Agricultural Investment Strategies: Prioritizing Land and Water, in FAO, *Investment in Land and Water,* RAP Publications 2002/09, March 2002, pp. 67-78

Alagh, J.K. and D.Buch, (1995), "The SSP and Sustainable Development", in W. Fisher, *Towards Sustainable Development*, Columbia University Seminars.

Ambler, J., (1991). *Bounding The System*, in IIIM, Colombo, pp.3-12.

Alexandratos, N., (Ed.), (1995), *World Agriculture Towards 2010, An FAO Study*, Chichister, Wiley.

Asian Development Bank (ADB), (1999). The Growth and Sustainability of Agriculture in Asia, Manila, ADB.

ADB (2000). Rural Asia: Beyond the Green Revolution, Manila, ADB.

ADB (2000). *Rural Asia: Companian Volumes* (I-III), Oxford University Press: Oxford.

Chopra, K., and G. Kadekodi (1993). Watershed Development, *Economic and Political Weekly*, June 26, pp. A61-A67.

Chambers, R., (1988). *Managing Canal Irrigation: Practical Analysis from South Asia*, Cambridge, University Press.

De Los Reyes, and S. Jopillo, (1987). An Evaluation of NIA's Participatory Communal Programme, in IIIM, Colombo.

F.A.O., 1978-81, (1982). *Agro-Economic Zoning Atlas*, Rome, FAO-UNESCO.

F.A.O., (1993). *Agriculture Towards 2010*, Rome, F.A.O., for final published version see, N. Alexandratos.

Fisher, W., (1995). *Towards Sustainable Development*, Columbia University Seminar Series, Sharpe: New York.

Gangrade, K.D. and N. Radhakrishnan, (Ed.), (2000), *Gandhi and The Changing Facets of India*, Gandhi Smruti: New Delhi.

Government of India, 1989, *Agro-Climatic Planning : An Overview*, New Delhi, Planning Commision (authorship, Y.K. Alagh, *et al.*)

———, (2000). Commission on Agricultural Costs and Prices, Reports *for the Crops Sown for the 1999-2000 Season*.

Hayami, Y., (1981). Understanding Village Communitty and the Direction of Agricultural Change in Asia, Delhi, IEG. Occasional Paper, New Series, No. 1.

International Institute of Irrigation Management (IIIM), 1987. *Public Interventions in Farmer Managed Irrigation Systems*, Colombo, IIIM.

———, (1991). *Performance Measurement in Farmer Managed Irrigation Systems*, Colombo, IIIM.

Joshi, Sharad, (2000). "What Price Alagh Economics", *Business Line*, December 20.

Jomo, K.S., (2000). Income Distribution in East Asia, IEG-New School of Social Science Research Seminar, New Delhi.

Korten, D., (Ed.), (1986). *Communitty Management : Asian Experience and Perspective*, Kumarian: New Haven.

Lele, U., N. Kumar, Y. Alagh, N. Saxena, and K. Mitra, (2000). *Forestry in India: An Evaluation*, Washington, World Bank.

Lindert, P., (1996). *The Bad Earth*, Davis, UCAL Agricultural History Center.

Manor, S., and J. Chambouleyron, (1991). in IIIM Colombo.

Mitra, Saumya Kanti, (2001). Ajanta and China, *Economic Times*, Jan. 19.

NABARD, (1991). Annual Report, Bombay, NABARD, Sec. On fianancing agroclimatic development.

Ostrom, I., (1994). *Neither Market Nor State: Governance of Common Pool Resources*, Washington, IFPRI.

Ostrom, I. *et al.*, (1996). *Rules, Games and Common Pool Resources*, University of Michigan Press: Ann Arbor.

Pronk, J., and M. Haq, (1992). *Sustainable Development: From Concept to Action*, UNCED and UNDP.

Sachs, I., (1991). Background Paper for the Hague Declaration, extensively reprinted in *Nature* and by RIS.

Siy, R., (1986). "Averting Bureucratization of a Community-Managed Resource", in D.C. Korten, *Community Management: Asian Experience and Perspective*, New Haven.

Small, L., and I. Carruthers, (1991). *Farmer Financed Irrigation*, Cambridge, University Press: Cambridge.

Sonntag, N., (1996). *Adaptive Management Policy Exercises : Two Methods for Integrated Resource Management*, in MRC, above, p. 27.

Taylor, Lance, (1994). *The Rocky Road to Reform*, Helsinki, WIDER.

Vaidya, C., (1999). Private Sector Participation in Financing and Management of Urban Water Supply and Sanitation Projects, Delhi USAID FIRE Project, reprinted in 31st Annual IWWA Technical Volume.

Vermillon, D., (1997). *Impacts of Irrigation Management Transfer: A Review of the Evidence*, IIIM: Colombo.

20

Systems Approach to Water Resource Management

M.S.V. PRASAD

I. Introduction

Water is so essential to human life that planning for its future use and protection is a necessity. Planning is both a means for ensuring that the need of future generations will be recognized and is moveover a mechanism for resolving conflicts of interest.

In the area of Water Resource Management (WRM), two new developments in the recent decades are becoming increasingly more important. These are the application of systems analysis techniques to improve the planning and decision-making process and the need for interdisciplinary team work.

Historically, water resource planning has been primarily the domain of engineers and economists, but with increasing complexities of the planning process it can no longer be so. Participation from other disciplines like biology, chemistry, ecology, sociology, law, mathematics, geography, or political science is not only desirable, it is now essential.

The systems approach has provided a new dimension to man's analytical capabilities in developing new and effective management strategies for a popular resource like water, the diversity of which and the intensity of demands on which have increased manifold in recent decades and are bound to further increase in the future.

Systems approach to Water Resource Management (WRM) is an attempt to look at the problem on an interdisciplinary basis by using systems techniques.

Even though the application of sophisticated systems analysis techniques to the planning, management and operation of Water Resources is of comparatively recent origin, this paper is an attempt to provide a broad and compendious account of various facets of the systems approach for effective Water Resource Management (WRM). The emphasis will be on systems analysis, programming and descriptive models, multi-objective planning and social science modelling techniques of the systems approach.

Thus, even though our water-resources planning process has become exceedingly complex at present, and will become more so in the future, it is becoming increasingly apparent that the average planner and decision-maker has been provided with a few fundamentally new tools and concepts in the past several decades. One of these few new tools is systems analysis; it is being used quite extensively in the operational phases, but the planners have so far used only a fraction of its total potential. The following are the broad techniques of systems approach that can enable water resource management to become effective and result oriented.

II. Systems Analysis

Systems analysis is an analytical study that helps a decision-maker to identify and select a preferred course of action among several feasible alternatives. It is a logical and systematic approach wherein assumption, objective and criteria are clearly defined and specified. It can significantly help the decision-maker to arrive at better decisions by broadening his information base, by providing a better understanding of the systems and interlinkages of the various subsystems, by predicting the consequences of several alternative courses of action, or by selecting a suitable course of action that will accomplish a prescribed result. Systems analysis has added totally new dimensions to the sciences of policy-planning and decision-making in Water Resource Management.

Systems analysis provides the answer by methods and techniques that are available to everyone for critical analysis and examination. These are not unique in the sense that everyone who has the necessary expertise and experience can exactly duplicate the analysis.

Basically, systems analysis is a problem-solving technique, wherein attempts are made to build a replica of real world systems or situations, with the objective of experimenting with the replica to gain some insight into the real world problem.

Quantitative methods are preferred in systems analysis, but qualitative analyses can also be incorporated in the process.

In contrast to other available decision-making tools that have the same limitations, systems analysis uses all the relevant information available and extracts the best components from different scientific methods from different disciplines on which the analyses are based. Thus, the virtues of systems analysis are also virtues of the methods and techniques on which it is based.

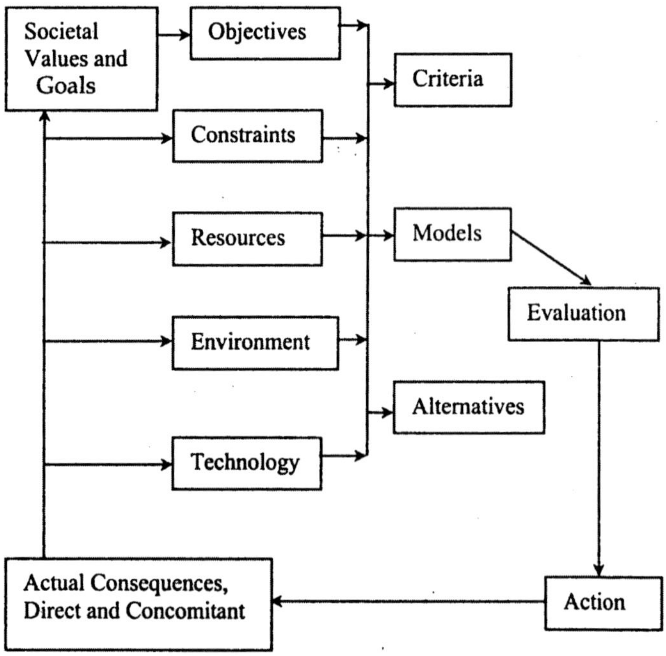

Fig. 20.1: Quantitative methods.

The system is represented by a series of mathematical expressions in such a way that the resulting relationship describes the phenomenon. The parameters that affect the system are included, as well as the factors that influence the parameters. Thus, in a real sense, it is implied that a good mathematical model of a system needs a thorough knowledge and understanding of that system. However, since in most cases of water-resources planning and management, all the factors affecting the system are not known, or, if known, often cannot be evaluated and quantified, the resulting model does not exactly describe the real-world situation, but can be fairly close to it for all practical purposes.

Broadly speaking, the analysis of a water-resources system goes through five related stages:

1. Identification and explicit statement of objectives.
2. Translation of objectives into measurable criteria.
3. Identification of alternative courses of action which will satisfy the criteria.
4. Determination of consequences that follow from each alternative.
5. Comparative evaluation of the consequences of the alternatives in terms of the criteria.

Fig. 20.1 shows the general steps involved in water-resources planning process. The starting point is the definition of national goals and values from which specific water-resources development objectives should be derived so that the programmes when completed will help to achieve those goals. The objectives then are translated into measurable criteria which can be used to appraise the degree to which the objectives are satisfied. The next stage is to develop a model of the system that will examine and evaluate the alternatives. The criteria used by the model will relate the alternatives to the objectives. During the process of development of the criteria, model, and alternative plans, due consideration should be given to the resources available, constraints to the system, and technological and environmental factors. There should be a good understanding of the interface between the system and the environment that it serves or alters and by which it is altered or constrained.

Finally, systems analysis cannot only help planners and decision-makers as an aid to prediction and planning processes, but it can also be used as a gaming or teaching tool. These types of models are generally less complex and are used to test assumptions, to explore consequences of different policies.

Alternative Models and their Relevance

In general, models can be divided into two categories — programming and descriptive — depending on the relationship of the model to problem-solving. Programming models, for a given objective function, attempt to derive the optimal policy. Descriptive models, on the other hand, attempt to predict possible future consequences due to a set of assumed exogenous variable and policy alternatives.

Theoretically, water-resources managers should find programming models more relevant as an aid to decision-making, since they are geared to obtain optimal policies, directly or indirectly. However, programming models are valid for rather simplified systems which can assume linearity of functional relationships. Also, often it is not possible to define objective functions, especially in the field of water-resources systems management, because of the many conceptual and empirical issues associated in constructing a truly compre-hensive social welfare function. Examples of this type of modelling effort are several linear programming models in the water-quality-management field, the interregional linear programming models of Henderson and Stevens and the interregional linear investment model of Rahman. Objective functions need not be defined for predictive models since they are not directly related to objectives. These types of models predict the values of endogenous variables for a given set of exogenous variables. Exogenous variables can be chosen by the model builder or the decision-maker and are often called policy or control variables. If it is assumed that the future consequences are functions of policy variables, the decision- maker can select an "optimal" policy by changing the policy variables, which will give conditional predictions of future stages.

Mera has pointed out the advantages and disadvantaged of this type of modelling efforts quite succinctly:

> The predictive model is inferior in the sense that each alternative examined must be chosen subjectively by the persons constructing the model and, therefore, the chance of missing a significant alternative cannot be eliminated. However, it is superior in many senses: the future state can be described in far more detail, the future state can be evaluated differently for different preferences, and the model can be used to test the sensitivity of response to any particular policy variable.

Multi-Objective Planning

The objective of water-resources planning and management, or of any other type of natural-resources planning and management for that matter, have drastically changed over the years. Broadly speaking, this can be attributed to our increasing awareness and understanding of our social needs and goals. Historically, the main objective of water-resources development has been economic efficiency, and the technique used for its evaluation has been benefit-cost analysis. Gradually, to this single objective of economic efficiency, other objectives have been added. These, in order of their emergence, are regional income redistribution, environmental quality, and social well-being. The addition of these three objectives has undoubtedly broadened the decision-making framework and is an improvement over the past practice, which was pretty much limited to economic efficiency analyses. The endeavour of the planners to simultaneously achieve these multi-objectives within the planning framework stems, to a great extent, from the needs and desires of the society; hence, if the planning analyses are properly conducted, they would presumably expedite social and political acceptance of the plans.

First, how should the different objectives be traded off against each other, and on what basis, especially when the subjectivity of some of the parameters is realized? Second, who should make a decision on the final mix of alternatives — planners, politicians, the public, or some mixture of the three?

The analyses of trade-offs between the various objectives can be quantitative, qualitative, or, as is usually the case, both. A system can be designed to perform optimally in terms of one objective, subject level of performance of the other, which in effect becomes a constraint. In other words, if both economic-efficiency and environmental-quality objective are to be considered for planning analyses, one can be included in the objective function and the other can be treated as a constraint.

It seems clear that water resource management will give increasing alternative to the multi-objective nature of planning. The explicit analysis of varied and often conflicting objective will be necessary and an alternative that give varying weight to the various objectives will be required. It is by no means clear, however, how this can be done; while economics has been added to the earlier engineering of water planning, it would appear that ecology, sociology and political science must be added to the planning effects. Somehow the various objective and the techniques for the plan formulated must be combined into a manageable interdisciplinary enterprise.

In fact, water-resources planning, project authorization, and level of funding are all essentially political processes. Hence, the planners decide the feasibility of the project, and politics decide the implementation of the plan. As Ogden has pointed out:

> No matter how sound a project may be physically, no matter how profitable it will be economically, it will come about only if effective political leaders can champion its cause in the right way at the right time.

Social Science Modelling

During the last decade or so, tremendous progress has been made on the development of physical models for water-resources planning and management, but commensurate progress on social sciences modelling is sadly lacking. Within the social sciences area, several models currently exist that consider some economic and demographic parameters, but very few, if any, include sociological and institutional factors. Biswas and Reynolds have presented a comprehensive review of the current status of socio-economic modelling in water management.

Since modelling is a scientific method used for analyses of known parameter relationships and observe data to understand and explain different phenomena and is a predictive tool, its role in social sciences should be similar to that in physical sciences. Social scientists are becoming increasingly aware of the potential of the mathematical models to analyse social phenomena. Rex, for example, categorically states that the "important question for the sociologist is not whether he should interpret observed human behaviour in terms of models, but what sort of models he should apply".

It is often argued that social phenomena cannot be modelled since it involves consideration of human behaviour, which is often random and unpredictable. This argument is not really valid.

Potential for meaningful analysis and prediction of social phenomena are there; it has only to be developed. And it can only be developed by realizing the tremendous contributions that the social scientists can make to the planning process and by making special efforts to get the social scientists really involved as equal partners in that process. Without a concerted effort by social and physical scientists, it is difficult to visualize how this potential can be exploited

There is also another important aspect of the interdisciplinary considerations of water-resources planning and management. In the initial stages of analysis and planning, it is often difficult, if not impossible, to predict which disciplinary viewpoints will turn out to be the most important. In other words, we do not know *a priori* what the major issues and problems are going to be and who will be the most appropriate people to analyse and solve them.

III. Conclusion

It is clear that systems approach to water resource management is an essential technique to make water resource management more efficient and goal oriented.

It is desirable that the members of a planning team should have different disciplinary backgrounds and should at least be familiar with the techniques and goals of other disciplines. The planner should not only be able to understand the technical terms commonly used in other disciplines but should also be familiar with their methodologies and techniques.

Individuals trained in different disciplines could also look at a common problem to be analysed and solved in totally different fashions, depending on their education, background, and experience.

The point I am trying to make here is that we should have an interdisciplinary group for water resource planning and management because different disciplines could perceive the same problem in a different light as a result of their training and background.

Thus, unless we make a determined efforts to develop a truly interdisciplinary planning team, the water resource policies and programmes that are developed are highly unlikely to be the best possible decision for society as a whole.

References

Biswas, A.K. (1973). Social: Economic Consideration in Water Resources Planning, *Water Resour, Bull.*, 9: 4.

Biswas, A.K. (1973). Mathematical Model for the Planning and Management of the Saint John River Systems, *Proceedings of the International Symposium on River Mechanics*, Bangkok, Thailand.

Biswas, A.K. (1974). "A short History of Hydrology", in A.K. Biswas (ed), *Selected Works in Water Resource.*

——— (1974). *The Role of Mathematics and Computers in Resources and Environmental Management*, Education Media Press: Costa Mesa, California.

Biswas, A.K. and P.J. Reynolds (1969). Socio-Economic Simulation in Water Resource Systems Planning, Vol. 1, proceedings of the International Association for Hydraulic Research, Kyoto, Japan.

Biswas, A.K., P.J. Reynolds, and R.W. Durie, (1972). Water Resource System Planning, *Proceeding of the Symposium on Water Resource Systems Planning*, Varma, Bulgaria, International Commission on Irrigation and Drainage, New Delhi, India.

Hamilton, H.R., S.W. Goldstone, J.W. Milliman, A.L. Pugh, E. Roberts, and A. Zellner, (1969). *Systems Simulation for Regional Analysis: An Application to River Basin Planning*. The M.I.T. Press: Cambridge.

Mera, K., *Survey of Model Building for Regional Economics*, Discussion Paper No. 55, Programme on Regional and Urban Economics, Harvard University, Cambridge, Mass, 1969.

Odgen, G.M., *Politics of Water Resources Development in Social and Ecological Aspects of Irrigation and Drainage*, pp. 353-75, American Society of Civil Engineers, New York, 1970.

Productivity, Water Resource Management, National Productivity Council, October-December, 2000.

Rakesh Hooja, "Administering Water in India in 2050", *The Indian Journal of Public Administration*, April-June 2000.

Rex, J., *Key Problem in Sociological Theory*, p. 60, Routledge and Kegan Paul Ltd, London, 1961.

Index